Total Harmonic

The Healing Power of Nature's Elements

Case Adams, Ph.D.

Total Harmonic: The Healing Power of Nature's Elements
Copyright © 2012, 2008 Case Adams
LOGICAL BOOKS
Wilmington, Delaware
http://www.logicalbooks.org
All rights reserved.
Printed in USA

Publishers Cataloging in Publication Data
Adams, Case
Total Harmonic: The Healing Power of Nature's Elements
First Edition
1. Science. 2. Health
   Bibliography and References; Index

Library of Congress Control Number: 2008903728

ISBN 978-0-9816045-3-4

For my Teachers, Professors and Mentors

Table of Contents

Introduction

My awakening to the harmonic nature of life came as a young surfer probing the ocean's wave rhythms for a few good rides to the beach. I quickly discovered that not all waves were alike. They might look repetitive from the beach, but if I waited for just the right wave, I could catch one that crested just right, and allowed my board and body to tuck into it. Once my board was positioned within its fluid belly, I found I could turn and cutback variously, rendering the wave face a canvas for expression. Over time, I found this experience of wave riding triggered a surge of physical elation. I soon discovered this feeling was universally felt amongst almost all capable surfers. Although the wave ride itself might have lasted only for a few seconds, the feeling it created lasted for days if not years afterwards.

I thought: *What caused this surge of elation?* There is something about the force of a large ocean wave breaking in such a precise way—allowing one to 'plug' in to it—that seemed to stimulate this feeling. It seemed there was something about slotting into an overhead wave with its crest lifting and curving over—the "barrel" or "tube"—that intensified the surge. It is truly an awe-inspiring experience, and most every surfer will agree. In fact, there is an oft-used expression used in surfing literature: *"Only a surfer knows the feeling".*

The evidence of the effect of this surge of elation explains why surfers will journey out into the deep, cold ocean waters where numb appendages and a hard rock or reef bottom below combine with the potential of sharp-toothed visitors from the depths. It is just to achieve this feeling. To achieve the feeling, surfers will sit for hours in the harsh ocean environment to catch ten or fifteen few-second rides on a good day. As this surfing 'affliction' continued into my adulthood, I found it difficult to remove myself from the undulations of the ocean for long periods. Like a migrating bird or tidal mollusk, I had to return to the sea with some sort of periodicity.

The education offered by surfing rivaled my college and graduate studies in many ways. The ocean is quite the laboratory for studying first hand how the combination of weather, water, wind and sun move together in synchronicity, paced by nature's plotting sense of time. Thirty-plus years of surfing, with its cycles of good offshore waves, blown-out waves and big winter swells have allowed me to observe how living organisms and the elements cycle together in utter harmony.

My studies and eventual doctorate in integrative health sciences reflected not only my interest to understand in the more subtle realms of the human body, but how the cyclic harmonies of nature relate to the health of the human body. My studies led me to the various disciplines of traditional medicines including Chinese medicine; Ayurvedic medicine;

Western herbology; North American Indian holism; the healing arts of the Egyptians; the physicians of the Greeks; and the healing traditions of the Arabs, Polynesians, Aborigines of Australia and Fijians among others. These studies revealed to me a strangely similar common view of a conscious natural world expressed in rhythmic behavior. While modern science has quickly brushed aside this view of a conscious world, the evidence of it is nonetheless illustrated through the congruity between these ancient traditions and the latest findings of modern science and technology.

I began to gradually become aware that the wisdom revealed through humanity's intimate relationship with nature has been become blinded by an overconfidence in sterile double-blind studies and rigorous research protocols. Ironically, these very same double-blind studies also expose the very nature they were attempting to dismiss. While our scientific community seems appropriately intent to find answers to many of the mysteries of our existence, it is also mired within a culture that has lost touch with nature.

Our researchers—like the rest of our society—primarily live and work in isolated apartments and laboratories, thoroughly removed from nature. While attached to blinking screens of virtual reality, we have created for ourselves a two-dimensional view of the world.

For thousands of years prior to this past century, humans were intimately connected to nature. We lived with the tides, the moon, the rivers, the oceans, the mountains, the snow, the rain, the sun, and the plants. Before our self-bound slavery to the electrical light show of modern civilization, humankind survived and paced with the balanced rhythms of nature. Humans were thus intimate with the harmonies existing within nature. With that perspective, humankind could better view the harmonic nature of life with a sense of wisdom and balance.

Those days are gone. Now our only concern with nature seems to be what our local road and airport conditions are. We quickly dismiss our ancestors who worked harmoniously with nature as crude and simple. We dismiss their views of nature as uneducated and naive. Yet those very ancestors were somehow smart enough to hand their next generation a sustainable world and lifestyle. By comparison, our current generation hands the next a planet in chaos—teetering on the verge of destruction.

As we look at the conclusions of modern researchers, we can clearly see a problem exists with regard to how the universe is viewed. The track that modern science has laid down, from the beginning of the industrial age forward, has been towards a path of removal from nature. Our very survival on this planet is now in danger because of the many technological inventions created over the past 200 years. Not only do we find our environment poisoned by the effects of our modern science and technology, but we find our bodies increasingly robbed of their natural powers and potencies; replaced by autoimmune diseases and cancers.

The vicious irony of cancer, autoimmune disease and superbug infections tells us there exists a better strategy: A strategy that works with nature rather than against it: A strategy of harmony rather than adversity.

In our rush for chemical solutions and genetic manipulations, we are losing sight of the very rhythms of nature that give our bodies life, vitality and survival. This book expresses the need to tie the technological advances modern man has made back to the fundamental rules of life and nature. With it I hope we might learn the lessons we need to, before we destroy our environment and our next generous with it.

While the predominant thesis of this journey through nature's elements is to show how the elements are synchronically and consciously organized, our journey also bears witness to many the healing properties nature's elements provide when applied correctly. Much of the knowledge of the application of nature's elements is unavailable from modern research however, simply because researchers tend to focus towards pharmaceutical and technical solutions. Alternatively, much of the knowledge of the application of nature's elements has been developed and handed down over thousands of years among cultures and traditions that have developed and prospered within nature's laboratory. Some of these strategies are documented here, though by no means exhaustively..

As I explore this vast topic, I am humbly reminded of the story of the frog in the small pond:

The frog sat in his little pond and conjectured on the width, breadth, and depth of the ocean. In order to properly understand it, the frog figured that if he could just puff himself up a little, he could be large enough to understand the ocean. Not realizing just how expansive the ocean was compared to his tiny pond, the frog puffed himself up so much that he burst and died.

I want to thank my teachers, professors, mentors, fellow students, friends, and the many ancient and modern researchers who have provided the evidence, information, and encouragement for this exploration into the harmonic nature around us and within us. I hope they will excuse my errors and omissions.

Chapter One

The Harmonic Universe

Our universe is pulsing with rhythm. Throughout nature, we see repeating rhythmic occurrences. Each day we observe the sun's rise and set, establishing a cycle that is repetitious, adjusting slightly with every cycle. Seasonal changes with the rotation of the earth with respect to its orbit are apparent. We see this seasonal rhythmic rise and fall reflected in plant-life—waxing in the spring and waning in the fall. We see birds and other migratory animals move with similar periodicity, traveling synchronized with the seasons to amazingly exacting locations.

As we analyze these various pulses of nature, we notice a behavior of precise waves. We see a distinct and precise rhythm repeating and oscillating through the media of our environment. We see nature's oscillations pulsing through the oceans, causing waves and weather conditions. We see larger periods of ocean tidal rhythms bringing an exchange of ocean creatures and their food to and from the seashore. We see the rhythmic upwelling of cold waters from the ocean depths rotating and recycling the ocean's various biochemicals and marine life. Meanwhile, these surface waters are spun and rotated by the wind through recycling temperature gradients. We also see a similar rhythmic pulsing throughout our atmosphere; recycling temperature, water vapor, and various gas mixtures with periodic precision. We see rhythmic oscillations pulsing through space, sending us the radio, light and other radiation waves from twinkling stars billions and trillions of light years away. We see our closest star, the sun, periodically radiating heat and light along with cyclic solar storms that apparently affect behavior.

Through our own bodies, we see similar rhythmic behavior, from the periodic beating of our hearts to the pace of our breath. As we look deeper into the body, we see a myriad of subtle rhythmic pacing mechanism, with every cell and every organ involved in its own orchestration of timed behavior. We see hormones and neurotransmitters pacing through the blood and nervous system, stimulating our various physical activities. We see a rhythmic immune system pulsing through our lymphatic ducts and nodes. As we measure frequencies, we find that amazingly, each cell pulses with a distinct and unique frequency. Even as we peer into the finer matrix of the smallest parts of nature, we find every molecule and each atom vibrating rhythmically at frequencies unique to their composition and position within the environment.

As we observe the movement of the range of physical elements of the universe, we see a macrocosm of rhythmic oscillation. The various changes of the season, the revolution of the magnetic poles, the shape of the eye, and the various cycles of rhythm around our natural environment are all circular in shape, repeating their oscillations with precise intervals of rising and falling activity. Meanwhile the planets revolve around the sun in unique yet consistent orbitals, creating a variety of cyclic circumstances for each, rotating in relation to seemingly stationary objects. Each planet also tilts with a particular axis, creating cyclic seasonal changes to augment its periodic orbital revolution around its sun.

On the grand scale, our solar system is only one small solar system among many, revolving around a thermal core of our spiral-shaped Milky Way galaxy at an estimated rate of 230 million years per revolution. Furthermore, cosmological observations now indicate the Milky Way is but one of hundreds of billions of other galaxies within our observable universe. This universal megacosm appears to consist of a collection of larger and smaller galaxies, each rotating with their own precise cyclic procession, and likely rotating within a grander scheme. Most galaxies have been observed rotating with a predominantly spiral shaped array: Their arms each uniquely reflecting the same rhythmic parameters of the oscillations we observe at nature's quantum levels.

Over the past century, science has unveiled a human physiology that appears to oscillate harmoniously with the greater rhythms of the sun, the moon, and the greater magnetic rhythms circulating through our galaxy. We have determined many organisms, from the smallest single-celled organisms to the largest of mammals, cycle with the beat of the solar cycles and to a lesser degree the paths of the constellations and lunar cycles. Increasingly, researchers are discovering that solar storm cycles and lunar cycles impact human and animal activity on various levels including moods, metabolism, and hormone secretion.

The world around us is moving with sequence, pace and timing. In a word, *rhythm.* The rhythms of nature are also *harmonic:* They cycle together, grounded to the same fundamental beats and sequences. Thus the pulse of nature is not a single rhythm, but a collection of rhythms that resonate together to create harmony.

The most relevant physical example of harmonic nature is the cycling conducted by planet earth. Because the sun is tilted about 23.5 degrees from the axis of north and south, she cycles with seasonality, magnetic flow, temperature gradient and sunlight. In June, the northern portion of the planet is tilted towards the sun, giving that part of the planet summertime, warmer temperatures, and longer days. During December, the southern part of the planet is tilted towards the sun, giving that part of the planet warmer temperatures and longer days. Summers on each pole have no nighttime and winters have no daytime. Of course, those winters are extremely cold. The moments when the earth's tilt is at its greatest on each side is called the *solstice.* There are two periodic solstices each year— June 21 and December 21. This means there are two days each year when daylight and nighttime have equivalent periods: When the sun is at an even angle with the equator. This is called the *equinox.* Like the solstice, the equinox occurs precisely twice a year.

The rotation of the earth with its slight tilt and recurring extremes also means the other movements of the earth are periodic. It rotates through cycles that precisely and periodically alter its environment. This cyclic nature also therefore can be measured periodically and rhythmically. One complete season from solstice to solstice would be called one period. The combination of the timing of this period with the amplitude or aspect (23.5 degrees) create a precise waveform, which when charted over its period of twelve months, comes together to form one complete *rhythm* of the earth.

We observe the same type of rhythmic movement throughout the physical universe. Some have proposed the various movements occurring around us are chaotic and random—meaning they are unrelated. However, as researchers have increasingly measured and tracked the seemingly unrelated movements of nature, we have found a continuum of related motion with periodic cycling. As this library of scientific data has grown, an increasing trend is being observed: Nature's multi-dimensional movements are illustrating an expanded relativity between the motion of one rhythm and the motions of others. As these movements are being related to other rhythmic motions through scaling, charting and measurement; the inter-related rhythmic behavior of the universe is becoming obvious: A total harmonic is becoming evident.

The mysteries involving the relationship between movement, space, and time have been plaguing humankind for thousands of years. From the most ancient philosophical queries to the most recent, we have been fascinated—no, obsessed—with the moving, temporary relationships between these elements. Just about every ancient civilization has come to realize the various relationships between the elements of nature. Increasingly, we are realizing the mysteries of life are not abstract and separate problems: They are inter-related: They are to be found among the magical precision between the rhythms pulsing within us and around us.

From the earliest of ancient records, humankind has considered the pacing of time and the rhythms of life an indication of meaning and purpose within the universe. Early practices of the ancients reflected this adoration of harmonic pacing. We cite the ceremonial beating of drums to announce celebration, worship, and warfare. We point to the rhythmic singing and dancing that became a prominent part of every ancient culture. Even our own culture is grounded upon rhythmic vibration, as we communicate through the radiowaves and the oscillating appliances of telephones and computers.

From the timbre and meter of prose to the elaborate compositions of instrumentation and song, humans have found ways to harmonize with the rhythms of nature. With these orchestrations of harmony, we have expressed our inner emotion and character. We have used song to communicate our desire to love and be loved. We have expressed with song our need for purpose and meaning.

In a variety of ancient civilizations—most notably the peoples of the Mayan, Incan, Indian, Tibetan, Japanese, Chinese, Egyptian, Arabian, Mesopotamian and Polynesian cultures—language was considered more than communication. Speaking was considered an art form. Many of these ancient languages were spoken with precise cadence and intonation. They were spoken with a rhythm that matched the intention to transfer particular concepts and thoughts. Even today, many honor this heritage with poetry, songs and chants. Each observes their own standards of tempo and rhythm. Each has its own roots to the ceremonial harmonizing between our inner selves and our physical exterior.

The rhythms of living chemistry began to be translated into scientific knowledge some time ago. The rhythmic behavior of the universe gave rise to the use of mathematics. This was evidenced in the texts of Mesopotamia from the third and fourth centuries B.C. The math within nature was apparent to the Greek philosophers of the sixth and seventh centuries B.C. Mathematics were used by the ancient Egyptians as early as 3000 B.C. Earlier we find mathematical formulation recorded in the texts of the Chinese monarchs of 3500 B.C. They set forth various measurements and calculations, on subjects ranging from building to the stars. Earlier than this, we find the ancient Sanskrit texts of the 4500 B.C. Indus valley civilization of ancient India brought forth a mathematical and scientific view of the earth's rhythms. These early mathematical insights came to influence the scientific culture of later societies as we have investigated the precise nature of motion and movement. From the earliest of these recordings we find a common acceptance that the physical world moves with a synchronized pacing or pulse. This pacing came to be understood to be measurable and repetitively precise. These ancient cultures also observed these relationships with a reverence toward a larger conscious mechanism—an elegant oversight with purpose.

The measurement of nature's rhythmic order descended onto western science through the capable doorway of the ancient Greeks. The famed

Greek Pythagoras of the sixth century B.C. was considered by many in the West to be instrumental in developing a mathematical basis behind the rhythms of nature. Pythagoras and his students utilized reason, logic, and the observation of nature to find an affinity between the natural rhythms of life, the harmony of music and the mathematical relationships between integers and their ratios. Though it is understood that Pythagoras guided many of those principles directly, credit for many insights presented as Pythagorean concepts should be spread among a number of philosopher/scientists who learned and taught cooperatively within the renowned Pythagorean community.

Through the works of Philolaus of Tarentum, we learned Pythagoras expressed the rhythms of song and instrumentation within a context of between pitch, scale, octaves and harmony. All of these were brought into rationale through mathematical relationships. This same approach led the Pythagoreans to perceive various other connected rhythms within nature. While Pythagoras may or may not be responsible for the famous *Pythagorean Theorem*, he is still considered as the "father of numbers." He is also credited with the form of philosophical and logical reasoning that blossomed later through the teachings of Plato, Aristotle, Socrates, and Ptolemy.

Second century Greek Claudius Ptolemaeus was also known as Ptolemy. Ptolemy was a famed mathematician, astronomer, and natural scientist from Alexandria. He was responsible for a number of treatises that influenced natural scientists over the next 1500 years. His book *Harmonics* focused on the rhythmic qualities of music theory. His *Optics* treatise covered the realms of light rays and vision, and his book *Geography* established many of the principles utilized by geographers and cartographers in mapping and quantifying spatial relationships.

Ptolemy's *Harmonics* repositioned the Pythagoras' approach of relationships between harmony and music within a stricter definition of mathematical ratios. His work delivered music theory from the clutches of the fourth century B.C. Greek Aristoxenus of Tarentum, who proposed music harmony to be relative to what he termed the *"irrational exercise of perception."* While Ptolemy broadened music harmonies into various other relationships using the concords such as the fifth (3:2) and the fourth (4:3), his mathematical proofs—grounded in the scientific method—became a fundamental process of relating to the physical world using the precise relationships of mathematics.

The ancient Greek philosophers proposed that the synergies between the pulses of nature and mathematical logic were tied to a common bond with elements of a transcendent nature. This greatly influenced the progression of natural science for many centuries to come. The scientific contributions of Hippocrates, Pythagoras, Socrates, Plato, Aristotle, Copernicus, and others established the groundwork for centuries of progressive scientific endeavor and a reverence for mathematical modeling. The fundamental understanding that nature followed the relationships of numbers and mathematical formulae set into motion a codification of nature's movement. This codification led us to relationships such as $F=ma$ and $E=mc^2$, which became the basis for further correlation between rhythmic order and the natural world. This logic of the natural world's mathematical relationships provides us with our first proof that physical existence is pervaded by rhythm.

Using early mathematics, modern humankind began to divide time and movement into smaller and smaller increments. We began to split nature's motions through time into tiny fractions of periodicity. As we have developed instrumentation to access these deeper fractional elements of nature, we have continued to find more intricate, deeper rhythmic cycles evident in even these tiny units of nature. We have found that matter down to the smallest parts—atoms, electrons and other subatomic units—all appear to be moving with orchestrated and precise rhythmic periodicity. We have found that radiation, light and heat are cyclic and periodic: pulsing in waveforms of precise frequency. We have found that all our basic sensory reception is oscillating in frequency and wavelength. We have found that the only way to "see" the basic elements of nature through instrumentation is not to look for a particular unit, but to find smaller portions of periodicity. This, unfortunately, has led us to confuse the rhythms of our world with units of "solid" particles.

The Pulse

Most every occurrence in the natural world is complete with repeating rhythms and sequences. The famed *Fibonacci sequence,* giving way to the *Golden Rectangle,* the *golden mean* and the *golden spiral,* illustrates the rhythmic behavior being exhibited throughout the natural world.

The Fibonacci sequence is a series of numbers, 0, 1, 2, 3, 5, 8, 13, 21, 34, 55.... observed throughout nature. A *Fibonacci number* is found by adding the two preceding Fibonacci numbers together: 1+2=3, 2+3=5, 3+5=8 and so on. First observed by Italian Leonardo Pisano Fibonacci in the thirteenth century while tracing a family tree of rabbits, the *Fibonacci sequence* can be seen all around us and throughout nature. It can be seen in plants, fish, insects, animals, and humans, both from a perspective of dimension and appendage. For example, the angles of outward projection of branches and leaves from trees and plant stalks are always assembled in precise Fibonacci fractions: one-half in grasses, lime and elm; one-third in sedges, beech, hazel and blackberry; two-fifths in roses, oak, cherry, apple and holly; three-eighths in bananas, poplar, willow and pear; five-thirteenths in leeks, almond and pussy willow; and eight-twenty-firsts in pine cones and cactus to name a few. Some plants are aligned in a related sequence, the *Lucas sequence,* named after nineteenth century Frenchman Edouard Lucas. Lucas numbers are also assembled by adding two con-secutive numbers to get the third. However, the Lucas sequence begins with two and one rather than zero and one as in the Fibonacci sequence: 2+1=3, 1+3=4, 3+4=7, 7+4=11, and so on, to arrive at 2, 1, 3, 4, 7, 11, 18....

When sequential Fibonacci measurements are arranged into polygons, they form rectangles. When one of these rectangles is laid against a square of the next Fibonacci number, it becomes the famous *golden rectangle* or *Phi proportion.* The golden rectangle is made from two adjacent 1x1 squares, which become a 1x2 rectangle. This can be laid against a 2x2 square, becoming a 2x3 rectangle. If laid against a 3x3 square, this be-comes a 3x5 rectangle and so on. The Fibonacci rectangle is observed throughout nature, including the outside dimensions and inner segment measurements of plant, animal and human organisms and their various appendages.

Another natural pattern observed throughout nature is revealed when golden rectangles are assembled around each other into a spiral. The *golden spiral* is formed concentrically outward by the *golden section* dimen-sions of 1:1.618. The golden spiral is seen repeatedly throughout our natural world. It is seen in the nautilus shell. It is seen in storm systems such as hurricanes and tornados as observed from space. It is seen in the swirl of water down a drain. It is seen among the tops of plant florets like cauliflower and broccoli, and within the cross-section of an ocean wave and within a three-dimensional perspective of various natural frequencies.

The Fibonacci sequence can be compared fundamentally with the musical *harmonic sequence.* The harmonic sequence is established with a fundamental pace factor for each step. An example of a *harmonic sequence* would be 5, 10, 15, 20, 25, which has a fundamental pace of five. This type of sequence produces a wave harmonic. In a function of time, the sequence of a wave harmonic is then considered *rhythmic.* Fibonacci and Lucas sequences might be considered a progressive harmonic because the fundamental pace is progressively being updated by the previous two factors of the sequence. We might then refer to the typical music har-monic as a *static harmonic,* and the sequencing in nature as *living harmonics.*

The cyclical nature of the living universe has also become evident to early mathematicians. As the geometry of the circle was broken down and charted over a two dimensional horizontal plane, it converted to a waveform now known as the *sine wave*. As the early mathematicians studied this further, another sequence or series became evident, called the *Fourier series*. We will discuss this further as we delve into the wavelike nature of the universe. For now, we note that the circle is simply a repetitive sine wave.

The Currency

The world around us is pulsing with electricity. While we might be amazed at the tremendous surge of electricity given off during a lightning storm, in reality everything in the physical universe has an electrical nature. While we delight in humanity's abilities to alter and utilize electrical current, the discovery of electricity quite simply illustrates the reality that electricity is a baseline fundamental of all physical matter.

Through the eighteenth century, research by classical scientists such as Alessandro Volta, Georg Simon Ohm and Andre Marie Ampere—hence the *volt,* the *amp,* and the *ohm*—humankind began its quest to understand the nature of electricity and how electricity passed through physical substances. How does current conduct differently through different mediums, they wondered? Copper was a better conductor of electric current than nickel, for example. Why is that? What was it about current that distinguished one type of substance from another?

In the early nineteenth century, the connection between electricity and chemistry become evident, primarily due to the electrolysis work of eighteenth century physicist Sir Humphrey Davy. Sir Davy's work led to the discovery of a number of new elements, which included conductive elements such as sodium, potassium, calcium, magnesium, strontium, and barium. This work was continued by another Brit and early assistant to Davy, Michael Faraday. Faraday formulated the famous *Faraday's laws of electrolysis.* These laws concluded the conductivity of solutions of particular electrolytes, leading to the development of voltaic cells like lead batteries and dry cell batteries.

Early research on the electric nature of chemicals led to the assumption that the ability of a substance to attract electric current had something to do with its electric *potential.* This seems rather obvious, but the fact is, different substances have different abilities to carry currents and have different *resistance* capability. A substance's conductivity and resistance were eventually linked to its inherent lack of charge balance (electrons versus protons) and atomic number. A substance with a net negative charge (more electrons than protons) tends to attract positive charge and vice versa. It was observed this could be due to the substance being able to transport either positive or negative charges. While positive charge flow was the original concept of electricity, later the concept assumed the direction and path of current was due to the direction of electron flow.

This was born from the simple observation that a substance with a net positive charge tends to attract a substance with a negative charge. Observations led early researchers to believe these positive or negative potentials were balanced by small units dubbed electrons. These tiny parts—ascribed as particles—must be traveling from negatively charged substances to positively charged substances. Current, then, must be the passage of negative particles from one substance to another, they theorized. This theory created a nice picture for textbook drawings and chemistry class. There was a problem however: No one could locate these particles.

This movement of atomic negativity was found to be typical of the direct flow of electricity—called *direct current*. In a direct current, negative charge tends to flow in one direction only at a particular pace or rhythm. In an 1820 public lecture, Hans Christian Oersted placed a magnet on a battery current. The magnet moved around as the current flowed. This indicated a direct current not only had a flow or current direction, but also apparently had a magnetic quality. On that day, *electromagnetism* was realized.

The original flow of direct current was found to be in one direction only, from a "negative" source to a "positive" source. To set up a battery—a source of direct current flow—two cells or chambers are filled with electrolyte fluid. One is inserted with a negatively-charged metal *anode* and the other with a positively-charged metal *cathode*. The chambers are tied together by a bridge allowing electrons to flow in between. Each solution retains an opposite charge potential, which draws electrons from the anode of one chamber to the cathode through a connected metal conductor. This basic loop allowing the flow of electrons is called a *circuit*.

Now if the loop were wired to travel through a light bulb, the electrons would move from one side of the cell to the other through the wire and bulb. When the electricity current (or electron flow) reaches the light bulb—should the bulb be set up with a little *resistance* to the current—the light bulb will convert the resisted electricity into visible light. Of course this requires an isolated filament made of a special metal (tungsten filaments are typical) inside a bulb with either a vacuum or gas—as early bulbs were made—or under heightened pressure. The flow of these negatively charged units—the current of electrons—excites the orbital energy levels of the filament and the gas molecules surrounding it, emitting the theoretical unit called the *photon*.

As electricity gets drawn into the resistance of the light, the flow of electricity may exhaust the battery because the flow of electrons now are leaking into the energy excitement process and photon creation, leaving fewer to reach the cathode portion of the cell.

The amount of potential (or pressure) within a circuit of current is the *voltage*. The current itself is the rate of flow. The *ampere* or *amp* is the amount of current flowing through a particular point over a particular amount of time. A *watt* is a unit of power, obtained by multiplying the number of amps of flow times the voltage. A kilowatt is 1000 watts. The key relationship is that one watt is generated when a one amp of current flows from a source of one volt.

Assuming we had a 40-watt light and a 12-volt battery, we could calculate about 3.3 amps of current (40/12) flowing through the wire if the light was on. The bulb creates a resistance of about 3.6 ohms (12/3.3) in its utilization of that energy. This resistance converts the flow of electrons to light using a charged substance and filament within the bulb. However, a 12-volt battery in such a circuit would power a 40-watt light bulb with direct current for only so long before it drains the battery. The light bulb must be appropriately configured for a direct current as well. A small wattage is suggested because a 40-watt light bulb fired by a 12-volt battery current would wear out the battery within a little over three hours (12/3.3). A smaller bulb, say a 12 watt bulb, would be a better match because it would only draw about 1 amp of current (12/12). The battery could run this bulb for twelve straight hours without any changing. This matching of electrical current and voltage to appliance wattage would be better described as *calibration*.

We might say that nature is *preloaded,* because it never runs out of power. The required voltage is always available to run the various physical appliances—physical bodies—that inhabit the environment. How can nature be so precise?

Three centuries ago, the intrusion of magnets into the flow of direct electrical current illustrated a new type of current: A freak oscillating human-made rhythm called *alternating current* (or A/C). A/C was developed by first wrapping and then rotating magnets through a flow of direct current. The field from the magnets caused the current to pulse forward and then backward alternatively. An intertwining of current through wire wrapped around a magnet—an *alternator* or *generator*—created an *induced alternating current*. Another device dubbed the *inverter* was designed to convert natural direct current into alternating current, accomplished by locating magnets around coils.

As late-eighteenth and early-nineteenth century researchers began to develop and test this strange alternating current, it soon became evident electromagnetism was the result of an inter-relationship between two active vector motions—one moving horizontal to the flow of current and the other moving perpendicular. As magnets were rotated around direct current, these two vectors could be controlled and modulated. Humans developed the ability to harness the genie of electricity.

After perfecting rotor-generated electrical current, it became increasingly clear that alternating currents contained a number of qualities quite similar ocean waves: measurable frequency, amplitude, and wavelength. These wave shapes also turned out to be consistent with the sine wave motion of the circle.

Not only were the electrical pulse and magnetic pulse wavelike and circular, but they were found to produce a harmonic between the electronic sine waveform vector and the magnetic pulse vector moving perpendicular to the electronic one. The concept of electromagnetism became grounded in the reality that there existed a precise harmonic oscillation between these two forces. They were connected to each other harmonically. Once it became evident that electricity could be drawn from nature in the form of heat or combustion, humans worked to develop technologies that tapped nature's electromagnetic potential. After natural power is drawn from the earth or sun, we can now manipulate it through the use of magnets, rotors, transistors and resistors. A new entire industry has developed around the manipulation of natural energy: We call it power generation.

The basic model of electricity is to convert natural sources of power into alternating current. The energy created by the kinetic energy given off from the burning coal or oil, the releasing of pressure from water-filled damns, or the drawing of thermal heat from the earth's crust contains the *potential* for alternating current. The pulse of these power sources just needed to be converted into alternating current through the use large turbine generators. These generators modulated the inherent rhythms residing in natural energy with rotating magnetics. The rotation of magnets modulates the field around the flow of current, inducing alternating current. This mechanism was founded upon Hippolyte Pixxi's 1831 *dynamo*, along with the early *alternators* developed in part by a series of inventions during the nineteenth century. Nikola Tesla finally patented these technologies into a useful generator model that converted kinetic or mechanical energy directly into alternating current using this rotating magnet technology.

By inserting magnets around the flow of direct current, early electricians were able to insert a periodic change in polarity—or magnetic direction—within the current flow. This periodic change creates a cycle of back and forth current. The amount of time between the forward motion and the backward motion—the change in polarity—of the current is known as its frequency. So while an electric current has a voltage potential on either side of it—its potential draw—it also has a frequency of oscillation.

In order to utilize this oscillating current flow, electrical appliances were designed to *resist* and *transist* the alternating current. This process streamlines and channels current for the appliance's specific use, essentially recapturing these pulses as mechanical or kinetic energy to drive the appliance. The process essentially breaks down the cyclical nature of electricity into its smaller harmonic portions of capacity.

To move alternating current any respectable distance without losing too much voltage along the way, the voltage must be boosted. This means the wattage or the intensity of the current also has to be raised, and its oscillation speed or frequency has to increase. This means inducing more oscillation, and increasing the voltage potential on the far side of the conducting path. The power then has to be *transformed* from a weaker local current to a high-power current, capable of being conducted quickly through wires to distant destinations.

In order to facilitate this movement, transformers were developed to boost or reduce the rhythmic pulse of electricity. Again, magnets and coils are used for this purpose. Magnetic coil-based transformers boost the current before transmission and reduce it as it arrives at its destination. Typical voltage levels of 720 volts (or higher for transmitting over longer distances) are typically used at the beginning of the current journey. In the local area of the house or building, it is reduced to 240. Finally, as it enters the building, it is stepped down again to the 120 volts standard for the United States. As the voltage is disbursed through panel into the household, it is reduced again slightly, down to the 110-volt potential for which most appliances are set up. (Different countries use slightly different power voltage standards, however. Some use 115 and some use 120 standards for appliances.) Still, the current induction and transforming process of modern electrical plants are still based most closely upon the ingenuity of the Serbian-American Nikola Tesla.

Transformer boosting increases the sine wave amplitude and frequency of alternating current along with the increased voltage potential. When the electric waveform is reduced, the transformer will split the wave's voltage down from 720 to 240 to 120 volts. The harmonic fre-

TRANSISTORS CHANGE THE WAVE FORM

quency of this current is thus established at 120 volts, although it could theoretically be reduced to harmonics of 60, 30, or even 15 volts.

At the lower voltage of 110, the current is now ready to enter our house or business because it is *in phase* with the strength of the voltage and frequency requirements of the circuits within the facility or house. In the United States, most houses are wired for 240-volt and 120-volt reception, but the environment around the wiring may bring it in at 115 or even 118 volts. This range is moot as it is dampened to the 110 volts, calibrating to the 110-volt appliances. A 240-volt service current will usually drop in from the grid, allowing this higher voltage alternating current to be used for larger appliances and equipment to drive more kinetic energy.

As the waveform of 110-volts enters an appliance, the energy is transformed through its resistors. Resistors are points in the circuit where the current is opposed. This will typically lead to, depending upon the type of resistor, a drop down of the voltage from one level to another. Meanwhile a transistor will convert or modulate current from one type of pulse to another. By putting resistors and transistors within a circuit, voltage is converted to the power needs of the (often integrated) circuits of the smaller device.

Today we use the semiconductor "chip" as the ultimate transistor, converting alternating current to square negative or positive pulses—considered either 1s or 0s. Using a variety of increasingly complex semiconductor materials, our computer manufacturers push the envelope for speed and heat-resistance. For many years, the silicon semiconductor chip was the ultimate transistor for providing the gateway for the 0s and 1s of our computer machine languages. Over recent years, hafnium has been increasingly used. This semiconductor material provides a faster transisting conductance with the least heat production.

As energy moves through nature, there are a variety of natural semiconductors within the process, resisting and transisting natural power. The gap between our information-age technologies and nature's own semiconductance of information is illustrated with the introduction of semiconductor nanotechnology. Dr. James Fraser Stoddart, a Scottish nanoscience expert, received the 2007 King Faisal International Prize in Science for his work on transistor nanotechnology. Working at the California Nanosystems Institute at UCLA, Dr. Stoddart developed a process for utilizing natural elements combined into complex molecules as information semiconductors.

Dr. Stoddart's process synthesizes molecular structures as small as four nanometers thick with gateway structures such as chemical rings and rotaxanes to provide mechanisms that recognize and respond to chemical bonding parameters. The semiconductance of the chemical bonds provides switches that can transist and resist the flow of electricity. These have been named *Nanoelectromechanical Systems*, or NEMS. This technology underlines the reality that nature contains complex transistor and resistor systems. This illustrates that we are only just becoming aware of the depth of nature's harmonic technology.

We can easily observe a number of natural resistors and transistors throughout our universe. The layers of atmosphere act as a natural resisting and transisting system. Cosmic rays such as gamma rays and ultraviolet rays are all channeled, modulated, or reduced as they hit the atmosphere. When we sit under the shade of a tree during a hot day, the tree provides a resistor from the hot rays of the sun. Reefs and beaches are resistors and transistors for the waves of the ocean, converting tidal and deep ocean waves to beautiful beach breaks. Our body's sense organs are resistors and transistors: They convert the rhythmic pulses of light, sound and so on from one waveform to another—providing complex information to the brain and mind. The entire biochemistry of our bodies—from DNA to nerve cells to hormones—all modulate and transist information semiconductance throughout the body.

While nature's elements provide natural forms of electricity and information conductance, human society has developed a massive enterprise surrounding the use of altered electricity. Our lives now revolve around the use of the electrical alternating harmonic derived from the naturally electric universe. While we revel in our seeming intelligence over this feat, we must not forget that both the power and the harmonic derived from the magnets' interaction with current are both produced by nature. Our power plants simply modulate the natural electricity produced by burning coal or petroleum; converting the sun's rays; drawing the electromagnetic power from the atom; or mining the thermal energy of the earth. Our oil- and coal-burning power plants draw the natural energy contained within hydrocarbon bonds found within the earth. Solar panels draw the natural thermal radiation of the sun using semiconductor conversion. Wind energy systems capture the kinetic movement of the atmosphere through foiled blades that mimic the wings of birds to turn turbines. A nuclear power plant uses the radioactive bonding activity from uranium dug up from the earth to ignite a chain reaction of nuclear fission (when nuclei split) to heat water. All of these natural forms of electricity are being transisted from one type of energy currency to another. Their natural alternating rhythms are being modified to suit our insatiable desire for increased physical power, communication, light and sound. In the end, we should know that diverting and translating energy does not create it. The energy was already contained within nature's harmonic. We simply have mined it, translated it into alternating current, and conducted it around through wires.

Thus we can say with confidence that no energy shortage exists.

The Magnetic

When we look at a magnet, we see a needle pointing virtually in one direction. This is assumed to be the North Pole—a stationary object—or so it seems.

Magnets were named after the lodestone, a rock found by the Greeks in the province of Magnesia. This was a curious stone, and it was immediately thought to have healing properties. The Greek philosopher Aristophanes began to explore this mysterious rock. Hippocrates was a great fan of the magnet for healing purposes. The Chinese had also used magnets for healing centuries earlier. Through the centuries, everything from heart disease to gout was treated with magnets in Chinese, Greek, medieval, and Ayurvedic therapies. In 1175, English monk Alexander Neckam experimented with and eventually described how a magnetic could be used in a compass. In 1269, Petrus Peregrinus de Marincourt described the pivot compass. William Gilbert's sixteenth century *De Magnete* described many uses for the magnet, and described the pointing of a magnetized needle to not only the north direction of the compass but also downward. From this, he proposed that the earth must have a *"magnetic soul."*

For the next three hundred years, magnets and electricity underwent greater focus and exposure. However, their combined effects were not considered until the fateful day in 1820 when physicist Hans Christian Oersted noticed that electric current moving through a wire moved a compass needle next to it. Over the next 80 years, a flurry of discoveries and inventions by the likes of Andre Ampere, Joseph Fourier, Georg Ohm, Joseph Henry, Jan Daniel, Michael Faraday and others delivered the concepts of electromagnetism from theoretical to application.

A static magnet is a unique material. Most of its atoms theoretically line up with their polarity pointing in the same direction. Every molecule and atom has a specific polarity. This means that one side of the atom or molecule will have a greater negative charge and the other side will have a greater positive charge. While we might be tempted to compare this to a ball with a north and south pole, the reality is bit more mysterious. Researchers have thus far described this effect within the theoretical context of quantum mechanics: The angular momentum of their spins and orbits appear to create a *magnetic moment* of sorts. This theory proposes that a substance's magnetic moment is proportional to the amount of angular momentum needed to create the polarity of the atom or molecule.

Magnetism in nature presents itself primarily in the form of the fields that form around electrical currents. Whether these currents are observed through motion or polarity, magnetism appears to be interconnected in nature and unable to be isolated from the motion of the living universe. When electricity moves in the form of a current, the angular-natured magnetic field spreads out in directions perpendicular to the flow of the current, effectively wrapping the current in a sheath of interactivity. This effect can be seen by wrapping an electrical wire around an unpolarized metal: As the current begins to flow, the metal will become magnetized. A softer ferromagnetic metal will produce even a greater magnetic field. While magnetic fields are not visible to the naked eye, they are observed by their ability to attract other metals, and pass that ability on to other metals.

The realization of magnetism's inter-relationship with electricity donned on Michael Faraday in 1831 when he coiled wire around an iron ring and demonstrated *induction* by passing a magnet through the ring. Faraday followed these demonstrations by calculating the relationships and proclaiming four calculations as the basis for what is now known as the *Field theory*. Faraday's proposals for a combined electromagnetic effect were not well received, however. As many discoveries are, they were not accepted for many years later.

The famous *homopolar generator* followed these theories. The electric generator remains even today as the primary method of advancing direct current through a circuit—the most basic form of electrical generation—still referred to as induction. Should that induction be directed through a circulating magnetic disk, the induction will begin to alternate. This circular rotating disk is often called *Faraday's Disk*. It is the predecessor to today's generator motors.

Within a few years of Faraday's work, Heinrich Ruhmkorff developed a higher-voltage pulse from DC current. The *Ruhmkorff coil* consisted of copper wires coiled around an iron core—very similar to Faraday's disk. As a DC current was passed through one of his coils, the current potential increased. When the pulse was shorted or interrupted, the immediate magnetic field decrease drove the voltage to jump into high gear onto the second coil. This could be arced out to an outlet line, producing a large spike in the voltage with pulsed, alternating current flow.

Various demonstrations of these properties led to a maturity of the notion that Faraday initially proposed—the concept that electricity and magnetism were inter-connected: A change to one provoked a change in the other. While the notion of light and electricity having combined electromagnetic properties were proposed by Faraday some thirty years prior, they finally gained scientific acceptance following the writings of Scot James Maxwell in the 1860s. Maxwell not only proposed the cooperating nature of electricity and magnetic fields, but also proposed that light was a combination of these two fields, pulsing through space.

The distribution of alternating current took a big leap when physicist and fire alarm designer William Stanley conjured a crude AC electrical installation at a New York Fifth Avenue store. Prior to that, the distribution of direct current was dominated by the technical and marketing savvy of Thomas Edison.

Again, polarity is the key ingredient of magnetism. In a typical physical substance, the polarity of the various molecules and atoms making up the substance are arranged in such a way that the poles of each molecule balance each other. This is primarily because negative and positive poles tend to attract each other. Solid structures tend to have structured patterns founded upon this polarity balance. This of course creates stability between the various atoms and their polarity. As a result, most solids are either *paramagnetic* (attracted to a magnet) or *diamagnetic* (repelled by a magnet) depending upon which way their polarity balance is trending. In a lodestone or ferro-magnet, however, there is a little less balance in the structure. Groupings of atoms align together with their poles pointing in one direction or another. These aligned groupings tend to overwhelm unaligned atoms of the substance, rendering a large section (oftentimes one end) polar in one direction or another. The ability to attract or repel other substances works by this strong polarity on one side of the magnet. As a result, the groupings of polarity will reorganize themselves on one side and another group will form with the opposite polarity on the other side, forming the well known bar magnet. The strength of a field created by this polarity is typically measured in gauss or Tesla. A Tesla is 10,000 gauss.

When an electronic pulse moves through a magnetic field, the electron flow will be drawn either away from or towards the magnet, depending upon their polarity. The forward motion of the electronic current creates an arc. This arc is a representation or byproduct of angular momentum. As electronic currents and magnetic fields interact together in nature, this angular momentum effect creates a display of synchronized duality. The dynamics of induction and the interfacing of angular momentum between natural forces of electromagnetic current have presented us with profound mystery and beauty. As the quantum theory suggests, polarity may be broken down into tiny units. Their individual electronic waves may be described as moving with a particular orbital angle and tendency. This notion of electron angular momentum creates the concept of orbital motion around a central nucleus. This reveals a potential tendency that electronic current has both circular and cyclical characteristics, driven by the polarity forces inherent in magnetism.

In a simple sense, if we were to consider the trajectory of an arrow shot upwards, it would continue in a straight line until it was acted upon by another force. Since an arrows path typically curves upward, arcs, and then curves downward, we can say that the force of gravity was acting in a direction opposing the (upward) direction of the arrow. As we look at the perfect arc of the arrow as it heads back to earth, we can understand the direction of the force of gravity. In the same way, the circular and angular momentum of the theoretical electron indicates its interactivity with an opposing force: magnetism.

When we say that the earth is a giant magnet (which by all observations it is—but so are all living organisms) we must again respect the relationship between electronic wave movement and magnetism. Geologists propose that the earth is magnetic due to the motion of magnetic metals within the surface—a sort of liquefied magnetic core. The concept of a core in motion stems from the fact that the earth's magnetic fields are not static, but are changing. Findings of mountain and desert core samples show that the earth's magnetic north and south poles have varied from the current poles over the past few million years. Now the magnetic North Pole is near Bathurst Island—equidistant from the north geographic pole and the Canada's Arctic Circle. The magnetic South Pole is closer to Hobart, Tasmania. In the late sixteenth century, William Gilbert measured the north magnetic declination at 10 degrees east. By the early nineteenth century it was 25 degrees west, and now lies about 6 degrees west.

Core samples with magnetometer readings have confirmed that the earth's magnetic flow has maintained the same approximate direction over the past 700,000 years. Before that, it appears the magnetic field direction of the earth has changed periodically. More than once the poles have completely reversed. Several times in the planet's history, the earth's major magnetic field even traveled from east to west. The current phase of the weakening of the field—some 16% since 1670—indicates a reversal or another abrupt change is in motion. Some estimate a complete phase change could come as soon as 2000 years from now.

The moving core theory is not the only theory that tries to explain the phenomenon of the moving magnetic earth. Some have proposed that the magnetic direction changes are caused by the movement of theoretical magnetic field loops circulating from east to west within the earth's interior. This rhythmic field loop has been compared to the looping magnetism that takes place between each side of a bar magnet as it changes polarity. Still others have proposed that an impact from a large meteor might be strong enough to change the earth's magnetic field. Whether these theories or another cause completely, the bottom line is that the magnetic fields of the earth appear to be rhythmic in nature.

The study of animal migration has revealed that migratory movement is connected to the earth's magnetic fields. After much debate regarding migration, 1974 studies at Cornell disclosed migration's link with magnetism. Researchers tied magnets on bird's heads and let them fly. The birds became disoriented. They had little or no directional basis for flight until the magnets were taken off. Further testing with birds and other migrating species such as lobsters have since confirmed that migrating organisms find their direction using tiny magnetic elements within certain cells, which coordinate somehow with subtle steering mechanisms. Certain nerve cells appear to contain magnetite metals, which orient with the earth's polarity to guide their migratory path—much the same way a compass might be used by a ship's navigator to steer a course over the sea.

Amazingly, the existence of magnetic cells has also been found in the smallest of species, and magnetic components have been found even within bacteria. Many bacteria could very well be compared to bar magnets, in fact. Tiny pieces of magnetite material will line up within the center of a bacterium—approximating a rudimentary spine. This observation has led some researchers to speculate that the human and mammal spine is also magnetic. This theory seems to be supported by the various clinical successes with magnetic spinal therapy. Indeed, researchers have recently discovered magnetic molecules such as iron oxide within certain human brain and spinal nerve cells. Thus we might conclude that our bodies also contain a magnetic compass of sorts.

Ironically, humankind utilizes magnetization as our central means for information storage. We magnetize various types of oxide films with special magnetic heads. Our computer hard drives, CD players, DVD players, and tape recorders all use magnetic heads to store information as digital polar molecules. As we increasingly move away from the paper society, we find our society beholden to these fleeting magnetic moments of data storage. One large magnetic pulse could easily wipe out gigantic chunks of critical data across our society—a magnetic tsunami of sorts.

Recent discoveries involving magnets and data include the finding of *single molecule magnets*. Professor George Christou and his research group at Indiana University have isolated several of these SSMs, allowing even more data to be recorded onto magnetic media. In the late 1990s, IBM accomplished a new record for hard-drive magnetic data storage, obtaining three gigabits—or three billion bits—of data storage onto one square centimeter of area of cobalt magnetic material. The SMM technology has since enabled the expansion of capacity to 30,000 billion bits—10,000 times the IBM record (Soler *et al.* 2000).

In considering our use of magnetized information, the magnetic material within brain cells, the pulsing of magnetically driven brain waves, and the magnetic nature of migratory travel, we can safely conclude that living organisms are magnetically aligned with nature.

The Thermodynamic

The first law of thermodynamics—a concept born in the middle nineteenth century by English physicist Dr. James Joule—concludes that regardless of the internal work involved in a system, there is never a net change in energy within that system. In other words, total energy is always conserved. Dr. Joule's law also reveals the rhythmic nature of energy: Energy can be cycled through various states yet its basic fundamental potency remains intact. While we consider thermal change to be tantamount within the discussion of energy, in reality heat can be converted to other forms of energy. The potential of the energy remains stable despite the eventual transmutation of the energy: Whether it is thermal, kinetic, electromagnetic or seismic.

When electricity is fully engaged within a circuit, heat is often one of the byproducts. This indicates a current's ability to conduct through a wide range of materials—either within or outside the intended circuit. Every form of energy—whether it be thermal, conductive electricity, or radiation of any type—has the ability to transfer to or create heat. This points to a unity of energy: All energy has the same functional basis. All energy has the same fundamental abilities to conduct, transfer, convert, and crossover through different states into different forms. This is because at its baseline, all energy is harmonic.

The second law of thermodynamics, put forth by nineteenth century German physicist Dr. Rudolf Clausius, presents a more philosophical stance towards energy and heat. This theory theorizes that spontaneous processes should yield an increase in entropy. In other words, accidental processes should always yield increased chaos. The inverse proof of this states that organization within an inherently chaotic system must have an outside source of that organization. As we observe complex and precise (even harmonic) relationships between matter in motion, a rather glaring question regarding certain thermodynamic law assumptions arises: *Does chaos actually exist?* In other words, *do spontaneous or accidental processes exist within our physical environment?*

While a modern scientist might scoff at such a question, it is certainly worth asking. The assumption of an ultimately chaotic universe necessarily puts the thinker of such a premise above it all. It presupposes the ability to peer beyond the universe into the realm of meaning and purpose. It assumes we already know there is no ultimate meaning or order within the universe. A truly scientific approach would not be so hasty.

The central fulcrum to this question is whether accidents truly exist. If accidents exist, then at least we could assume that chaos at least exists. One way to test an assumption of accidental behavior within our microcosm is to isolate one particular event we already assume is a random event, and test it for ultimate randomness. If an isolated event can be random, then we could at least accept the notion that accidents do exist.

Enter the *random event generator:*

For at least hundreds of years, researchers and philosophers have counted flipped coins, dice throws, and card deals in an attempt to reveal the underlying principles of random events. It was assumed that these events were ultimately random because there seemed no way to influence them. Each event seemed to be thoroughly disconnected and isolated from the other. Each throw of the dice appeared to be affected only by the potential options on each die. The previous throw seemed thoroughly disconnected from the current throw. This also held true for coin tosses. Despite only two options, each throw was seemingly isolated from the previous throw. Hundreds of coin tosses would indicate a growing trend toward a 50/50 split of the tosses. At least until the numbers started getting larger. As the coin tosses increased into the thousands, the differential between the theoretical 50/50 remained intact and even sometimes grew larger. For some reason, the tosses were not conclusively arriving at what most thought would be the random result: That as the number of tosses increased, the percentage of heads and the percentage of tails would increasingly approach 50/50.

In an attempt to resolve this seeming riddle of nature, Dr. Helmut Schmidt—then a physicist at Boeing—invented a machine he called the random event generator (or REG) in 1969. This device utilized a mechanical basis to produce a theoretically random flashing of one of four lights, using the decay emissions from the strontium-90 isotope. This produced a theoretically natural random event, as this decay was considered ultimately chaotic. The light system was set up to enable an observer to predict which light would come on by pressing a button under one of the lights. With a choice of four selections, the statistical average over a large number of guesses should be no more than 25%. However, large trial numbers resulted in levels closer to 27%, indicating some sort of inherent human ability to either predict or influence the result (Schmidt 1969, Palmer 1997).

The research became somewhat controversial due to this unknown: Were the observers predicting the results or affecting the results? Were they involved at all? In an attempt to isolate these factors, Dr. Schmidt refined the methodology and instrumentation to reflect the early coin toss experiments. His new random number generator (RNG) performed random calculations, resulting in either even or odd results. Using computer programming, large numbers of results could be compiled quickly and accurately.

Dr. Schmidt's series of studies with the RNG resulted in the same curious results as the coin tosses: As the numbers got increasingly higher, significant variances between 50/50 remained, staying within the 1-4% range. Could a person consciously will a coin to land on the heads or tails side? What about a large range of tosses? Could a person's conscious intention influence more even or odd results over a huge number of lightning-fast computerized runs?

Enter 1970s Princeton Professor Dr. Robert Jahn, who refined the random number generator research. Dr. Jahn improved upon the machine, increased the number of controls in the protocol, and expanded the range of its study. Like Dr. Schmidt's machine, Dr. Jahn's machines would randomly produce either a one or a zero in a random sequence. Dr. Jahn's software was careful to remove any possible source of bias removed. Additionally, Dr. Jahn's software could be run on smaller personal computers, while Dr. Schmidt's programs required a computer the size of an office to run.

As hundreds of RNG studies were conducted and compiled by Dr. Jahn and other researchers, a consistent result again emerged. RNG variances from 50/50 continued with larger runs. Trying to find association, Dr. Jahn began investigating outside events or environmental conditions to find a possible correlation with these variances. One of the first relationships Dr. Jahn discovered was related to the personnel attending or observing the RNG runs. Variances trended differently for female observers than for male observers, for example. Seeing a possible human influence, Dr. Jahn's trials began to ask observers to wish for one result or another. This nearly always resulted in larger variances: While some results trended towards the wished result, some trended in the opposite direction from those wished.

In other words, some observers could influence the RNG results more than other observers could. Note that observers were not physically able to affect the results. They were merely observing the results—primarily from within the room or near the room. As we will discuss more later, incredibly the RNG results were found to be connected with not only local human influence, but also with seemingly unrelated global events taking place thousands of miles away (Jahn and Dunne 2007; Dunne and Jahn 2005; Jahn and Dunn, 1987; Jahn et al. 2007; Dunne et al. 1983).

While the RNG research initially focused on the ability of humans to influence theoretically random events, another conclusion began to become evident: A theoretically random event—supposedly isolated and thus seemingly unattached to any other event—appears to be connected to apparently unrelated events, after isolating all known forms of bias. If any machine could prove the chaos theory it would certainly be the peer-reviewed, controlled RNG machine. Yet this was not the case. RNG statistics continue to reveal that its seemingly isolated events are connected to other events, as will be described further later.

The assumption of even a partially chaotic universe may well be false.

Associated with this somewhat questionable and philosophical second law of thermodynamics is the unproved assumption of the vacuum. The mystery of the vacuum is an old one: the Greek philosophers discussed it thousands of years ago, as did other scientists over the centuries. The concept of a space void was even considered taboo for centuries following Aristotle's *horro vacui* declaration, which stated that empty spaces were abhorred by nature. It was not until the seventeenth century German scientist Otto von Guericke invented the vacuum pump that the simple relationship between work and the gas-pressure-vacuum was bridged. Guericke's vacuum pump was accomplished by a cylinder and piston, which pumped out the air from two copper hemispheres. Still the mystery of the vacuum remains: What is left within the space of a vacuum? Certainly no gas contents or molecules should be left. As many have proposed, should not a pure vacuum collapse its container?

John Dalton recognized gas absorption by fluids in the early nineteenth century. This became part of the *Law of Partial Pressures*, and eventually *Dalton's Law*. The key component in Dalton's experiments was the pressure created when fluids were heated and steam was produced. He found that steam produced greater pressure due to the thermal effect. The relationship between pressure and temperature entered the lexicon of science.

Heat conduction can be very closely analogous to electrical conductivity. While silver, copper and aluminum are considered good conductors of thermal energy; glass, plaster, rubber, masonry, and wool are considered poor thermal conductors. Thermal conductors relate to the ability to transfer heat. Those substances without good conduction are considered good insulators. For this reason, wool is known as one of the warmer types of clothing and masonry one of the best insulating building materials. The body also carries on heat conduction and insulation. Fat is a poor conductor and therefore a good insulator, for example. For this reason, a layer of fat will keep the body warmer.

This concept of conduction in thermodynamics brings us to the point of *metabolism*. We often consider metabolism in the context of the biological organism. However, as we will discuss further, the carefully arranged movements of thermal conduction and convection—where thermal energy is transferred through airflow—indicate a larger cyclic motion of thermal flows moving within our environment. These thermal flows are precise, keeping migrating and non-migrating organisms nourished, hydrated and just warm enough to maintain life. Noting our inability to prove randomness among isolated events, we must consider the possibility that nature is arranged by design.

We also know that thermal energy is fundamentally rhythmic. This is illustrated by the fact that thermal radiation has consistent periodic waveforms. Thermal energy is typically radiated in the wavelength of *infrared*. As we consider the first law of thermodynamics in this context, the transfer of energy from one type to another within our environment indicates the cyclical nature of the energy moving within the universe. It indicates the harmonic relationship between the cyclic nature of heat, radiation, electricity, and kinetic motion.

The Periodic

Chemists have long stimulated our warrior mentalities with notions of violent chemical reactions. Most of us who took high school chemistry were introduced to the subject on that first day of class when the instructor demonstrated a spectacularly violent chemical reaction bursting into blue flames or a cloud of smoke. As our indoctrination to violent chemistry proceeded, we were told of the embattled reactions between the forces of *oxidation* and *reduction*. While oxidizers were introduced as those elements deficient in electrons or with negative charge, reducers were accused of being bloated with electrons, possibly in waiting to aggravate some poor oxidant. While oxidizers seemed noble enough, including chemicals like oxygen, nitrates, and chlorine derivatives, reducing compounds included rough riders like metals and fuels. When these two substance types clashed, the violence ensued. They might explode, or at the very least generate some significant heat.

The existence of chemical violence is certain should the scheming chemist get in the middle. In reality, the reciprocal reduction and oxidation of various substances occur rhythmically within nature's course. They are perfectly timed, balanced, and integrated within the larger ecosystem of cyclic behavior. Oxidation reactions lend negativity to needy compounds while reduction appropriately utilizes nature's positivity to compose a precise balance. Plants reduce carbon dioxide to carbons and sugars in the presence of sunlight, and oxidize water into oxygen using the sun's rays. That same oxygen becomes involved in respiration in mammals, converting sugars and oxygen into a reduction of nicotinamide adenine dinucleotide, triggering the famed ATP/ADP electron transport energy cycle. Alternating biochemical reduction and oxidation (sometimes shortened to *redox*) within nature is complementary and strategic: One triggers the other in a revolving, cyclical orchestration of congruent activity. The two reactions are not battling each other in a meaningless chaotic struggle for supremacy: They work in unison within a rhythmic array of balance and design to harmonically transmute energy.

When most of us look around, we see *objects*. Some objects are moving, while some are stationary. Objects like rocks, mountains, desks, and buildings seemingly stand still. Yet if we were to carefully watch those objects over time, we would undoubtedly see dramatic change to their shape and structure. If we were to take a snapshot of any stationary object, and compare it to the object one hundred years later we would see substantial change in the object if it were even still there. Did the object move? No, but the molecules and atoms of the object have. In this way, buildings collapse; desks rust; and mountains are worn away by the elements.

The atoms making up the objects around us are constantly being recycled. Electrons are being exchanged and atoms are being displaced or replaced. An atom may be part of one object one day and part of another object the next day.

Thus the objects around us are not so solid after all. An electron microscope journey through smallness reveals "solid" objects are made up of rhythmically gyrating molecules and atoms aligned structurally in spatially aligned blocks of resonating waves. Upon closer inspection of even the most "solid" of objects, these blocks of matter undergo tremendous multi-dimensional microscopic movement in fact: They each spin, gyrate, tumble, and orient in different formats to a particular yet regular rhythm.

As we look deeper, we find every "solid" in the physical world involved in a balanced orchestration of structured motion.

Fluids are obviously less "solid." Molecules and atoms forming fluids flow within polarity and positioning—rendering surface area and structure within their unyielding motion. While we might think fluids are made up of chaotically moving molecules, the implications of surface tension and clustering puts these notions to rest. Watching waves of water pound the beach illustrates the fitting combination of fluid strength and motion, as water surges with incoming crests and backs away in rhythmic recession. When we see clouds taking shape and reforming, we witness water's orderly conversion into evaporated structure to enable remote redelivery—a synchronizing conductance and transistance into cyclic rainfall that nourishes the land and organisms with specific seasonal routine.

Dalton's atomic theory, put forth by John Dalton in the early nineteenth century, stated that the tiniest indivisible pieces of matter could be assigned the unit *atom,* and matter must be made up of these indivisible units. Furthermore, he suggested, each indivisible type of atom—each type of atom a specific element—must be unique in its weight and number of components. The later being more specifically its atomic number, theoretically determining that atom's ability to marry other specific types of atoms, thereby forming specific types of molecular compounds. This new approach envisioned molecular compounds made up of atoms brought together by an ability to share sub-atomic particles. While Isaac Newton also theorized the atom centuries before, Dalton's theories—with his notions of sub-atomic *electrons*—brought a mathematical basis to the aspect of these minute structures.

In the late nineteenth century, the Russian Dimitri Mendeleev and German Lothar Meyer each independently configured a periodic table of chemical elements. It is said that Mendeleev's configuration dawned upon him as he awoke from a dream. The table was based upon the notion that aligning the basic elements into groupings by number also grouped their relative chemical and physical properties. Atomic mass was used as the vertical basis for the chart, with atomic numbers providing the sequence. Their reactivity with hydrogen and oxygen provided the basis for their horizontal arrangement.

As atomic numbers of the different elements were compared, it became apparent that somehow their bonding properties and appearance were connected to their atomic weight. As measurement procedures were developed, due in part to the pioneering work of T.W. Richards, who was most known for his work in standardizing thermometers and his work with a radioactive transformation of lead, which he called "radio-lead." In fact, it was primarily through the reactivity of elements with other elements that allowed chemists to develop their relative weights and their relative attraction to each other, rendering the basis for the lining up of the periodic table—a two-dimensional chart screaming of rhythm and of course periodicity.

As the notion of atomic mass underwent further query, it was proposed that the atomic number of an element is equivalent to the number of protons inside the atom. The mass was figured to be the number of protons plus the number of neutrons in the atom together, however. While each element has a unique atomic number, that same element could also theoretically have a different neutron count, depending on the circumstance. Thus, the mass theoretically varies to the neutron count. When an element's proton count differs from its neutron count, it is called an *isotope*. The isotope is considered theoretically unstable because it is ready and waiting to combine with other atoms to enable a sharing of sub-atomic particles. This sharing of sub-atomic particles, early atomic scientists presumed, created that stability.

Amazingly, the crude periodic table design of Mendeleev and Meyer created a standard still used and still very relevant to chemists. While it arranged the known elements into a congruent order, numerous new elements discovered over the years have all fit nicely within the same table logic. Furthermore, holes in the table have predicted a number of elements found later.

The chemical periodic table naturally arranges the alkali metals into the first column, and the alkaline earths into the second column, and as successive columns move right, transition metals gradually give way to metalloids, halogen gases, and finally noble gases in the last and eighteenth column. Furthermore, as polarities of each element are closely studied, it becomes obvious that as the columns move left, the electronegativity of each element increases consistently. This electronegativity gives the respective element a greater tendency of attracting electrons from associated elements. This characteristic renders it easier for that class becoming ions. Therefore, as the columns on the table move right, the likelihood of conversion to ions increases.

If such elements' characteristics are graphed on an X and Y-axis, the periodic table would take the form of a spiraling waveform. This is illustrated by the famed *Mayan periodic table,* in which the elements are arranged in circular shape with the smallest elements in the center.

As we step back and consider not only the circumstances for the periodic table's discovery, but nature's synchronistic organization of elements according to their sub-atomic numbers and weights, we might be surprised. Their characteristics and properties linked to column position provide additional consideration. Furthermore, within this periodic arrangement we find an unquestionable cyclic rhythm between the different element types and their properties. The chemical periodic table is an astounding testament to orchestrated harmonic within the arrangement of chemical nature.

One essential take away point about the periodic table is that because of the limitations in the human senses to observe atomic elements directly, we have devised an indirect method of calculating a mathematical relationships between each element. Utilizing a basic fundamental unit—the hydrogen atom—provided us with a basis for comparison between the other elements. Once this basic unit had been established, the relationships played out with proportional congruity. Every other element had a fundamental relationship with the hydrogen atom. The fact that the hydrogen atom was given the arbitrary unit of one is besides the key point that hydrogen provides a harmonic basis for the rest of the elements. This simple point illustrates yet another confirmation of nature's harmonic design.

The unit concept within the atomic number also unfortunately allowed science's imagination to picture *little subatomic* planets—or particles—encircling a central nucleus or sun, which also contained hard units (protons and neutrons). This unrealistic rendering haunts atomic physics even to this day. It was not hard to imagine the fabric of the universe being made up of little tiny rocks encircling other rocks, as we see rocks all around us. This *unit-rock* concept—the *particle theory*—became a beguiling accomplice in both chemistry and physics theory for the next two centuries.

The Atomic

Once Dalton's early nineteenth century atomic theory began to help us visualize how compounds might form, other chemists and began proposing subsidiary laws of behavior. These included Frenchman Joseph Prousts' eighteenth century *law of definite proportion;* Antoine Lavoisier's eighteenth century *law of mass conservation;* and Dalton's own *law of multiple proportions.* These rules became the guidebook early atomic theorists used to connect the dots for a particle understanding of life.

Then in 1897 English scientist Sir Joseph Thomson (also known as Lord Kelvin)—who won the 1906 Nobel Prize for Physics—passed cathode rays through a slit within a vacuum tube. Using magnetic fields, Thomson was able to bend the rays. This indicated to Sir Thomson that this must be caused by a particle with inherent electronic and magnetic characteristics. Sir Thomson went on to project that an atom must be made up of the electron units—comparing them to plums sitting in plum pudding. This theory became the *plum pudding model,* eventually abandoned in favor of the image Japanese physicist Dr. Hantaro Nagaoka referred to as the *Saturnian model,* and finally the *Bohr model* named for the work of Dr. Niels Bohr. This led to the *Rutherford-Bohr model,* which included Ernest Rutherford's concept of electrons orbiting the nucleus in classical orbital motion.

The Bohr-Rutherford model of the atom eventually gave way to the quantum model. Through the progressive theorizing of Paul Dirac, John von Neumann, Max Planck, Louis de Broglie, Max Born, Niels Bohr, and Erwin Schrödinger, statistical probability quotients were eventually applied to *wave mechanics* models in an attempt to assess the quality, quantity, location, distribution, momentum, spin, and other behavior of sub-atomic particles.

Many of these atomic theories were the result of bombarding elements with various types of radiation—scattering their subatomic particles and recording the results. This process led to correlations between atomic number and the type of chemical or gas formed by the scattering. From these *mass accelerator* experiments came relationships between the number of individual *orbit shells* or *valences* within each particular atom. It was thought that atoms would combine into molecules as atoms shared orbital shells and electrons of other atoms. Sir Thomson proposed the fundamental reason for this is that each element has a unique *ground state,* where it has a balanced number of electron units and proton units. Should that balance change, an *ion* (an atom or molecule without this balance) would occur. As the theory went, this resulted in the ion seeking to arrive at back at its particular balanced ground state by either finding additional electrons or protons (whichever was their deficiency) or sharing them with another hungry ion. Molecules, then, were assumed to be multiple ions locked together in a sub-atomic particle orgy.

In the late nineteenth century, Swiss Nobel Lauriat Dr. Alfred Werner suggested some sort of sub-atomic coordination between atoms that came together to form molecules. Dr. Werner proposed the *coordination number* to be the number of direct links between the atoms already within a molecule. The German chemist Dr. Richard Abegg a few years later noticed that the number eight came up consistently in molecular combinations. This combined with the observation that oxidation states between elements also tended to differ by eight provided the grist for Dr. Abegg to suggest *Abegg's rule of eight.* Later the American chemist Dr. Gilbert Lewis provided the calculations to formulate the *octet rule,* and suggested electron pairs are shared to the extent of the capacity of their valance shells. As these shells fill up they will ultimately establish the most stable state—that of a noble gas. Later this combination recipe became known as the *valence bond theory*, a the *rule of eighteen* became the model of metal stability. Interestingly, Dr. Lewis also proposed the curious notion of cubical atoms sharing corners with other cubical atoms, something that has long since been dismissed.

As images of molecular orbits were plugged into the unit theory of chemistry, harmonic electron orbit characteristics began to become evident. Valence orbitals—sometimes referred to as *clouds*—were described as regions where electrons were most likely to exist. These cloud unit combinations were found to be consistent with specific energy emissions following mass accelerator bombardment. For this reason, the clouds came to be referred to as orbital energy levels. These valence energy levels elegantly fit into the periodic table. When an atom seemingly garnered enough electrons to fill up its outer shell, more stability resulted. On the other hand, as stable atoms or molecules were bombarded, often energy was released, creating less stable ions. While still short on completeness, together the octave-based periodic table and the orbital valence shell model illustrated a working harmonic amongst the tiniest parts of our universe.

Accelerator testing and orbital energy calculations gradually revealed the inaccuracy of the Rutherford-Bohr atomic model. The energy emissions simply did not correlate with this classical particle view of little electron planets nicely orbiting a nucleus-sun. Dr. Bohr's and Dr. Einstein's notions of theoretical trajectory or *angular momentum* essentially scrapped the single orbit. These shared electron orbits simply did not seem to function like a particular "orbits." An electron's location within the orbit at any particular point in time simply could not be established. At best, physicists were left with mathematical probabilities of the elusive electron's location at any particular time. Even then, the elusive particle escaped objective sense perception. While a planet circling the sun might have a predictable location at any point in time, the locations of electrons could not be established. Something was not right.

Throughout the twentieth century, research physicists were intent bombarding chemical solutions of gas and liquid with radiation, measuring sub-atomic emissions and recording their trajectories using photographic film. By observing the subsequent diffraction and interaction of sub-atomic emissions onto film, oscilloscope or computer graph, it gradually became obvious that electrons participated and interacted with light in ways not unlike light rays themselves. Following an impact, for example, some particles would apparently scatter. By all observations, it appeared that the radiation was exciting the atom, causing a release of electrons. In other circumstances, radiation bombardments were observed creating new ions and subsequent molecular combinations. These responses indicated that light stimulated the energy change from within the atomic structure.

Early twentieth century researchers like Dr. Max Plank and Dr. Niels Bohr eventually observed that the energy atoms consumed or emitted could be precisely correlated with oscillating energy frequencies. This *radiation* energy was quantified using the *Plank's constant* and the *Rydberg constant* (named after Swedish physicist Johannes Rydberg). These quantifying relationships allowed atomic physicists to calculate a number of mathematical relationships inherent in different atomic emissions. These had wave character and little if any particle behavior. In other words, these tiny atomic bonds were not working quite like particles circling like planets—they pulsed more like waveforms of light and electricity.

The twentieth-century technologies of spectroscopy, chromatography, crystallography and electron microscopes afforded the indirect perception of an elegant myriad of rotational orbital clouds existing within the atomic world. Radiation bombardment using better technology allowed chemists and physicists to observe the existence of orbital clouds arranged within relationships of magnetism and polarity. The probable orientation of any particular electron shell or cloud appeared to relate to the relative polarity between the other parts of the molecule. The relative polarity between orbitals created various complex yet precise angles of positioning. While some held orbital shell angles of 90 degrees, others formed 120 degree angles, 45 degree angles, or even 109.5-degree angles. Each type of atom or molecular combination, it turned out, displayed a unique arrangement of orbital clouds.

The interactions between these orbital clouds in molecular configurations also revealed fantastic geometric shapes. Molecular bonds projected images of linear, trigonal planar, trigonal bipyramidal, octahedral and many other uniformly balanced and symmetrical structures. For example, a methane molecule is apparently arranged in a perfect ring structure. Its tetrahedral carbon-hydrogen arrangement creates a repetitive orbital-sharing matrix that reconnects at the other side with utter symmetry. Methane is one of a family of hydrocarbons such as ethane, propane, butane, pentane, and hexane. Each of these hydrocarbons also has a unique polygon form caused by this carbon-hydrogen orbital sharing system. While propane forms a triangular molecular shape, pentane forms a pentagon shape, and of course, hexane forms a hexagon shape.

These fantastic atomic geometric structures are even overshadowed by the many beautiful crystalline or lattice structures that provide the architecture of the three-dimensional physical world. Even a facile imaging of the fabulous shapes of snowflakes, glaciers, and diamonds illustrates the graceful and elegant configuration of nature's harmonic assembly.

The harmonic wave nature of the elements is illustrated by the early nineteenth century research of Oxford University physics professor Henry Moseley. Professor Moseley followed Mendeleev's periodic table construction with radiation emission measurements. Using x-ray diffraction, he determined that each element emitted a unique frequency of radiation. These frequency relationships, he found, also correlated with the orbital valence relationships. As atomic number count increased among the elements, the wavelengths decreased and the frequencies increased. Moving along the elements of the periodic table, frequency measurements increased in a stepped fashion and tapered with elements with complete valence shells.

Dr. Alexander Beddoe illustrated in 2002 that these frequencies can be correlated with sound frequency. Each row of elements is corresponds with a stepping up of octaves. While the fourth row of the periodic table (potassium, calcium, and so on) corresponds with the seventh octave, the third row (sodium, magnesium, and so on) coincides with the sixth octave. Meanwhile the second row correlates with the fifth octave and the first row corresponds with the fourth octave. The rows under the fourth row fall into the seventh through ninth octaves. Dr. Beddoe also charted these frequency-octave correlations, illustrating a helical shaped graph.

Precision and order at the atomic level is also evident when considering the accuracy of the atomic clockworks. In today's standard for timekeeping—the atomic cesium clock—radioactive cesium provides a steady stream of waveforms passing through a magnetic field to routinely oscillate a crystal. The emission from cesium is so rhythmically accurate that the world now quite literally sets its clocks to it.

The relationships between the elements and their orbital energy levels illustrate they have a common mathematical basis. When we look at these relationships using different models of measurement, we find the same common basis among the elements, indicating arrangement around a common integral multiple. When a molecule reaches a point through atomic combination where its orbitals are filled out to a point where stability is found, that indicates the next link in the sequence. In real terms, this means a fulfillment of the next step in the sequence illustrates that the elements are isolated, but are joined within an *array*. This array is illustrated by not just the display of the periodic table, but the unique mathematical relationships between the elements, and the arrangements of their combinations. This might be comparable to an instrument having sequentially arranged strings, with the first tuned to a basic note. Each sequential string can step-up the tuning from this first string to reach a full complement of musical notes, enabling the strings to be played together in harmony.

The Optical

It was the seventeenth century Dutch Christiaan Huygens who first proposed that light was a wave. A peer of Sir Isaac Newton, Huygens is said to have arrived at his notion by observing wave fronts as they expanded outward and interfered with other wave fronts among the waters of a nearby canal. His observation of wave fronts expanding into their own wavelets seemed to Huygens to correlate nicely with how light might travel.

Sir Newton meanwhile established that light was composed not of one simple white ray, but of a spectrum of colors. He demonstrated this by observing light refracting through a prism. Sir Newton also espoused in his 1704 classic *Opticks* that physical objects do not in themselves contain color. Rather, he suggested, color was made of "corpuscles." Some objects absorbed them while other objects reflected them. Those reflected colors, he supposed, created the illusion of that object's color. Sir Newton and Hyugen's works were either largely ignored or refuted vigorously by the mainstream scientific establishment of the time. In Sir Newton's case, though the spectrum was quite visible to the naked eye, criticisms were raised about his research methods and various suppositions.

The light-wave concept was further advanced by Dr. Thomas Young in the late eighteenth century—a century after Huygen's work. Dr. Young observed that if light rays were passed through a slot within a barrier, they would expand outward from the slot. If the same light were shone through two slots in the barrier, the resulting light ray expansions would create both light areas and dark areas on the other side of the barrier—a *diffraction* of light waves. Dr. Young observed that the light rays acted just as water waves might under the circumstances. Two different types of interference patterns emerged as the light shone through the two slots. In some areas the light waves interacted negatively and dark areas were formed. In other areas the light waves interacted in such a way that brighter areas resulted.

A few decades later, using calculations based upon earlier discoveries by Faraday and Oersted on the relationship between electric currents and magnetic fields, nineteenth century scientist James Maxwell created a formula implying light not only travels in waves, but also consists of dual electromagnetic waveforms. Using the velocity of light as a measurement basis, his new formula fit the observations of light waves as they exhibited the alternating duality of a wave oscillation. Maxwell and his peers proposed that light's pulsing dual waveform comprised of both electronic and magnetic components, moving perpendicular to each other.

Albert Einstein further proposed light traveling in units with wave-like properties in 1905 with his Nobel Prize-winning paper, _Corpuscular Theory of Light._ Later these same "corpuscular units" as espoused by both Newton and Einstein would be referred to as _photons_ or even _quanta._ Dr. Einstein proposed these photon units carried their energy potential until encountering atoms within the atmosphere (as he thought space was a vacuum). This encounter would either raise electron orbit energy levels or knock electrons out of atoms' orbits. Either way, the theoretical photon—racing at a theoretically consistent speed—would alter the atom's energy levels following such a collision. Einstein further proposed that photons were of a nature consistent with electron theory of the day: These photons were assumed as being both waves and particles simultaneously. This later became known as the famous _wave-particle theory of light._

The assumption of a consistent speed of light—one of the fundamental assumptions of the quantum view—has hit a snag. For over a century physicists have assumed light travels unchanged at close to 300,000,000 meters per second or about 186,000 miles per second. This speed is supposed to be regardless of the frame of reference or location of observation. Collaborative research led by Texas A&M University physics professor Dr. Dimitri Nanopoulos, Dr. Nikolaos Mavromatos of King's College in London, and Dr. John Ellis of the European Center for Particle Physics in Geneva confirmed in 2001 additional influences that alter the speed of light. Their calculations showed that the speed of light actually varies to frequency. Furthermore, in 1999 University of Toronto professor Dr. John Moffat showed evidence that the speed of light has slowed down over time.

Today the visible spectra is seen as composed of red, orange, yellow, green, blue, violet, and ultra violet waveforms. Each of these waveforms has a distinctive wavelength and frequency, which gives it a unique perception of color. The rate of oscillation is different between each color. In essence, each color beats to its own drum. The smallest wavelengths of light have been observed to have the highest energy potentials. Violet for example, has a wavelength of about 375-450 nanometers. Red has one of the longer wavelengths at 625-750 nanometers.

When color radiation strength is measured, violet has the potential to create more energetic change than the red part of the spectrum. Thus we can say that its wavelength is inversely relative to its energy potential. This relationship between the various color wavelengths and their ability to affect distinct electron energy orbitals indicates there is a relationship between colors and the periodic table.

Sir Newton proposed the spectrum of color could be arranged within a circle, with each color relating to a particular musical note and planet within the solar system. Although he missed several planets we now recognize, and his red range failed to reveal purple as it ranges to black, the concept certainly made sense to Sir Newton and his colleagues. As we broaden our view of the color spectrum with our awareness of polarity, we can reconsider this cyclical view as we ponder the aspect of polarity. While light's alternating magnetic and electrical properties seem to make it impervious to the effects of the earth's magnetism, color is perceived as light interacts with the elements in our atmosphere. Rainbows are a good example of this. As light refracts through water vapor, displays of majestically brilliant color result. The aurora borealis also illustrates interactive effect between solar influence and atmospheric elements, as solar waveforms are trapped within the magnetosphere of the earth. As the energy levels of these atmospheric particles become excited, fantastic shapes and colors are seen in the skies.

Light has been shown to bend via magnetic/gravitational influences. Light reflecting from mercury has been seen bending around the sun during a solar eclipse, for example.

Different substances refract and diffract light differently. This is because light will travel with different speeds and vectors through different mediums. As such, each medium has a unique *refractive index*. This relates directly to the substance's molecular makeup. For example, a diamond will refract light differently than a piece of glass might. Water refracts light differently than does air.

Sending light through various substances and then through diffraction gratings based on the principles of Dr. Young have became one of the standard techniques to determine chemical composition or compound purity. A molecule's ability to interfere with the path of light or other radiation as it passes through, it renders a means for identification. Because light interacts with each atom and molecular structure in a distinct manner, it enables us to quantify this characteristic. This is performed using a *refractometer*.

Visible light waves, for example, have particular polarity, depending upon their source, media and history. Light with different polarity can be *polarized* if its electronic waves and magnetic waves are standardized so that they are consistently at the same angle with the direction of the light. This can be accomplished by refracting the light at specific angles, or simply by viewing the light through a filtering mechanism that screens out waves that do not have the same polarity. These methods became the basis for the Polaroid camera and polarized sunglasses, which both screen out light waves that are not polarized. This is not a new concept to nature, however. Various animals and insects see with dramatic polarization. Examples include bees, octopus, and squid, among others. Humans can learn to distinguish light of different polarity with a little training apparently. Still, because we can see light of particular polarity, our eyes also conduct some polarization filtration.

Even with all our technology and discovery, our understanding of light is still unfolding. While we might assume our observation of light forms the standard, we may well be missing dramatic pieces of the puzzle, just as Sir Newton missed purple on his color chart.

The absolute harmonic of the visible spectrum is implied by the model of the sine wave: A congruency found not only in visible light, but throughout nature.

The Quanta

No one has exactly seen an atom face to face. Stumped by this reality, science is still debating: *What is an atom?* Newtonian physics models illustrated atoms not unlike little solar systems, with spinning particles circling a central nucleus. Eventually physicists like Bohr, Einstein, and Rutherford saw the shortcomings of this mechanical view. Radiation bombardment with particle accelerators revealed a link between light, radiation and the electron. This new perspective also brought into focus an element common among all these basic elements: the electromagnetic waveform.

At the turn of the twentieth century, the realities of electromagnetic oscillation and the atom began to unfold. Physicists could not help but correlate atomic theory with the same energies being observed within alternating electrical currents and radiation. The paradigm of electromagnetic oscillation was recognized as the common feature of every atom, every color, every chemical reaction, and every biochemical function. Combining observations with calculations, Dr. Max Planck's theory characterized the smallest unit of energy absorbed or emitted at the atomic level was characteristic of light emission. He captured this within a fundamental unit called the *quantum*. From that moment on, the connection between wave mechanics and the electron orbit of the atom became inescapable.

As particle accelerator instrumentation developed, physicists accumulated a complete array of characteristics linking sub-atomic activity with the elements of wave mechanics. While experiences with magnetism, polarity, and conductivity awoke nineteenth century physicists to the possibility of a tie between electricity and magnetism, radiation bombardment awoke twentieth century physicists to electromagnetic oscillation. As physicists like Bohr, Einstein, Planck, and de Broglie began to consider sub-atomic units with waveform properties, the perception of particle electrons and protons became increasingly fuzzy and theoretical. The location of theoretical particle became a probability rather than an object with a specific position. Meanwhile, the theoretical orbit of the electron expanded from a single line to an electron cloud.

The early twentieth-century proposal that atoms had electromagnetic characteristics eventually developed into the famous wave-particle theory of atomic structure. Dr. Albert Einstein's theories of *general relativity* and *special relativity* finally abandoned the Newtonian notion of movement. Newton proposed the forces of motion were relative to the positioning and movement of the observer. Dr. Einstein's relativity assumed that the universe was in constant motion. Therefore, every observer in motion will perceive the physical characteristics—such as speed, distance and time—differently. That is, unless the observers are moving in uniform motion (the same speed and direction). Furthermore, his special relativity theory proposed space to ultimately consist of a vacuum, and the speed of light was consistent for all observers. These assumptions concluded a rather fluid picture of the universe. The black and white of mechanical motion gave way to a universe in fluidic, rhythmic motion.

With the turn of the twentieth century came the publishing of Dr. Planck's famous quantum proposal. This theory proposed that every atom radiated a number of distinct energy components. He labeled these energy components *"quanta"* (Latin for "measurable unit") because he thought each sub-atomic particle should have a measurable and quantifiable amount of energy. Dr. Planck also defined a formula that arrived at a numerical value upon each quantum. This was called *Planck's constant*.

The constant proved important because it allowed Dr. Niels Bohr to arrive at his expanded *quantum theory*. Bohr's theory focused upon the notion that the electron's orbit was essentially electromagnetic. Dr. Bohr's proposal focused upon an atom's ability to release light energy or absorb energy as orbiting electrons were theoretically released, bumped up to a new level, or bumped down to a lower level. He also proposed each atom had a natural *grounded energy state*, where its energy level was balanced. He suggested that chemical reactions occurred primarily because electrons became either excited to higher quantum states or reduced to lower quantum states. Bohr's quantum model illustrated how these quanta were consistent with Lewis' orbital shell orbital models, as balance was reached when atoms shared electrons to fill out their shell potential. Dr. Bohr received the Nobel Prize in physics in 1922 for his theories.

In 1924, Louis de Broglie wrote a doctoral dissertation that stunned the physics world: He proposed that all sub-atomic matter was better described as composed of waves than orbiting particles. Using sub-atomic momentum and velocity measurements, de Broglie created a formula that calculated the wavelength of an electron.

The wavelength of an electron? This was hard to conceive or perceive in the particle view. Despite the dramatic research of physicists over the previous decades, the Newtonian model of the electron particle orbiting the nucleus was firmly implanted into the imagination of both the scientific community and the media. The particle image provided a solid approach for the chemical element table with its atomic numbers. The particle could be successfully used to describe the combination of atoms into molecules. Yet the impossible to perceive became the only reasonable explanation for the results. Research using diffraction among sodium chloride crystals cinched de Broglie's hypothesis. The *wave mechanics theory* of sub-atomic matter stubbornly rose into solid scientific view.

The quantum view of the atom took a turn in the mid 1920s when the work of Werner Heisenberg, Dr. Max Born and Pascual Jordan resulted in the *matrix mechanics* version of the quantum theory. This matrix version approached the combination of wave-based elements as time-dependent operators within a rotating reference frame to accommodate change. This was also called the *Heisenberg picture*, eventually rejected in favor of the *Schrödinger picture*, which assumed a system evolving with time and independent operators. Dr. Erwin Schrödinger proposed this new wave mechanic vision by unveiling the now-famous *Schrödinger equation*. His vision seemed to provide the solution to the mechanical problems existing within the quantum perspective. The wave and the particle could exist simultaneously, he figured.

Professor Heisenberg unleashed a little bomb that got physicists thinking about the problems of quantum perception. In attempts to determine the position of the theoretical electron, Heisenberg illustrated the mathematical impossibility of simultaneously determining the position and momentum of a sub-atomic particle. This famous proposition was termed the *Uncertainty Principle*. The principle was later ascribed as the problem of the major quantum numbers—driven by the electron's apparent dualistic wave and particle qualities.

The uncertainty principle became a call to reason for the new wave-particle theory. It was mathematically inconceivable to use the quantum theory to locate the particle, and it was inconceivable to attempt to measure the various energy components of sub-atomic wave probabilities using classical mechanics. The quantum theory was delegated as the opening of a new dimension of matter very unlike the "solid" physical world.

Nonetheless, on a practical basis, as has been proposed by a number physicists, that this may simply be an observer problem: In order to make a physical measurement or an attempt to perceive the sub-atomic particle in a certain position, the particle's motion becomes altered. While this can be surmised from the Heisenberg's proposition, the ability to ascertain the mathematical relationship between a particle's location and quantum component is outside of the observer's conditional range. Because a mechanical element is required for the momentum calculation, the two components (probability-wave positioning and particle momentum) cannot be sequentially abstracted. The wave-like quantum element must be collapsed to achieve the classical mechanical qualities of particle matter.

The mechanical particle concept was a difficult one to lose, however. The premise of the Newtonian particle has not been abandoned, even as today's teenagers take their first chemistry and physics courses. That hat of Young's famous double-slit experiment was hung onto the unlikely perception that electrons could be both waves and particles.

Yes, electrons acted like waves when they were shot through the two-gap opening. Their waves collided and refracted with each other, forming reflected images similar to ocean waves moving through two side-by-side openings. The confusing thing occurred when one of the gaps was blocked. The movement through the single gap was consistent with mechanical particles when the interaction with the medium on the other side of the gap was observed. In atomic particle bombardment research, sub-atomic behavior has revealed a similar incongruence: bombardments indeed indicate both particle and wave characteristics. The obvious characteristics of sub-atomic waves—interference, diffraction, and refraction— are nonetheless unavoidable. But so are the particle characteristics.

Although the wave-particle theory has been frankly described by physicists as *inconceivable,* it was has been the accepted theory for sub-atomic matter over the last half-century. Curiously, some physicists have explained this inconceivability with a twist of incomprehensibility: Sub-atomic elements *are particles sometimes and waves at other times.*

The proposal submitted by Dr. David Bohm in the early 1950s was that the particle aspect is simply *guided* by wave motion. This required, he theorized, a hidden *local effect* at work within the electron. Alternatively, he figured this also required a *non-local effect* guiding information between all electrons throughout the universe.

This possibly resolved the inconceivability of the quantum wave-particle theory proposed decades earlier. The four basic quantum numbers proposed a unit baseline for particle existence. The first quantum number is the *principal quantum,* described as the sub-atomic *singular* quality. Without first defining an electron or other sub-atomic particles as distinct units with singularity, no further quantum characteristics could be possible. While this singularity is given wave qualities, the first quantum supplants the particle aspect of that wave. The second quantum number is based upon the orbital qualities of the particle-wave, notably its *angular momentum.* This is also sometimes referred to as the second sub shell quantum. The next and third quantum number is the *magnetic projection of that angular momentum.* This represents the magnetic field projecting from the electron wave motion, being perpendicular to its aspect in electromagnetic tradition. The final of the four principle quantum numbers is the *spin quantum.* This spin characteristic provides the third-dimensional view of the orbital qualities of the wave-particle.

These primary quantum numbers provided a mechanical picture to the waveform characteristics consistent with the appearance of the oscillating, orbital, rotational, magnetic, and centrifugal electron.

As mass accelerator research progressed into the 1960s, Caltech physicist Dr. Murray Gell-Mann proposed the existence of several even smaller particle parts, theoretically contained within protons and neutrons. He supposed that if protons and neutrons could be quantified as particles, their scattering upon collision should enable a quantification of sub-components. These tiny particle parts were successively labeled *quarks, leptons, gluons, bosons,* and even *antiquarks.* Furthermore, these quantifying characterizations were expanded with other theoretical qualities such as *charm, upness, downness, strangeness, top, bottom, hyperchange,* and of course *flavor.*

Using linear accelerator collisions and subsequent formulae it was first proposed that quarks were thoroughly disconnected from each other. They apparently contained no charge. This circumstance presented a problem, however. Since quarks were conceived as elemental units without charge, they should also be able to separate from the proton quite easily. What force was keeping these theoretical units apart?

It was later proposed the quarks must rather be confined within the proton. This meant another force must exist within the *hadon* (a proton or neutron) that keeps the quarks inside the hadon separated. Otherwise, there would be a loss of singularity. This proposed force was called the *strong force:* The force holding the protons and neutrons together. This strong force had to be mediated, it was supposed. The *gluon* was proposed as the mediator or transfer agent for this strong force within the hadon. It was subsequently proposed that quarks could be *colored* (or given these strong charges) by these gluons—themselves colored particles.

The puzzle this presented was a picture where quarks are moving around freely yet contained within the proton. There must be another, weaker force keeping these quarks confined, these physicists figured. This force was labeled the *weak force*. It was suggested that these weak forces were mediated by the existence of yet another sub-atomic unit, the *boson*. Bosons were assigned activities such as nuclear decay and neutrino function. Instead of exerting a *color* as in the case of the gluon, bosons theoretically transfer *flavor* to these sub-atomic parts through the passing of weak forces.

Through this effort to quantify wave-like behavior using abstract and primarily symbolic unit representation, physicists have constructed a theoretical basis for measuring and predicting various nuclear activities. Whether these units actually exist is presumably beside the point. Regardless of not being directly observed, these quantifying virtual objects fit observation—allowing physicists to quantify previously undefined forces of nature. The quantification of quarks, gluons, gravitons and even photons and electrons allow a semblance of congruity for theoretical physicists. Thus we might conclude that quantum mechanics is essentially an attempt to physically explain something that is frankly beyond our comprehension.

In the early days of quantum mechanic formulation, there was heady disagreement between theoretical physicists on how atomic quantum characteristics were cast and retained, especially in describing particle-waves from different perspectives. One of the most famous arguments set forth by Albert Einstein was the concept that *"God doesn't play dice...."* This comment was characterized in a 1935 paper presented by Einstein, Boris Podolsky, and Nathan Rosen as the *"EPF Paradox."* This position argued that either the theory of quantum mechanics was missing critical elements (termed *"local hidden variables"*) or there was a broader *"non-local effect"* allowing particles to act and respond consistently—even at great distances from the observer. A non-local effect was presumed to be a force acting outside of the system, while a local hidden variable was presumed to be some hidden variable that filled the gap between the probabilities quantum mechanics provided.

This argument was embraced by Irish physicist Dr. John Bell in the early 1960s, who attempted to explain the existence of a non-local effect pragmatically. The basic assumption of the now-famous *Bell's Theorem* states that the gap between the probabilities of quantum mechanics and reality could not be due to local hidden variables. As this theorem was tested, *Bell's inequalities* were applied to two particles splitting off from each other. While these two particles may be independently tested to show no local hidden variables, they also may end up either neutralizing each other or otherwise reflecting each other. This implied their relationship continued beyond any effect of quantum mechanics. This led to the possibility that particles are able to instantaneously exchange information. This is often referred to as the *Bohm Interpretation,* and sometimes as the *seamless whole.*

Over the last decade, physicists have discovered other anomalies not envisioned in the classical quantum mechanic view. One example is *g-forces*. G-forces are theoretical wave clouds containing protons that apparently move about as electrons do, rather than remain in the nuclei as the Bohr model assumed. As their definite existence and location is merely probable, the possibility served to fill gaps in the quantum matrix.

While congruent with mathematical and linear accelerator application, the quantum mechanical view of the universe has yet to offer clear answers regarding the makeup of the universe. The parade of inventive particle parts and abstract jaunts such as the *membrane theory* and the various *string theories* provide creative yet unproven models in an obvious attempt to bridge the gap between the unknown and the observed.

On a more practical basis, what we have discovered from the study of sub-atomic acceleration and collision is that physical matter is inseparable from the wave mechanics of electromagnetism. Thus we can now take the characteristics uncovered regarding atomic arrangement and apply them more appropriately to waveform language: Atomic weight is a function of the element's frequency. The electron shells or orbitals are frequency levels. The reason we perceive particular valences among the shells within an atom is because that the waveform frequencies within a stable atom or molecule are harmonic: In other words, they have a basic underlying integral frequency multiple whereupon they become balanced or tuned to the rest of the environment. For this reason we find atoms or molecules with frequency levels outside of these valence shell multiples as imbalanced— and seeking balance among nature's sea of harmonic waveforms.

The Spectrographic

The oscillating nature of sub-atomic forces and electronic orbits became obvious with the development of the *crystal field theory* of the 1930s. When light travels through a substance, a portion will be absorbed by the atoms and a portion will be reflected back—depending upon the substance. A ruby looks red because the chromium in the ruby absorbs some of the blue-green wavelengths (around 490 nanometers) while reflecting back a greater amount of red wavelengths (around 650 nanometers). As these 650 nm wavelengths strike the retina, we perceive the color red. This technology is basically the same process of spectroscopy used today by chemists to determine the atomic makeup of a particular molecule or substance. Because atomic particles making up molecules interact distinctively with light, the molecular configuration of a substance can simply be identified by the wavelengths absorbed, reflected, and/or diffracted.

All imaging for atomic identification or structure is performed through the interaction between radiation and substance-matter. One such method is called *x-ray crystallography*. X-rays are shot into a crystallized version of a particular substance glued onto the glass of a diffractometer tube. The x-rays react with atoms within the molecular substance and the waveforms of these diffracted rays are recorded onto film or into a computer. These diffraction recordings are measured for amplitude and waveform to yield the theoretical atomic structure. Because x-rays are short-wavelength electromagnetic waves, they interact with the electromagnetic waves within the electron clouds. As these interactions occur, they are absorbed or diffracted in a variety of different angles. These angles can be plotted out onto photographic film or computer imagery to display the shape and probable location of the electron clouds.

The resulting crystallographs can indicate any number of angles of wave diffraction. Using diffraction measurements and a formula created by William Bragg and his son in 1913, these plotted angles measure the level of constructive or destructive interference between the x-rays and the sub-atomic particles of the substance. Constructive interference creates a resulting larger wave while destructive interference creates a smaller wave result. The type of interference is often a factor of the extent the waves are in-phase or out-of-phase with each other. This in turn relates to polarity, which indicates the possible orientation of the electron cloud.

X-ray diffractions have revealed numerous molecular structures over several decades. X-ray crystallography has identified various proteins and enzyme molecules, and revealed the fantastic DNA double-helix structure. These ghost-like reflective crystallographic molecular images reveal atomic structures quite different from the ball and stick models we saw on our chemistry professors' desks.

Another tool physicists and chemists use to perceive structure at the atomic level is the *electron microscope*. The electron microscope is an instrument that fires thin magnetically directed cathode rays into a substance. Their interaction with the substance is then reflected and recorded onto an electronically charged screen. Here a series of magnifications allow the scientist to observe what we could call an electromagnetic shadow of the substance's atomic response to the ray. Over the years of comparing and analyzing this interactive viewing of thousands of substance reactions, the electron microscope has enabled us to image atomic structure in extremely minute detail.

These techniques are similar in concept to spectrometry, which again is the process of analyzing substances by their ability to either absorb or react to certain radiation. It is not unlike the process of visual reception the eyes perform. While one substance will absorb sunlight's radiation, another substance will reflect parts of that radiation, producing the perception of colors and shapes.

Waveforms used in spectrometry include sound waves, radiowaves, microwaves, and x-rays. The spectrometer has a waveform sensor plate like the electron microscope or x-ray crystallographic equipment might have, but the sensitivity of the plate or scope (sometimes oscilloscopes are used as viewing media) will be set to the waveforms of the radiation being used.

A specialized form of spectrometry emerged after World War II called nuclear-magnetic resonance (NMR). This technique is actually a form of spectroscopy. Instead of firing continuous streams of radiation at a substance to elicit electromagnetic responses, NMR beams are magnetized and polarized during the firing process. This increases the likelihood of the beam *resonating* with the electromagnetic wave character of the electrons. Once a resonance has been established, the scientist can make a number of assumptions regarding the qualities of that particular substance's atomic structure.

Yet another development in the area of visualization in the sub-atomic landscape is *electron spin resonance* (or ESR) spectroscopy. By comparing the radiation response of materials with assumed spin characteristics, the spin of a new substance can be approximated.

While these observational techniques appear quite technical, they still do not give us a clear understanding of electron orbits and other sub-atomic matter. The complex quantum mechanics view of these media results tend to govern our perception. Because sub-atomic structures are moving in the *ultra-fast* range of motion, these visualization techniques are abstract approximations. Some physicists have suggested that quantum electrons move in the *femtosecond* (quadrillionth of a second) range, while atoms move in *picoseconds* (trillionth of a second).

Recent attempts to capture the elusive electron include short-burst technology developed at the Brookhaven U.S. Department of Energy

National Laboratory. The short bursting technology was developed by Dr. Yuzhen Shen, Dr. Larry Carry Carr and a team of technicians (2007). These electromagnetic pulses beam in the teraherz range (between microwave and infrared). As the short bursts are fired at substances, the researchers are discovering a cross-phase modulation of the electron waves. This produces a *nonlinear optical effect*. A nonlinear optical effect is a theoretical breaking apart of the various charges composing the atom. This effect creates the theoretical potential to reveal still unresolved aspects of electron composition.

An even newer technique of viewing sub-atomic elements is the *scanning transmission electron microscope,* or STEM. The STEM system fires electronic beams through a sample substance on a thin film. As the beam moves through the substance, the resulting spectrographic response is measured. While some parts of the atoms will absorb the energy from the beam, other parts create specific emissions, which are collected through a detector. The detector picks up and measures the energy signatures of each emission and absorption.

In 1995, respected physicist Dr. Milo Wolff said, *"It is now seen that the electron consists entirely of a structure of spherical waves whose behavior creates their particle-like appearance. The correctness of this structure is supported by the physical laws originating from this wave structure, including quantum theory, special relativity, electric force, gravity, and magnetism. This type of structure is termed a Space Resonance."* Dr. Wolff also pointed out that while the wave-structure of light and electrons is elegant and consistent with wave mechanics, the particle portion of the wave-particle theory presents puzzles, paradoxes, and complexity: *"Clifford, Mach, Einstein, Wyle, Schrödinger all pointed out that only a wave structure of particles (matter) can conform with experimental data and fulfill the logic of reality and cosmology."*

Dr. Wolff's statement certainly captures the essence of logic. Newtonian particle mechanics are simply not consistent with these observation-reflections made at the sub-atomic level. At its fundamental essence, sub-atomic matter becomes available for indirect imaging only when it interacts with radiation. What does this tell us about the composition of sub-atomic matter then? While we might think the hand blocking the light to form a shadow on the wall is solid, we should realize that the only reason that light cannot pass through the hand is because the particles of the hand are interacting and absorbing the light's radiation. This absorption prevents the light from emerging from the other side of the hand as it might in the case of glass, for example. As we come to understand the elements of wave mechanics, we realize that in order for waves to interact and interfere, they must have coherence. As we will discuss later and throughout this text, coherence requires first that both energies are ultimately waveforms. Secondly, the two energies must have some common baseline or harmonic wave structure in order to either constructively or destructively interfere. Thus we can safely arrive at the logical conclusion that in order to obtain duplicatable and precise absorption/emission results using the various forms of spectroscopy and crystallization techniques, we must be dealing with sub-atomic matter composed of the same waveform structure as the radiation we aim at the sub-atomic.

We might compare the reflective technology of spectrometry as detective work performed well after the crime was committed. We did not see it happen, and we can only make assumptions based upon various clues. The problem is that we did not see who was responsible for the crime in the first place, so we could easily be led astray by misleading evidence or missing evidence. We could well be sniffing the evidence with incorrect assumptions.

We might also compare these electron visualization techniques to looking at a shadow of a hand puppet on the wall. As the light moves from its source onto the wall, part of it is blocked by the hand. The light blocked by the hand allows for a dark shadow on the wall. Should we only look at the wall we might not know a hand is creating the shapes. We might presume the character we see on the wall is the real thing rather than simply being a shadow of a hand. Even if we guessed the shape was a shadow of a hand, we still would not know just by looking at the shadow whose hand it was or how the hand was contorted to made the shadow's shape on the wall. We'd have to turn the lights on for that information.

The Relativity

The basic force of gravity has perplexed humankind for thousands of years. While easily observed, gravity still evades comprehension. This is precisely why Dr. Albert Einstein spent nearly thirty years of his life trying to connect gravity with electromagnetism. While he established a number of *field equations of general relativity* in an attempt to establish mathematical cohesiveness, the single unification between gravity and electromagnetism evaded him.

After centuries of ancient scientific endeavor, including Galileo Galilei's reputed sixteenth century ball droppings in Pisa; Sir Isaac Newton proposed the *universal gravitation theory*, composed of the inverse-square law and planetary gravitational orbit hypothesis. This proposal became the foundation for Einstein's field equation work, which described massive orbits and predicted the existence of black holes.

Through the last part of the twentieth century, Einstein's field equations and Newton's gravitation theory are loosely tied together and packaged within a theoretical mechanical particle model called the *graviton*. Still, Einstein departed from Newton's model substantially. Einstein's theory of *general relativity* states that gravity is not due to forces pulling two masses together as hypothesized by Newton and Galileo. Einstein's general relativity postulation proposed that gravity is caused by the four-dimensional space-time system curvature responding to motion through time. The field equations and the theory were considered so complex Sir Arthur Eddington joked that no more than three people in the world understood the theory. This probably holds fairly true today.

Recently the assumptions of general relativity are being challenged. Louisiana State University professor Dr. Subhash Kak has published findings introducing an element not previously considered in the relativity model: *Time relative to a distant star*. In Dr. Einstein's equations, the relative ages of two people were compared: One relative to a traveling spaceship and the other relative to the earth. By introducing aging relative to a distant star, the relativity model approaches a broader perspective for time and motion.

Though little understood, we know gravity synchronizes with the orbital motion of the solar system. In turn, the solar system's gravitational forces coordinate with the movement of the solar system within the galaxy. As such, we can assess gravity as a force working in conjunction with various other planetary forces. Gravity also synchronizes with the magnetic fields circling the planet. Thus, we can safely conclude that gravity is connected harmonically with these other forces. They are synchronized, because they work conjunctively and effectively. Therefore, these forces all follow some common baseline fundamental.

The Rhythms

When most of us think about waves, we think of the ocean. We think of waves pounding onto the beach. Stirred up by the forces of wind and weather, large waves will march onto the reefs and beaches, standing up with ferocious crests. The beauty and power of a large wave lifting and crashing onto the rocks or beach is often the subject of popular photography and film. What we may not realize is that each single wave is communicating an event that took place thousands of miles away: A particular mix of wind, temperature, atmospheric pressure and moisture combining in just the right way to instigate a *weather system*. This weather system converts its potential into waveforms in the surrounding ocean waters. Should we look at this confluence of elements from space, we will see nature's characteristic spiral. Harmonically, we see this same spiral shape within a cross-sectional view an ocean wave.

A wave is a repeating *oscillation of energy:* A translation of information through a particular medium. Waves can travel through solids, fluids, gases, thermals, or electromagnetics. Waves are not restricted to a particular medium, either. Most waves will move through one medium and as that medium connects with another medium, continue on within the next medium. A sound, for example may vibrate a drum skin first. Where the drum skin meets the air, it oscillates the gases in the air, carrying the information through the medium of air. Where the air connects with the *tympanic membrane*, the information waveform is translated to the *malleus, incus* and *stapes* of the middle ear. After vibrating through to the *round window*, the oscillation is translated through the cochlea into nerve pulse oscillations. During its journey, the original wave of the drum beat transverses each of these mediums before being sensed by us.

A repeated oscillation or waveform through a medium against the backdrop of time is a *rhythm*. This tendency of to move with repetitive pace translates to a continuous waveform and rhythm. As a particular waveform is associated with an event, we can say that rhythm is the repetitive cycling of information waves through the physical world.

Every movement in nature has a signature rhythm: The earth oscillates in specific types of seismic waves—some causing damage but most hardly noticeable. We each walk with a signature pace as our feet meet the ground. Our vocal cords oscillate to the reflection of our thoughts with a unique pace and timing. Our heart valves oscillate with the needs of circulation. Our lungs oscillate as we breathe in and out—unique to our lung size and cells' needs for oxygen. Even rugged, seemingly solid structures like rocks oscillate—depending upon their position, size, shape, and composition. A cliff by the seashore will oscillate with each pounding wave. A building in a windy city will uniquely oscillate with the movement of the wind through the streets. Each building will oscillate slightly differently, depending upon its architecture and location.

All of these movements—and all movements in nature for that matter—provide recurring oscillations that can be charted in waveform structure. Moreover, the various events within nature come complete with recurring cycles. While many cycles obviously repeat during our range of observation, many cycles have only recently become evident, indicating that many of nature's cycles are beyond our current observation range.

Natural oscillations balance between a particular pivot point and an *axis*. The axis is typically a frame of reference between two media or quanta. An axis showing quantification may illustrate time in reference to height, time versus temperature, time versus activity or time versus other quantifying points of reference. Waves will also transist between media. The ocean wave is the transisting of waveforms between the atmospheric storm system. The water's surface tension gives rise to the ocean wave as it refracts the pressure of the storm system. The storm system's waveform will eventually transist through the ocean to the rocks and beach.

Nature's waves are relational to the rhythms of planets and galaxies. These rhythms translate to electromagnetic energy and kinetic energy, which translate to the elements of speed, distance, and mass. Momentum, inertia, gravity, and other natural phenomena are thus examples of the cyclical activities that directly relate with nature's wave rhythms. Every rhythm in nature is interconnected with other rhythms. Like a house built with interconnected beams of framing, the universe's rhythms are all interconnected with a design of pacing within the element of time.

The most prevalent waveform found in nature is the sinusoidal wave. The sinusoidal wave is the manifestation of circular motion related to time. Thus the sine wave repeats through nature's processes defined by time. For example, the rotating positions of the hands of a clock translate to a sinusoidal wave should the angles of the hand positions be charted on one axis with the time on the other axis.

Sinusoidal waveforms are thus the typical waveform structures of light, sound, electromagnetic waves and ocean waves. Late eighteenth and early nineteenth century French physicist Jean Fourier found that just about every motion could be broken down into sinusoidal components. This phenomenon has become known as the *Fourier series*.

The cycle of a sine wave, moving from midline to a peak, then back to midline then to a trough and back to the midline completes a full cycle. If we divide the wave into angles, the beginning is consistent with 0 degrees; the first peak is consistent with 45 degrees, the midline with 90 degrees and the trough with 270 degrees. The cycle repeats again, as we make another revolution around the circle.

Other wave types occurring in nature might not be strictly sine waves, yet they are often sinusoidal in essence. The cosine wave, for example, is sinusoidal because it has the same basic shape, but is simply *phase-shifted* from the sine. Other waves such as square waves or irregular sound waves can usually be connected to sinusoidal origin when their motion is broken down into composites.

We see so many circular activities within nature. We see the earth recycling molecular components. We see the recycling of water from earth to sea to clouds and back to earth. We see planetary bodies moving in cyclic fashion, repeating positions in periodic fashion. We see the seasons moving in cyclic repetition. We see organisms living cycles of repetitive physical activity.

While not every cycle in nature is precisely circular—the orbits of planets or electron energy shells for example—they are nonetheless linked within a grander cycle. Linked cycles often contain various alterations as they adapt to the other cyclic components. This modulation can be described as *adaptation*—a harmonic process observed among both matter and life.

This all should remind us of the notion of the *circle of life*, which has been repeatedly observed throughout nature in so many respects that it is generally assumed without fanfare. Circles recur in human and animal activity, social order, customs, and individual circumstances. We cycle emotionally and psychologically. The tribal circle is common among many ancient cultures—and for good reason. In modern society, we have circular conferences, round-table meetings, and cyclical ceremonies. The potter's wheel, the grinding wheel, and the circular clock are all examples of circular symbols in our attempt to synchronize with nature. Just about every form of communication and transportation is somehow connected to circular motion. For this reason it is no accident that the wheel provides our primary means for transportation. The motion of walking is also circular/sinusoidal, as the legs rise and fall forward, rotating the various joints.

In nature we observe two basic types of waves: *Mechanical* and *electromagnetic*. A mechanical wave moves through a particular medium: sound pressure waves as they move through air, for example. Mechanical waves can move over the surface of a medium. Ocean waves and certain earthquake (seismic) waves are examples of mechanical *surface waves*. Another type of mechanical wave is the *tortional wave:* This mechanical wave twists through a spiral or helix.

The electromagnetic wave is seemingly different because it theoretically does not move through a medium of any composition. Einstein physics assumes space is a vacuum and electromagnetic waves move through this vacuum. A heady debate regarding the content of the medium of space took place during the late nineteenth and early twentieth centuries. Dr. Einstein's proposal that space is a vacuum was made several times, including his *theories of special relativity*. According to the theory, time and space in this vacuum of space are collapsed: Instead of time and distance being separate, they were theorized as a singular element of *space-time*. However, now that a constant speed of light is in question, the notion of a wave being able to travel through a totally collapsed vacuum would be inconsistent with the wave mechanics within time.

Nature displays two basic waveform structures: *transverse* and *longitudinal*. Visible spectrum, radiowaves, microwaves, radar, infrared and x-rays are all transverse waveforms. As these waves move, there is a disruption moving at right angles to the vector of the wave. For example, should the wave move along a longitudinal x-axis, its disruption field would move in the perpendicular y-z axes. This might be compared to watching a duck floating in a lake strewn with tiny waves. The duck bobs up and down as the waves pass under the duck's body. In the case of the transverse electromagnetic wave, the disruption field is the magnetic field.

The other waveform is longitudinal. Here pressure gradients form regular alternating zones of *compression* and *rarefaction*. During the compression phase, the medium is pressed together, and during the rarefaction, the medium is expanded outward. This might be illustrated by the alternating expansion and compression of a spring. Instead of the wave disturbing the medium upward and downward as in the case of a transverse wave, the medium is disturbed in a back and forth fashion, in the direction of the wave. Examples of longitudinal waves are seismic waves and sound waves. In the case of sound waves, air molecules compress and rarefy in the direction of the sound projection.

These two types of waves may also combine in nature. An ocean wave is a good example of a combination of transverse and longitudinal waveforms. Water may be disturbed up and down as it transmits an ocean wave, and it may convey alternating compressions and rarefaction as it progresses tidal currents.

Waves are typically referred to as radiation when the waveform can translate its energy information from one type of medium to another. In this respect, seismic waves and ocean waves can be considered radiating as they translate their energy onto the sand in the case of ocean waves, or through buildings in the case of seismic waves. The classic type of radiation comes from electromagnetic waves such as x-rays or ultraviolet rays, which can travel through skin or other molecular mediums after transversing space.

Waves are typically measured by their wave height from trough to crest (*amplitude*), rate of speed through time (*frequency*) and the distance from one repeating peak to another (*wavelength*). Waves also may have specific types of specific wave shapes such as sinusoidal, square, or otherwise, as we've mentioned.

The frequency of a wave is typically measured by how many wave cycles (one complete revolution of the wave) pass a particular point within a period of time. Therefore, waves are often measured in CPS, or *cycles per second*. The *hertz*, named after nineteenth century German physicist Dr.

Heinrich Hertz, who is said to have discovered radio frequency electromagnetic waves. Note that hertz and cps are identical: Both the number of complete waves passing a given point every second. Other frequency measurements used include machinery's RPM *(revolutions per second)*, special radiation's RAD/S *(radians per second)*, and the heart's BPM *(beats per minute)*.

Wavelength is frequently measured in meters, centimeters, or nanometers to comply with international standards. Each radiation type is classified by its wavelength. A wave's wavelength has an inverse relationship to its frequency. This is because a shorter wave's length will travel faster through a particular point than a longer length will. Note also that speed is the rate measured from one point to another, while frequency is the rate of one full repetition to another past a particular point. Therefore, a wave's wavelength can be determined by dividing its speed by its frequency. Of course, a wave's speed can also be divided by its wavelength to obtain the frequency.

Despite popular science literature's penchant for naming only one aspect of a particular wave (often either wavelength or frequency), we must consider the various other specifications of a particular wave to have a useful understanding of it. When we describe a sinusoidal waveform, however, we can state either its frequency or wavelength, since the two will be inversely related. Otherwise, the wave's amplitude is an important consideration, as this relates to the height of the wave from peak to trough. Among sinusoidal waves, larger amplitude will accompany a larger wavelength. We also might consider the specific *phase* of a wave, its medium of travel, and again its wave shape. These together will help us arrive at a more precise set of specifications to more accurately describe the nature of a particular wave. While one or more of these specifications might be used to describe a particular wave in popular media, in reality all should be considered. Here we will refer to a unique combination of these specifications with the term *waveform*.

Waves travel with some form of repetition or periodicity. The very definition of a wave describes a repeating motion of some type. This repetition, occurring with a particular pace and particular time reference, together forms a rhythm. All around us we see wave rhythms. Can waves be chaotic? To the contrary, it is their very consistent, non-chaotic rhythm that allows us to interpret light, color, sound, or warmth with duplicatable precision. All of these waveforms connect with the senses because they have consistent and congruent oscillations. In sensing the world around us, we do not perceive each wave individually. Rather, we perceive the information the wave is carrying.

When an information-containing waveform collides or *interferes* with another information waveform, the result is often a more complex form of information: A combination of the two. As waveforms collide throughout our universe, they comprehensively present a myriad of complex information conductance. This information is only available to us to the extent we can sense and interpret those interference patterns, however.

We can thus surmise that nature is composed of various combinations and interactions of these two types of waves (longitudinal and transverse) with various waveforms. The classic waveforms vibrating through space and radiating through physical molecules may be fairly easy for us to isolate, chart and measure. Within nature however, waves collide and interfere with each other to create various effects we simply cannot measure. Colliding waves interfere with each other's continuing motion in some way, forming a multitude of complex interactions.

The more precise reason why not all waves are obviously sinusoidal is that nature is complicated by these different types of interactions between different types of waveforms. When dissimilar waveforms collide, there is a resulting disturbance or *interference pattern*. Depending upon the characteristics of the two colliding waveforms, this interference could result in a larger, complex waveform—or *constructive interference pattern*. Alternatively, should the waveforms contrast each other; their meeting could cause a resultant reduction of waveforms—a *destructive interference pattern*. The ability of two waves to interact to form a greater waveform lies within their similarity of wave *phase*. If one wave is cycling in positive territory while the other is cycling in negative territory as the two collide, they will most likely destructively interfere in each other, canceling some or all of their effects. However, if the two waves move in the same phase—where both cycle with the same points on the curve—then they will most likely constructively interfere with each other.

As a result, interference between waves can be *in phase* or they can be *out of phase*. In phase waveforms will often meet with superposition to form larger, more complex waveforms. Out of phase waveforms will often conflict, reducing, and canceling part of their effective rhythms in one or many ways. This canceling or reduction of interfering waves is not necessarily bad, however. Destructive interference can also communicate various types of information.

The degree that two or more waves will interfere with each other—either constructively or destructively—is their *coherence*. In other words, if two waves are coherent, they will greatly affect each other, creating either a greater resulting pattern or a canceling and reducing pattern between them. Waves that are too different to create any significant interference are said to be *incoherent*. This term usage is very similar to how we describe comprehension in language. If someone considers another person's language or sounds to be coherent, it means the sounds are better understood by the listener. Whether the communication is interpreted by the listener positively or negatively, the clarity of the communication is indicative of its coherence. This is analogous to wave mechanics coherence. Coherent waves interfere either constructively or destructively as they interact.

Resonance occurs when individual waves are expanded to a balanced state—one where the amplitude and period is the largest for that waveform system. Thus, *resonating waves* typically occur when waves come together in constructive interference. This results in a maximization of their respective wave periods and amplitudes. This is illustrated when two tuned instruments play the same note or song together. Their strings will resonate together, creating a convergence with greater amplitude, which will typically (depending of course upon the surrounding environment) result in a louder, clearer sound. We also see (or hear) this when we create the familiar whistling sound accomplished by blowing into a bottle spout: To get the loudest sound, we must blow with a certain angle and airspeed—positioning our lips with the shape of the bottle. Once we find the right positioning, angle and speed, we have established a resonance.

As waves move from one media to the next, they will partially *reflect* or *refract*. *Reflected waves* will bounce off the new medium, while *refracted waves* will move through a new medium with a different vector and speed, depending upon the density and molecular makeup of the medium. The ability of a particular medium to provoke these changes is referred to as its *index of refraction*. Some mediums will reflect certain waveforms while refracting others. Most mediums will also absorb certain types of waveforms, as we will discuss further. The type of waveforms reflected and absorbed will usually determine the medium's perceived color and clarity.

As we have mentioned, *surface wave* is typically seen at the surface of a particular medium where that medium interfaces with another medium. Surface waves are seen on the surface of oceans and lakes, for example. Surface waves are mechanical in nature and thus tend to respond to surface pressures from the colliding medium. In the case of surface seismic waves, the collision of the wave with a building may mute the waveform with one type of building material and exaggerate the waveform with another type of building material.

Surface waves are divided into two basic types—the first being a *capillary wave*. The capillary wave is a smaller wave that forms during the beginning of the build-up process. Therefore, it is considered a smaller wave—often seen as smaller ripples on the water as the wind freshens. The second type of surface wave is a *gravity wave*, typically having a larger wavelength and speed than the capillary wave has. It is the gravity-type *rogue waves* seafarers respect for their shipwrecking abilities.

In deeper water, a combination of transverse and longitudinal wave motions combine to form *monochromatic linear plane waves*. This is a type of wave called an *inertial wave*. Inertial waves are typically moving within rotating fluid mediums. Inertial waves are common in not only the ocean and lakes, but also within the atmosphere and presumably within the earth's core. The various currents and winds within the atmosphere all travel in varying length inertial waves. In a medium, the surface waves will interact with these inertial waves to move energy over the surface of the medium. This energy movement allows surfers to ride a wave from the outside to the inside of the tidal region, for example.

A *simple harmonic* is a recurring wave (usually sinusoidal) that repeats its own rhythmic frequency. When different waveforms converge and their frequencies are aligned—they are multiples or integers of each other—their waveform combination becomes *harmonized:* There is a mathematical integral multiple between them. In other words, harmony is based upon waveforms having a multiple of the same fundamental base. For example, waveforms with frequencies at multiples of a particular waveform will harmonize. Other waveform aspects of a wave can also create harmony. Waveforms having the same amplitudes or wave shapes—or multiples thereof—will be harmonic on different levels. Though their resulting pattern might not appear the same as a frequency-harmonic combination, we can still recognize their interaction as harmonic.

As forward-moving waves interact with returning waves, both waves will become compressed and dilated. This effect is known as the *Doppler effect*—named after nineteenth century Austrian physicist Johann Christian Doppler. If the incoming waves have the same waveform, frequency, and amplitude, this will create a *standing wave*. If they do not, either the incoming or the outgoing wave will divert the waves it meets, and distort those waves in one respect or another. This distortion would be analogous to the oblong or parabolic orbitals of solar systems and electrons, which are distorted from sinusoidal by oncoming waveforms or magnetic fields.

Standing waveforms will typically have the same frequency, wavelength, amplitude, and shape as they oscillate. This creates a balance and resonance that gives the perceiver (using retinal perception) the illusion that those standing waves are solid objects.

As suggested by Zhang *et al.* in 1996, and confirmed by multiple physicists over the last decade, multiple electrons within shared orbitals among multiple atoms situated within a close-range matrix are best described as *multiple standing waves:* These are standing waveforms within minute space. They create some of the strongest forces in nature, as they compile the illusion of physical reality. The convergence of these multiple waveforms standing together in a harmonic resonating pattern for a unique period of time is best described as *architecture.*

Harmonic synchronized and resonating waveforms also continually undergo *displacement,* thus transferring their energy through convergence or interference with other waveforms. This convergence or interference displacement conveys the rhythmic energy and information traveling through these waveforms on to others, mixing to create the resulting cyclic nature of our universe. These conveyances would also render the perception of *movement* within the universe.

As researchers have probed deeper into the nature of energy and matter, we have found that everything around us is oscillating at a particular rhythm. All information and movement moves within waveforms. Even nonmoving 'solid' matter has the slower, standing waveform movement as mentioned above, even while it undergoes larger life cycle waveform phases. For example, we might see a mountain as non-moving, but in actuality, it is not only rotating with the earth through the pacing of time, but the mountain is undergoing a hardly-noticed cycle of growth due to the earth's motion of heat and lava. This growth inter-cycles with the earth's recycling of molecular components, creating gradual reformation due to water, winds and weather. Finally, this 'solid' mountain may undergo a dissolution phase during a volcanic event, with its molecular waveforms cycling into fluid and gas states.

When we see an object with a consistent color, we recognize this color consistency because the frequency of the reflective color waveform is oscillating repeatedly. It has a particular consistency or pace through time. If we were to look at a mirror with a mirror inside the reflection view, we will see a repeating pattern of whatever is inside that second mirror. This repetition forms a visual wave pattern. It may not appear oscillating to our vision, yet its repetitious nature creates a waveform. We will also note in this array, the mirrors become increasingly smaller as they repeat. The vision of the repeating mirrors is not simply repetitive: It is rhythmically reductive—starting from the larger to the smaller with a precise pattern of reduction. This reducing pattern is also rhythmic and harmonic—allowing us the opportunity to see many mirrors in array within the one mirror view.

In 2002, Gabriel LaFreniere unleashed upon the internet a series of wave models incorporating Doppler effects and interacting standing wave patterns. Through his models and calculations, he was able to present a visual case for the increasingly accepted hypothesis that electrons making up matter are standing waves, oscillating within a magnetic field of *aether*. His proof is called the *LaFreniere's wave*, a single basis wave pattern of standing waves in accordance with the Lorentz equations. His visual images are quite stimulating (LeFreniere 2002).

The Spirals

As coherent waveforms of nature interact and interfere, spirals develop. For example, when we see the rhythmic spiraling growth of leaves or branches in Fibonacci sequence around the trunk of a tree we are presented with an interference of Fibonacci waveforms and the vectors of growth coming from the living plant. This interference creates a spiral or helix pattern of leaves growing upward and/or outward. Should we spread out the spiral orientation of a plant into two dimensions—x and y coordinates—we would find that the branching reflects a sinusoidal wave pattern. Should we look down at the plant from its apex, we would see this spiraling or helical effect, depending upon the size and nature of the plant. Looking at a younger plant—where we could see the top shoots with respect to the bottom trunk—we might perceive a spiral. Should we look at a larger tree with a large trunk at the bottom with its branches swirling and widening to the top, we might perceive a helix.

These helical and spiraling forms provide the basic structures for life within the physical world. We see these structures present from the smallest elements of life to the largest elements, from the double helixed DNA molecule to the spiraling galaxies of the universe. We see the spiral within all types of anatomical shapes. The nautilus shell is most famous, but just about every shell formation also reflects this spiral. The swirling of tornadoes, hurricanes, and weather systems mirror this spiral as well. Our senses are also tied to rotational spheres and spirals. Our cochlear hearing system utilizes a spiral to convert air pressure waves to neuron impulses. Our eyes are circular, spiraling through the pupils to the retina, bending light with filtration before they hit the retinal cells. When we consider the twisting and bending path of light through our visual senses, we ultimately arrive at the trigonometric sinusoidal and spiral. Other displays of nature's spiraling energies include the magnetic fields of the earth; animal displays of spiraling structures such as claws, teeth, horns, irises, ear pinea, and fingerprints.

Just as the sinusoid wave is derived from the circle, the classic spiral may be derived from the sphere. Beginning at any one of a sphere's apexes or poles, a spiral is formed if we move around the curvature of the sphere towards the opposite poles. Hence this most basic type of spiral is known as the *spherical spiral*. The spherical spiral is also known as the *arithmetic* or *Archimedean spiral*, named after the third century B.C.E. Greek mathematician Archivedes of Syracuse. In this spiral, the distance between each layer (or spiral arm) is held equidistant. This creates an angular moment that is consistent throughout.

There are various other types of spirals in nature. The Fibonacci sequence is often displayed within either helical or spiral forms. *Fibonacci spirals* are close relatives to the *logarithm spiral*. *Fermat's spiral*—named after sixteenth century Frenchman Pierre de Fermat is related to the arithmetic spiral. In 1979 Helmut Vogel proposed variant of Fermat's spiral as a better approximation of nature's Fibonacci spiral. This is the spiral observed within *phyllotaxis* forms, which include sunflowers, daisies and certain spiraling universes. Around 1638, Rene Descartes revealed the *equiangular spiral*. This spiral reflects geometrical radii outward as polar angles increase. The relationship Descartes discussed (S=AR, which Evangelista Torricelli also developed independently during that era) has also been called the *geometrical spiral*. Edmond Halley's seventeenth and eighteenth century work revealed the *proportional spiral*. Jacob Bernoulli developed its logarithmic basis, revealing the *logarithmic spiral* shortly thereafter. Bernoulli gave it the namesake of *spira mirabilis*, meaning "wonderful spiral." It is said Bernoulli's fascination of the spiral led to his request it be engraved on his tombstone.

As pointed out by Giuseppe Bertin and C.C. Lin in their 1996 book <u>Spiral Structure in the Galaxies</u>, the spiraling galaxies may well be generated through a combination of density waves that rotate in a slower rhythm than the rest of the galaxy's various stars, planets and gases. The *density wave theory*, first proposed in 1964 by C.C. Lin and Frank Shu, explains that the harmonization of the angular paths and the mutual gravitational attraction of the galaxy's components form areas of greater density. This allows the spiral arm formation without a *winding problem*.

Much of nature is arranged in helix or spiral shape. What may not appear to be is most likely should we peer through its cross-section. For example, an ocean wave breaking over a reef may appear to be a half waveform as it is looked at straight on from the beach. However, a cross-sectional view of the same wave reveals its spiral motion: As we watch the water falling from one side of the crest—crashing into the trough from the side, the spiral effect comes into view. The combination of the forward movement of the wave to the beach and the sideways movement of water along the crest creates the surfer's classic spiraling *"tube"* or *"barrel."*

To ride the tube or barrel requires the surfer to stay just ahead of the final eclipsing of the water with the trough. Should the surfer lapse into the center point of the spiral, the surfer will most likely be separated from the surfboard and experience the *"wipe out."*

The hurricane is a similar example of this effect. Waves from two different pressure and temperature fronts interact to form the classic cyclone effect seen from satellite. The hurricane's spiral is only visible from above. This means for thousands of years humankind had no direct awareness of this spiraled form. Looking at a hurricane front from the land renders a view of the coming front 'wave' of rain and wind from the storm. While many speculated about weather systems as they experienced the 'eye of a hurricane,' and compared this with that other classic spiraling interaction of waveforms—the tornado—it was only as humans began to take to the air that these beautiful spiraling images unveiled themselves.

D'Arcy Wentworth Thompson's 1917 classic *On Growth and Form,* and Sir T.A. Cook's 1903 *Spirals in Nature and Art* both illustrated the many examples of nature's tremendously precise spiraling waveform elements. Mr. Thompson details how various elements in nature have a tendency to coil, such as hair, tails, elephant trunks and cordiform leaves among others. Other interesting helix and spiral movements include the spiraling burrowing of rodents and the spiral swimming of dolphins and whales.

Let us not forget nature's molecular spirals and helixes. The most famous of which is DNA and RNA—storing and processing the programming of metabolism.

In 1973, Dr. Michael Rossmann reported the finding of a protein structure where multiple coiling strands are linked together with two helical structures. The connection between the strands and the helices were found to be alternating, forming an available structure for nucleotide bonding. This structure proved to be an important helical structure: Nicotinamide adenine dinucleotide (NAD) is a critical coenzyme involved in cellular energy production and genetic transcriptional processing within every living cell.

We have since found that numerous other biochemical molecular structures are actually helical when we are able to observe their *tertiary* structure. Various polysaccharides, polypeptides, hormones, neurotransmitters and fatty acids produced by the various living species have helical molecular structures. Many electron clouds have also proved to have helical or spiral shape. Certainly, this spiraling micro universe of atoms and molecules would have astounded the generations of thoughtful scientists over the centuries.

At the same time, physicists have been disturbed by the *paradox of the spiral*. This issue was noted by Dr. Einstein, who concluded this in discussion of the related *Faraday disc problem*, as apparently a radial conductance: *"It is known that Maxwell's electrodynamics—as usually understood at the present time—when applied to moving bodies, leads to asymmetries that do not appear to be inherent in the phenomena."* He went on to propose *"asymmetry"* arises when the currents are produced without a *"seat"* of forces. As we have previously discussed, the magnetic field tends to exert a force vector moving perpendicular to that of electrical current. As this happens, angular momentum is inferred from the induction. When the torque of angular momentum arising from the conducting Faraday Disc is considered together coherently interacting currents and fields, the dynamic of the spiral becomes evident (Serra-Valls 2007).

If we overlay nature's spiraling motion within the cyclical structure of the molecular world, a tremendous symphony of alignment becomes apparent. As we observe this mysterious collection of movement and symmetry, we naturally seek to find the basis for the underlying tendency of nature to harmonically move into a twisting wave motion concluding in rhythmic spirals and helixes. Dr. Einstein proposed that we accept that many of the movements of nature are either moving relatively too fast or too slow for us to observe. We can apply this to our understanding of the universe's wavelike, spiraling oscillations. For this reason, we find our research only gives us momentary glimpses of the complex harmonic existing within the universe.

As we peer deeper into the electromagnetic interactivity between electronic current and its reciprocating magnetic fields, we unveil—through an analysis of induction—the potential disturbance they create. This implies a designed functionality between conduction and induction. All of nature's wave and spirals oscillate within a cooperative context. The fields of one electromagnetic rhythm will affect surrounding electromagnetic rhythms. The inter-relationship between these rhythm motions indicates that their ability to continue their cyclical harmonic oscillation in each other's presence is ultimately organized on a grander level. A chaotic arrangement with this level of replicating depth of organization would simply be an oxymoron.

We can ponder more deeply the reality that a sphere is really a three-dimensional circle. When we measure radiation on oscilloscopes or computerized data we are essentially measuring a two-dimensional view of a force that likely has a third dimension. Playing out this perspective illustrates that the magnetic field moving away perpendicular to the electronic vector in any electromagnetic wave would create a helical and spiraling arrangement.

Regarding the apparent differences between the spiral and the helix, we point out that when cross-referenced with the axes of time and space, a helix will convert to a spiraled helix. This may require a cross-sectional view to complete the image, however—just as we explained with a beach break wave. If we are looking at the wave fully breaking (say a *"close out"* wave) at once, we see a cylinder from the beach. It is the view from the wave, looking down the length of the wave from the side that brings the spiral into view. In the same way, as time and space accumulates helical motion along one axis or another, we observe that one end of the helix will be more relevant than the other end. This progression may also be perceived as the helical arms expanding outward as the helix approaches us. Even without this relativity of motion towards the perceiver, we can understand this effect. We can illustrate the effect of time and space in motion by observing a train approaching us. The front of the train as it approaches us is closer and thus appears larger than the rest of the train. Though the train size is consistent through all cars, our perception at that point of time is of a large locomotive and small caboose. In the same way, would we be able to travel through time within the motion of the spiral we would arrive at a helix from our perspective, and vice versa in the inverse. Likewise, a person caught in the eye of a hurricane or tornado will not perceive the funneling shape of the storm. The helical or spiral view would only be perceivable from a distance or from above, respectively.

If oscillating waveforms are information carriers, spirals imply a conglomeration of waveform interference and thus a complexity of information. We might compare a waveform to a word and a spiral to a paragraph or even a book. We could pull a word out of a book but it would have no meaning without the rest of the arranged words of the book. While combined waveforms form interference patterns, spirals are the standing representations of harmonious waveform interference.

Information is conducted through waveforms. The medium or space-time may be disturbed by information waveforms, but the information does not move the medium. The information-wave *passes through*. The analogy that Dr. Einstein and Dr. Infeld used in *The Evolution of Physics* was to compare a wave to gossip that travels from one person to another, even over large distances like city to city, yet the people who communicate the gossip move very little. In the same way, we can understand that the spiraling waveforms of nature are information carriers.

And those *electron orbit clouds* with their various amazing geometric wave shapes? As we mentioned, these are illustrative of zones where nature's rhythmic influences interact to form resonating and sometimes standing waves. Like spirals, orbitals are created by wave interference and confluence. The shape and angle of the orbital depends upon the confluence of reciprocating waveform events on a harmonic scale. How do we know orbitals occur on a harmonic scale? The valence shells of twos and eights are specifically that: Harmonic frequencies of repeating integral multiple waveforms. As these standing harmonic wave orbitals are bombarded with radiation as demonstrated in mass accelerators, their resonation is disturbed, releasing radiation emission of precise measure.

Consider observing standing waves within a small pond of water caused by dropping a couple of small pebbles into the pond. While we see waves appearing to be standing in the same place as they meet each other, their motion is still reflecting the original pebble drops that disturbed the water. In other words, whatever motion is on the water reflects the original cause of the motion. If we were to drive a car over a cliff into the pond below, the resulting large waves within the pond will also reflect the weight and volume of the car. Those waves may reflect onto the basin at the side of the pond and possibly also form a standing wave pattern. This will be quite different from the pattern formed by the pebbles, however.

The oscillating elements of nature are all spiraling waveforms; each having a particular rhythm, along with a unique combination of specifications like wavelength, frequency, amplitude, field, and phase—as well as spiral characteristics such as radius, loci and angular velocity. As spiraling waveforms interfere, the more complex interference patterns among molecular orbitals and physical life forms form.

Most consider rhythms to be exclusively a characteristic of music. Certainly a gathering of interference patterns between harmonious rhythms creates a melodious song. However, is this not what is occurring all around us? All of the various waveforms of color, temperature, wind, motion, sound, heat and light are radiating and interfering harmoniously to form the song of the physical world.

We might ask ourselves what information is being communicated through these harmonic waves of the natural world. Put another way, what song is playing?

The Coherency

The discovery and utilization of electromagnetic frequency has enabled humans to successfully communicate sounds and images with precision over long distances. Contrary to popular thinking, radiowaves, microwaves, x-rays and others were not invented: They were discovered. Nature produces these types of radiation, and the discovery of them allowed humans to utilize radiation to communicate over longer distances.

Pictures and sounds are converted to electronic pulses and broadcast to distant receivers. Once received, the pulses are converted back to sound or picture images through resistance and transistance using semiconductors and integrated circuits. At first glace these communications may seem incredibly technical. Once we stumbled upon these different forms of natural radiation after tinkering with electricity for a couple of centuries, we were able to utilize and manipulate nature's technology. For this reason, we note the various contributions of many scientists in establishing these communication systems. Hundreds, even thousands of researchers have tapped into these natural rhythms in one respect or another. Eventually a few ambitious scientist-businessmen exploited the collection of discoveries to commercialize the technology.

We should note that the discovery of radiated communications and our ability to use it has not necessarily been the result of our understanding it. The development of radio wave signaling by Gugielmo Marconi and Ferdinand Braun during the late nineteenth and early twentieth centuries led to the 1909 Nobel Prize in Physics for the wireless telegraph. The telegraph was developed over a hundred years earlier through the combined tinkering of Francisco de Salva, Alessandro Volta, Samuel von Soemmering and Johan Schweigger. These efforts began with simple electrostatics and further developed into the manipulation of currents and voltage. Heinrich Hertz introduced the existence and detection of electromagnetic radiation in the late 1880s. An Italian physicist and university professor Augusto Righi investigated their optical properties in the years following. He published no less than 200 papers on the subject. While Marconi developed methods of electric wave propagation, it was Righi who improved upon the technology. Righi was able to increase wave stability and reception clarity.

Edouard Branly, utilizing some of the work of Italian physicist Temistocle Calzecchi-Onesti, eventually assembled a crude radio transmitter. This became referred to as the *Branly coherer*. It utilized current resistance to transmit alternating radiowaves across an electrode. It was Englishman Dr. Oliver Lodge who coined the term *coherer*, postulating that the medium through which the radiowaves transmitted was the *aether*. Dr. Lodge is thought to have demonstrated wireless transmission prior to Marconi; and quite possibly Nikola Tesla may have demonstrated the first wireless radio wave communication in 1893.

The key element was added around 1898 by German Ferdinand Braun, however. Braun invented the *cat's whisker-crystal diode rectifier*. This formed the basis for the *crystal radio receiver*, which utilized a natural crystal as a semiconductor. The natural crystal was positioned to receive and conduct radiowaves through contact with a thin bronze wire—the *cat's whisker*.

Eventually a tuner was installed to fix the radio crystal and whisker upon a particular station emitting radiowave pulses. The crystal semiconductor converted these waveforms into electrical pulses, driving a speaker. Marconi, the ultimate businessperson, assembled the various equipment—much of it under patent by the original inventors—and combined them with existing telegraph technology to send and translate real-time communication.

It should be noted that the signaling system on both ends must be grounded to the earth: Nature provides not only the facility for semiconductance but also the grounding for the electrical pulses to provide the right polarity. Early semiconducting devices were made from crystals of natural minerals such as galena or pyrite. Prior to the cat's whisker crystal radio, other minerals like silicon carbide and vitreous silicon were also used as crude semiconductors. All of these of course preceded the use of elemental silicon for the semiconduction of integrated circuit and microchip technology.

The ability to broadcast communications riding on the bandwidths of nature's radiowaves was an invention of equipment only. The equipment enabled humans to utilize natural radiation as a carrier for information communication. With various experiments and mechanisms developed through trial and error, inventors and physicists have been able to build upon this natural technology. The technology itself is utilized by nature through the transmission of voice, energy, and heat. Moreover, if we examine the timeline of the inventions and some of the theories proposed by Bohr, Einstein, de Broglie, and others, the mechanical ability thoroughly preceded humankind's understanding of radiation technologies. As we have discussed, physicists are still debating many of the basic theories regarding radiation and communications. In other words, we still do not completely understand these natural waveform technologies.

Radiation transmission is really an innate rhythmic process natural to all living beings. Radiation communication can be compared to the barking of a dog, or the tapping of Morse code. The sending of the signal is the act of the tapping or barking, and the receiving of the signal is the act of sensing or hearing those rhythmic waveforms, followed by a translation of the information communicated. This allows the sender and receiver to have an intended communication, filtering out other communications with a *handshaking* protocol. As long as each party agrees on how the rhythm is to be converted, communication can be extensive and personal.

A television camera converts visual data into a series of digital pulses. Those pulses are converted through electricity into broadcast radiowaves. A television set is the receiver and converter of those radiowaves, translating the pulses back into digital mode. As long as the television receiver is set up with the same conversion coding (or handshaking) used in the camera to broadcast signal conversion, it can convert the pulses into the right images.

If we were to analyze voice or even Morse code on a two-dimensional oscilloscope, we will see the same thing we see when we look at an electronic broadcast transmission: A series of waveforms. As a signal is converted from electronic broadcast pulses to voice and pictures, the information contained in the waveforms are converted from one carrier system (radiowaves) to another (digital integrated circuitry). During this conversion the information is conserved yet the waveform may look altogether different as it changes from radio wave to electronic pulse to projected pictures and sound waves. This is the *transmission* of information through radiation. The information is the energy being transmitted and the waveform media are its carriers.

Photosynthesis is another example of rhythmic energy transmission and conversion. Light wave pulses are converted by photosynthesis into the nutritional building blocks needed by the plant utilizing a natural technology to convert the sun's radiation into energy. Radiation translation mechanisms are common among all living organisms. Each living organism has the capability of converting radiation from one form to another. This conversion process occurs during protein synthesis and energy metabolism, for example.

All living entities code, transmit and receive signals in a rhythmic information exchange. Dolphins and many whales, for example, can not only code and transmit though sound, but they can utilize rhythmic waveform signals to *echo-locate*, obtaining three-dimensional pictures of targeted objects or creatures. This sense is typically referred to as *sonar*, which stands for *S*Ound *N*avigation *A*nd *R*anging. Sonar allows these intelligent creatures the ability to analyze an object's shape, movement, and location from very long distances. While research has long confirmed that dolphins and whales use sonar, it now appears they may have the ability to sense the feelings and emotions of other animals during these complex sound wave transmissions.

Even ants communicate through a complex coding system of touching each other's antennae. Other animals can broadcast reports and emotions over many miles. They can announce their proprietary territories along with their state of affairs with complex sounds. When a dog's domain is faced with a threatening situation, for example, he can broadcast that situation out to other dogs in that area. Those dogs can in turn broadcast the information to other regions if necessary. Theoretically, remote dog populations can almost instantly know a single dangerous situation through a relay of sound transmission.

This illustrates how broadcasting and reception technologies are simply an extension of natural processes. Just as the ears are equipped with a converting mechanism in the form of the bones of the ear and the cochlear hair that translate sound frequencies into nervous impulses, our gigantic radio telescopes receive transmissions from deep space and attempt to convert this into useful data. The same operation is taking place, except that the telescope technology has not been able to translate much meaning from the many space signals. While both are simply conversion instruments, the translation technology of telescope computers is no match for the complex technologies of our natural sense mechanisms.

Different sense organs translate light, sound and tactile waveforms. These waveforms allow us to receive information from a variety of energy sources. These information waveforms provide the basic platform for structure within our universe. The information carried through waveforms of various types connects everything together with resonation and coherence—aligning molecular waveforms into sequential progression. This provides an environment designed for learning.

The ability to broadcast *intended information* through waveform radiation is well established. We can communicate intention within sound waves. We can broadcast intention within radiowaves. We can deliver healing intentions or destructive intentions within lasers, x-rays and atomic energy. Just as the waves of the ocean can carry a surfer or a boat, we can ride the airwaves with intention—transmitting consciousness through radiation to intended receivers. The effects of intention were illustrated by thirty years of research by Cleve Backster. Dr. Backster happened upon intentional communications in 1966 when he connected a dracaena plant to a polygraph machine. To his surprise, *galvanic skin response* readings from the equipment revealed that the plant responded to the potential of harm, specifically when there was an *intention to harm the plant*. This intention-sensing was found to also take place among fertilized chicken eggs and human leukocytes separated from their hosts. Human white blood cells kept *in vitro* sensed their hosts' emotion episodes (Backster, 2003).

This brings us back to the development and implications of the *random number generator* studies we discussed earlier. Emeritus Princeton professor Dr. Roger Nelson took over the research from Professor Jahn. Dr. Nelson began taking his smaller, more compact RNG machines to group events and discovered that group intentions could significantly influence the RNG results. At one point in the 1990s, Dr. Nelson brought together a team of seventy-five researchers from around the world and connected forty RNGs placed in different parts of the world through the internet. This essentially delivered RNG data from all over the planet into a central computer for analysis. At first, the data did not seem to reveal anything of great significance.

Everything changed on September 6, 1997. On this day, billions of people throughout the world watched the funeral of the once Princess of Wales Diana. Also on this day, RNG results from around the world made a massive shift, illustrating an effect from mass population consciousness. After this extraordinary event, the *Global Consciousness Project* went into high gear. Dr. Nelson and his associates began watching other events that affected the masses. Consistent with the Diana funeral, events like the Super Bowl, the Olympics, the O.J. Simpson verdict, and the Academy Awards all produced spikes in the RNG results. Major catastrophes such as earthquakes produced RNG spikes. In other words, events involving greater levels of consciousness among large populations would affect the RNG results more significantly than events nearby. It became clear that events once considered to be disconnected and even random were actually significantly affected by mass consciousness (Radin 1997).

A stirring RNG effect took place on September 11, 2001. Of course, the RNG results were significant after the plane crashes, associated with the world's massive reaction. However, something else, even more mysterious took place: The shift in RNG results *began four hours before* the first plane hit (Radin 1997).

We might also consider research on intentional thoughts that took place in the late 50s. Dr. Bernard Grad of the McGill University School of Medicine in Montreal studied depressed and non-depressed patients' effects on growth rates of barley plants. Subjects equally watered barley sprouts with sterile glasses of water. Barley plants watered by the depressed patients grew less then those watered by control and normal groups (Brad 1964).

In 1932, Dr. John von Neumann introduced the concept of the *wave function collapse*. Some also feel that Dr. Neils Bohr's Copenhagen interpretation of the wave theory also assumed the wave function collapse. Initially the wave collapse theory was introduced as an accommodation to quantum mechanics theory in order to cover the issues of continuity and non-local effects of electromagnetic waves through time. As the theory went, in order to measure and discuss the functional nature of sub-atomic reality, it is necessary to *collapse* the wave character and presume a particle characterization and locality. As this notion has evolved within theoretical physics, it has been used as explanations for other theories, including vast suppositions like the *string theory* and the *theory of everything*.

However, another perspective has also been suggested. Perhaps *consciousness* collapses the wave function. This *consciousness-collapse* theory was first proposed by Dr. George Berkeley, a divinity professor and well-known empiricist, not to mention the namesake of University of California's Berkeley campus and now thriving borough. Dr. Berkeley's notion was that physical perception had a lot to do with consciousness. Dr. Berkeley proposed that physical reality could thus not objectively be known by observation alone, because perception was faulty by definition.

Dr. Neumann hinted at the role of consciousness in his collapse theory as well. While the emerging quantum perception took hold, other prominent scientists like professor and prominent radiation researcher Dr. Walter Heitler proposed conscious perception affected the wave-particle atomic model. He proposed that the particle or object-based constitution is a perception illusion created by the conscious observer. Like a snap shot of a waterfall, the static picture of the particle is provided by the observer, but it is not necessarily the reality. A number of other prominent theoretical physicists such as Dr. Russell Targ, Dr. Jeffrey Schwartz, Dr. William Tiller, Dr. Stuart Hameroff and Dr. Bernard Baars have built upon this concept over the past few decades. As to why it has not entered the mainstream of physics thought—we are not quite sure. Perhaps the theory of *consciousness collapse* is simply a little difficult to break down into easily digestible mass media metaphors.

Consciousness gives us the impression of particles and solid-ness because through perception, waveforms can act in a singly responsive manner. For example, our voices travel seamlessly within sound waves, but because our ear drums, limbic systems and minds are able to translate these waves into specific intentional communications; these waves appear to have subunits such as words, sentences, or musical chords. The reason for the illusion is the speaker's and listener's common conscious intentions to communicate and perceive purposeful expressions. This intention manipulates the realm of continuous sound waves. Though the means for communication did not contain any actual units, the translation renders that perception.

In the same way, the entry of light through Young's two slits gives the illusion that light contains particles because light contains rhythmic information. When those waveforms interact, the interference pattern renders a perception of particles rather than waves. This would be a combined factor of waveform interference and perception. The interference created by light's pulsed waveforms, together with our pulsing brainwaves creates the illusion of particles and solid-ness. This effect has also been described as being caused by *nodes* and *anti-nodes*—points where waves intersect and interfere either positively or negatively.

We might also illustrate this effect by observing the interaction between waves traveling on the surface of the ocean. The actual water molecules are not moving with the wave. The wave is moving through the water. The shape of the waveform gives us the perception that the water is moving, but this is an incorrect assumption. The water itself stays put within the confines of the ocean and the waveforms are translated through it.

This same ocean wave carries waveform information from a storm thousands of miles away, and interacts with other waveforms along the way. As the waveform hits the shallowness of the shore, the wave stands up and sends that storm's information onto the beach. While the storm's information is continuously being translated through a series of waves that hit the beach, we only see one wave at a time. This gives the perception of units rather than a continuous effect.

As we begin to grasp the reality that the waveform rhythms of nature are information communications, we may want to widen our perspective regarding communication: Are we limited to communicating through sound, sight, movement, and touch? We know now there other, more subtle means of communication, as we have discovered through television, radio and the internet. The appropriate follow-up question might be, are there communications taking place within the waveforms around us that we have yet to perceive?

The Memory

In 1982, a physics research team led by Professor Alain Aspect at the University of Paris determined that subatomic particles exhibited correlating waveforms despite being separated by long distances. This contradicted *Bell's theorem*, which effectively eliminated non-local hidden variables (independent from perception and outside influences) from the quantum mechanics view of the universe. Einstein had issues with non-local influences. Einstein's *principle of locality* proposed that there could be no distant influences: each particle is influenced only by its immediate surroundings.

When two particles split from each other and continue the same waveform, vector and polarity though separated from each other alludes to the fact that either each molecule is continuing to be influenced identically from a distant force, or each particle somehow remembered its waveform activity following bombardment and separation (Aspect *et al.* 1982).

Either way, we have a contradiction between either or perhaps both of the rigid proposals by Bell and Einstein. It would be logical that there were independently local aspects influencing the memory of the particles' former union. It also appears that there is some distant influence maintaining the correlating activities of the estranged pair of particles.

The proposal of a memory of a substance once existing in solution long after the substance is diluted away has been clinically applied over the last 250 years of homeopathic medicine. Homeopaths and researchers have observed clinical success with dilution factors well-beyond one million parts to one: A level at which theoretically no molecule of the substance could remain. Furthermore, homeopathy has documented successful clinical applications with these diluted substances with not lower response with dilution but with deeper and more lasting healing responses. With millions of case histories and hundreds of clinical trials illustrating the effectiveness of diluted homeopathic dosing, researchers have entered the controversy to settle the case. Hundreds of studies provide clear evidence of efficaciousness among homeopathy, while still other studies have brought it into question. Whether the positive study results are due to what homeopaths describe as molecular memory or something else is yet to be ascertained. Research on the subject continues to be controversial.

Some rather bold evidence for molecular memory has come from well-respected researchers with no prior acceptance of homeopathy. One of these was a well-known French medical doctor and researcher named Jacques Benveniste, M.D. At one time Dr. Benveniste was the research director at the French National Institute for Health and Medical Research (INSERM). Dr. Benveniste's career was very distinguished, having been credited with the discovery of the platelet-activating factor. Whilst performing research on the immune system—notably the action of basophils—Dr. Benveniste and his research technician Elisabeth Davenas inadvertently observed that basophil activity continued despite extremely low dilution levels: Dilution levels so low it was doubtful any molecules of the biochemical remained in the solution.

Over a four-year period of continual trials, showing repeated confirmation while instituting further controls, Dr. Benveniste and his research team concluded some sort of memory effect was taking place within a former solution following thorough dilution. It was suspected that water might have some faculty to retain and transmit an antibody's biological activity long the biochemical was diluted out of the solution.

Furthermore, as Dr. Benveniste and his team initially diluted a substance, the activity of the substance decreased, as would be expected. At least until the ninth dilution. After the ninth dilution, the activity of the substance began to increase with successive dilutions—as was experienced in the 250 years of clinical homeopathic success.

Dr. Benveniste's research effort was joined by five other research labs in four countries. All of these labs were able to independently replicate Dr. Benveniste's results. After conducting no less than 300 trials, the results were published in 1988 in *Nature* magazine, authored by thirteen of the researchers. The authors eventually concluded that, *"transmission of the biological information could be related to the molecular organization of water"* (Davenas *et al.* 1988).

The research became controversial to say the least. This *memory of water* conclusion had vast implications in the study of medicine and our knowledge of physics. The unintentional byproduct of the research was to inadvertently provide the evidence for the premise of homeopathy—something Dr. Benveniste initially had not agreed with. It challenged others too. *Nature* magazine's editor apparently assembled a team of outspoken "verifiers" who challenged Benveniste's results and protocol. Initially they observed while the lab confirmed the results. The "verifiers" then modified the protocols to theoretically remove any bias. With the change in protocol, the team could not duplicate the results. Dr. Benveniste and his associates responded to deaf ears by explaining that the protocol changes themselves eradicated the results. Until his demise in 2004, Dr. Benveniste and other researchers repeatedly confirmed his findings (Bastide *et al.* 1987; Youbicier-Simo *et al.* 1993; Endler *et al.* 1994; Smith 1994; Pongratz *et al.* 1995; Benveniste *et al.* 1992).

Ironically, the controversy appears to revolve around the premise of whether a chemical molecule had to be present in order to submit a particular biochemical action. This reminds us of the debate regarding the sub-atomic wave theory versus the classical particle theory. Determining with certainty whether there are any molecules of the original substance left in the water is calculated using probability. Using mathematical probability, the liquid content should be fully displaced with new liquid contents; the likelihood of molecules within that former solution existing in the new contents diminishes substantially, but absence is still only probable.

Viewing the liquid content's molecules and atoms as combinations of interfering waveforms creates a new paradigm. If matter is composed of waveform energy and those waveforms interfere with the solution's other waveforms, there would likely exist a residual memory of original waveforms within the remaining interference patterns. This might be compared with a pond's waves retaining the memory of what was dropped into the pond a few minutes previously.

Over the past decade and partially in response to the controversial nature of Benveniste's research, the scientific basis for homeopathy has undergone a flurry of research. Most of this research has occurred in Europe, where homeopathy practice is often practiced by conventional physicians. Hundreds of controlled and randomized studies assessing homeopathic treatments have now been accumulated. Over the past few years there have been four major independent meta-studies that have analyzed this volume of recent research. Three of these reviews concluded that the effects of homeopathy were more significant than the effects of a placebo, while one concluded homeopathy's effects were consistent with the effects of a placebo. However, this later review was also highly criticized for its elimination of studies (Jonas 2003; Chast 2005; Merrell and Shalts 2002).

The implication is simple: Contrary to classical chemistry and physics theory, chemical reactions would not require particles to physically touch within the waveform view. We can observe this because we can create chemical reactions simply by bombarding molecules with radiation. Subatomic waveform emissions often exert ionizing influences in much the same manner.

The inability of the "verifiers" to duplicate Dr. Benveniste's laboratory results appears to be as easily explainable as how the mixture was treated prior to and after dilution. In an interview shortly before his death in 2004, Dr. Benveniste explained his and the other labs' process of dilution and memory-testing. He described agitating the diluted solution for twenty seconds with a spinning motion, creating a spiral or funnel shape inside the beaker. He called this motion a *vortex*. Dr. Benveniste explained, *"Only then do you get the transmission of the information."*

Succussing has been standard practice of homeopathy since the father of modern homeopathy Dr. Samuel Hahnemann began his clinical provings. (*Ayurvedic* doctors practiced a form of homeopathy for many centuries before Hahnemann.) The process of homeopathic dilution as described by Dr. Hahnemann also required this process of *sucussion,* which was a swirling and knocking of the substance upon the heel of the hand in order to mix the memory components. This practice of sucussion is still widely practiced amongst homeopathic manufacturers and physicians conducting clinical dilutions for patients. This succussion process is quite consistent with the process of vortex shaking documented in Dr. Benveniste's research.

As we convert this process to waveform language, we would explain that succession is stimulating coherent interference between the substance's molecular waveforms and the water's molecular waveforms. This type of interference would logically imbed a sustained impact upon the waveforms and magnetic fields of the remaining water molecules.

As we will explore further later, water clustering illustrates one of water's many interesting attributes. Over the past four decades, chemists and physicists have been observing very organized yet weak-hydrogen-bond clusters forming and breaking up within water, seemingly on a spontaneous basis. Initially it was supposed that these structures were simply randomly forming these complex structures. However, upon further observation it became evident that once a cluster broke apart another cluster would form in its place. Many of the new clusters often replicated the shape of the previous clusters. It was also noticed that most of these clusters took on symmetrical shapes, such as icosahedrals. Is this not an indication of some sort of molecular memory among water molecules?

Controlled laboratory research has concluded that catalytic enzyme activity will vary greatly, depending upon the nature of the organic solvents in the solution (Zaks and Klibanox 1988; Lee and Dordick 2002). From this, we can determine that different solutes affect catalysts differently. The exact mechanisms for the catalytic activity are mysterious in many instances. As we consider the quantum nature of these types of molecules, we can appreciate that reactions do not simply take place between molecules or particles. Reactions take place through an interactive process between the electromagnetic bonding forces inherent within the reactive species. Still the surrounding environment is involved. The environment will either facilitate or buffer a particular reaction.

In a downgrading *allosteric regulation* for example, a molecule is bound with an effector atomic structure at the molecule's active bonding site—controlling its ability to continue that particular reaction. The mechanics of reactions such as these are not exclusive of water—water is surely a conductor and sometimes buffer of reactivity with its ever-changing, short-lived hydrogen-bond ion structures. Thus, water logically should have at least the same electromagnetic impact any other reactive substrates might have. As we will discuss later, water actually has an extraordinary capacity in this respect.

Consider the ability of an iron-oxide tape to memorize data or sounds through electromagnetic field manipulation. Our ability to tape-record a song or speech onto a magnetizing substance like iron oxide occurs simply by impinging a magnetic influence upon the surface with a magnetized head. Is this not the same occurrence as water's ability to memorize? Are we not manipulating iron oxide's polarity to memorize a particular electromagnetic contact? Not only can we read the magnetically charged iron oxide tape later, but the playback of the recording will be quite precise. We can then store magnetically recorded information for extended periods, erasing it to make way for new information.

When we press a bar magnet upon another magnetic metal we change the polarity of a majority of the molecules making up the metal. The polarity is changed through a restructuring of the electron wave orbital orientation, rendering an electron-heavy side and a proton-heavy side. This polarization causes an effective means for memorization. After removing the magnet, some molecules will revert to their original polarity. Others will remain in the same direction. In either case, there is a recollection of positioning and orientation long after the bar magnet encounter.

When a solution is diluted many times and the theoretical amount of molecules left in the solution is seemingly too small for probability equations, there are certainly still innumerable hydrogen ions left in the water that were once part of the original solute. If we envision these ions not as "particles" but as rhythmic waveforms, we can incorporate the dilution into an ongoing information wave technology: Coherent waveform interference patterns create the basis for memory.

As researchers have come to understand the placebo effect over the past century, double-blind studies have become the norm to isolate the treatment's success from various biases, notably expectations affecting the results. Following the results of thousands of trials, it has been commonly accepted that the placebo effect may skew results by as much as 33%. In other words, up to 33% of the test subjects will improve simply because they expected improvement from the therapy, or their doctors expected improvement. When we consider this is one-third of the population being tested, a placebo-range result has quite an influence on healing. What causes the placebo-result and why are results in the placebo range so frowned upon?

Placebo effects have gotten a bad rap. A study's results will be considered insignificant if the trends are under 33% of the study's population even if the study was double-blinded. This is because of the effect expectations have upon results.

Expectations are no more than conscious intentions for a particular result. These intentions can affect research results in either a positive or negative way. In medical research, it would be considered positive if the placebo effect increased the efficacy of the treatment. Perhaps the treatment required some personal interaction between the researcher or clinician and the patient. Perhaps this created a placebo effect and increased the effects of the therapy. Should we then say that clinicians should not have personal interaction with patients so we do not create any additional opportunities for healing? Certainly not.

The viability of the *non-local effects* and *local effects* of personal consciousness, clinician consciousness and group consciousness has significant implications toward healing. Certainly the placebo effect illustrates the non-local effect conscious expectations have upon healing. If consciousness can have up to a 33% positive effect upon treatment and the health of the body, then we must accept the notion that consciousness can somehow be retained, memorized, and transmitted within the waveforms exchanged during these studies (Leder 2005).

Whether or not we relate the clinical results of homeopathy to the effects of water retaining memory, we can illustrate through the placebo effect that intention and expectation can influence physical results. Therefore, should a substance be diluted down to the infinitesimal solute state by a therapist who intends for a particular therapeutic effect, there is some likelihood that the resulting solution will contain some molecular waveform memory of the substance. This will be accompanied by the effects of the conscious intentions of the healer. We could see the combination of these two effects taking place in homeopathy just as it takes place in pharmaceutical research.

We see this effect everyday: When we give a gift to someone, they might look upon and remember the gift decades later. As they look upon the gift, there will be a retrieval of the conscious intent of its giver. Does the gift itself physically contain that consciousness? Possibly not, but certainly the physical gift connects to the memory of the intention of its purchase. Certainly, the gift is a vehicle for the transmission of the consciousness of the giver. How did it arrive at that point? A connection between the giver's intent and the gift had to take place at some point— perhaps when the giver picked out the gift. Therefore, we would say that the gift *reflects* the consciousness of the giver. As soon as the giver interacted with the gift by taking possession of it, the gift has taken on a new character: It is now irrevocably tied to the consciousness of the giver. Does the gift physically contain the consciousness of the giver? By perception it certainly does. Perhaps it simply serves as a reminder or a trigger for the memory. In either case, the perceived effect is the same. The gift triggers the memory of the original conscious intent.

The Resonance

Every type of element and every molecular combination emits a unique frequency. These signature frequencies translate to each element, giving off a unique visual experience. This is why gold appears shiny yellowish and silver appears shiny whitish. We have discussed how specific waveform measurements and comparisons have been made possible over recent years with various instruments. X-ray crystallography and mass spectroscopy are two methods used by chemists to identify different compounds because these instruments read the particular frequencies given off by bombarding those compounds with radiation. After decades of looking at the frequency emissions of various compounds, chemists have cataloged the emission readings of each atom, allowing molecular structures to be identified by their frequencies. This of course indicates that each type of atom and each type of molecule emits a precise and unique *signature* waveform.

Signature waveforms can also be seen across space. For this reason, we have been able to identify a number of different types of elements existing on other planets and stars. Of course, the accuracy of these measurements is limited to the scope of the equipment and our perception. These observations nevertheless illustrate how precise nature can be.

(start here)Over the past few years astronomers at large array radio telescopes have been tracking the existence of persistent intergalactic rhythmic pulses. One of the most puzzling is the *gamma ray burst,* a consistent signature waveform originating from deep space. Gamma rays were discovered at the turn of the twentieth century by the French physicist Paul Villard while working with uranium. The cyclical yet persistent gamma ray bursts are puzzling to researchers because they stream through the universe from the most remote regions of space, yet maintain a consistent pulse strength throughout that range. Cosmologists have difficulty with these gamma ray bursts because their extreme energy pulses do not seem to be related to a known *physical mass.* This point is critical to these researchers because the famous postulation of $E=mc^2$—relating energy to mass and the speed of light—is contradicted should there be no physical mass related to these intergalactic oscillations. As a result, cosmologists cannot explain these tremendous rhythmic energies pulsing through the cosmos, opening the door to various speculations about their origin. Regardless of these speculations, the fact remains that gamma ray bursts are macrocosmic pulses out of our range of comprehension.

Humans have figured out how to electronically manipulate rhythmic energies for various purposes. Consider the sonar toothbrush, which sends rhythmic frequencies into the gums to deter certain bacteria. Consider electronic insect and rodent repellents, which plug into the electric circuit of a house, sending pest-irritating waveforms throughout the house. Lasers (**L**ight **A**mplification by **S**timulated **E**mission of **R**adiation) are now seen in just about every household, business, and hospital in one appliance or another. DVD and CD players utilize tiny lasers. Indeed, there are so many other examples of how we have learned to manipulate rhythmic energies using electronic technologies: Microwaves, x-ray equipment, radios, cordless devices, magnetic resonance imaging equipment, televisions, satellites, and so many other electronic devices utilize and manipulate waveform technology today.

Over recent years, we have determined that many of these electronically manipulated frequencies can be damaging to our physical health. Electromagnetic frequency (EMF) radiation emitted by microwave appliances has been intensely studied for their potential for causing cancers and other disease. For this reason, most wire is shielded by twisting and heavy gauge shields to protect our skin from damage. Hospital staff wear lead aprons to avoid negative health affects from excessive x-ray exposure. Shielding devices have been installed in cell phones to try to filter their radiation from the ear. All of these attempts have certainly created more safety amongst EMFs. Whether we are out of danger is questionable, as we will investigate later. In general, this research indicates that some waveforms are beneficial to the body while other waveforms may be harmful. For example, the radiation from sunlight is a beneficial waveform to most living organisms, while the radiation from uranium isotopes or x-rays can be very harmful. Still other waveforms appear to be neither harmful nor healthy as far as we can tell.

Let us consider these issues relative to the rhythmic theory. Why are some waveforms healthy while some are harmful? This question brings us back to the issue of *resonance*. If a particular waveform vibrates in such a way that synchronizes with the natural vibrations occurring within our bodies and around us in nature, this vibration would create a larger, stronger interference pattern. Assuming these stronger resulting waveforms are harmonious to the other vibrations pulsing through and around our physical bodies, these waveforms should lead to greater vitality.

Should the waveform be contradictory to the natural wavelengths occurring in our environment, those vibrations would be considered disruptive. We can easily see this phenomenon when we look at ripples or waves on the surface of water: If a pebble is dropped into a still pond, precisely concentric waves will expand outward from the point of contact on the water surface. If a larger rock were then dropped into the spreading pebble waves, larger waves from the rock drop will collide and overwhelm the waves from the small pebble. This will cause a disfiguring of the concentric design, leaving a host of angled wave collisions.

This illustrates how waves can collide and disrupt other waves, creating unique successive waveforms. While some waves will not interfere because they lack coherency, other waves may interact, leading to a cascade of either constructive or destructive interference patterns. As these waves continue to collide, they may form standing waves or simply create more collisions. Either way, they transmit information through their interference mapping.

The Holography

If we look at a very small wave break onto the sand or in our bathtub, we can make an interesting observation: *The tiniest wave looks exactly like a big wave.* This tiny wave has precisely the same waveform characteristics of even the largest monster wave. While it might have a different frequency, wavelength, and amplitude, the proportions between these measurements will be the same. This is the same principle of a harmonic—a proportionate reflection of a fundamental basis.

We might call this effect *waveform relativity.* To be more precise we might call this phenomenon *harmonic holography.*

This relative holographic effect is observed throughout nature. As we look around us, we see patterns of replication and duplication with size and proportion relativity. As we scale from the smallest to the largest, we see unified yet individual structure. We see multiple atomic structures cooperatively interacting to form molecular structures. We see multiple molecules interacting to form cellular structures. We see multiple cellular structures interacting to form organ and tissue structures. We see multiple organ and tissue structures interacting to form organism structures. We see multiple organism structures interacting to form family structures. We see multiple family structures interacting to form colonies, cities, villages, or tribes. We see multiple cities, villages, or tribes interacting to form nations. We see multiple nations interacting to form societies. We see multiple societies interacting with the environment as part of a living planet. We see multiple planets interacting to form galaxies. We see multiple galaxies interacting to form universes.

Many researchers may refer to this effect as *homuncular functionality.* This refers to the operations of sub-systems harmonizing with the purpose of a grander system—each part working with congruity and synchronization to the overall purpose of the larger system. We see in the largest and the smallest plants and animals this synchronized sub-system effect, tying together the larger ecosystems of the planet and even the universe with the smallest molecular activity. On a structural functionality, we see waves, crystals, spirals and their various interference patterns all forming *homunculi* with replicated architecture and motion throughout the natural universe.

Harmonic, holographical, and homuncular concepts have pervaded the ancient sciences of the Greeks, as found among the works of Pythagoras, Socrates, Aristotle, Hippocrates, and others. We also find these concepts documented among the works of ancient China (Tao, Buddhism), the ancient Vedas of India, the Egyptians, and various other cultures over the centuries. The term and theory of holography was formally put forth by Dennis Gabor in 1947 in an effort to improve electron microscope resolution. Gabor's general assumption was that a waveform crest would contain the complete information on the source of the waveform. His concept provoked the term from Greek meaning *holos* for "whole" and *gramma* for "message." In its current use, holography uses light (later laser) to separate an image into its third dimension with depth-of-field. This is a process called *wave front reconstruction*. The ability of light to present the parallax view—a view with a dimensional perspective—was contrived by splitting a light beam into two parts, each traveling to the object with different paths. The two beams arrive at a photographic plate from two angles—one referencing the object and the other reflecting the object. As they arrive, they create an interference pattern, which unveils a combined view with depth from two different perspectives.

A *homuncular hologram* then would exist when each of the parts, when separated, still reflect the whole. Often when we hear the word "hologram" we think of the space ships of a certain futuristic television show that use laser holograms reflecting a person or multi-dimensional object in its reproductive entirety. This is considered holographic in one respect because it reflects a three-dimensional physical shape or person into another theatre. This is not a true hologram however, because the anatomy or physiology of the subject is not carried through this reflection—only the exterior shell is.

Dr. Gabor's light holography images were crude estimations or modelings of true holography. True holography requires the interference of multiple coherent beams. An interference pattern produced by two laser beams can partially accomplish this same feat by reflecting the laser subparts into components of the whole image. Following the laser's invention (or discovery) in 1960, its use to reflect the wholeness of an image through coherent light beam splitting was accomplished through the efforts of Emmett Leith and Juris Upatnieks at the University of Michigan in 1962. Their laser applications produced realistic 3-D images with holographic clarity because the two beams interacted with coherence. While this is considered an image modeling rather than a full hologram, the ability of a laser to produce a hologram image illustrates this ability within nature. Since most if not all of nature's rhythmicity is coherent in one respect or another, holography is naturally resident throughout the physical world.

An outer image reflection can be holographic if each section of the image is reflects the entire form. In a real hologram, the image is duplicated throughout the smaller parts. To this is added the function ability and physiology of the whole reflected in each part. This is the case in nature. While we may not perceive the reflection of the image nor the physiology as perfect in the natural world, there are reasons for variances. Without variances in nature, there is no room for the conscious elements to make choices and learn lessons. A strict hologram with no flexibility would simply be a machine.

To picture this holographic effect in nature we could toss two rocks into a still pond a few feet from each other. As we watch the waves of concentric circles collide with each other, we notice a rather unusual event. As the waves from the first toss collide with the waves from the second rock toss, the resulting circular waves reflect the combination of both rock tosses: If there were conformity among the rock tosses—say the rocks were of the exact same weight and tossed simultaneously, we would see one type of interference pattern. If they were rocks of different sizes tossed at different times, the interference pattern would be altogether different. In both cases, the interference pattern reflects and retains the history of the two tosses.

One of the most obvious indications of nature's holography is the sheer breadth and size of the database of mathematical formulas accurately reflecting many of the specifications of nature's functionality. When we consider that one simple formula—such as mass-equals-force-times-acceleration ($F=ma$)—can be applied to physical movement in so many different applications—large and small, we can see the mathematical holographic. How could nature reflect this same simple property specification through all matter when it comes to mass, force and acceleration? *Bernoulli's principle of fluid*—another example—relates fluid pressure with velocity. This formula will apply to a large amount of water moving through a large pipe or channel, or a small amount of water moving through a small water pipe. It can also be applied to air and other gases and liquids. The pressure and velocities will relate every time with the same formula, but with a slight variance to allow uniqueness. This is because the elements and activities of nature transmit holographic reflection.

The ideal case for homucular holography is made with DNA. Each cell within the body contains a blueprint of the body's entire structure. The typical location of DNA is within the nucleus of each cell. Within this nucleus, DNA imparts instructional messaging throughout the cell using a process called transcription. This process requires RNA to copy the coding of DNA onto its own strand. Once copied, the RNA can execute or instruct the coding instructions of DNA by assembling specific proteins such as enzymes or hormones, which perform specific functions according to the cell's defined duties.

Each DNA molecule inside of each cell also reflects the entire body's structure and function. Each cell's particular function is differentiated by a repression of sections of genetic code not applicable to that cell. Even though repressed, those non-essential codons are still present in every nucleus of the trillions of cells of the body—giving each cell a reflection of the genetic code for the entire body. As biologists have broken down the process of fertilization via *mitosis* and *meiosis,* they have observed that genetic information is assembled together and shared between the sperm and the ovum. During the initial fertilization stages, the embryonic *blastocyst* duplicates a complete set of DNA into each initial cell as though each cell will perform all the functions of the entire body. Each cell has all the instructions for the eventual efforts of all the various cells.

Somehow, through a process biologists and medical researchers do not quite understand, the process of cellular differentiation will define and determine each cell's activity. This creates specialized cells within the body. Some cells become liver cells while others become bone cells, each with their specific function. Each cell has its own "brain" or nucleus, from whence it instructs the rest of the cell in its functioning. While each cell contains the blueprint for the entire body's structure—and therefore can synchronize its activities with the entire body—each cell has a predetermined specialty. Thus while the parts reflect the whole, the parts also engage in unique activities contributing to the whole function.

As we investigate this further, we can see that our bodies, this planet, and most of the organisms on it operate in the same manner: Each part contains a seed of the whole, while each part contributes in a special way to accomplish the functioning principles of the whole. For this reason, most of us have at some point imagined an atom to be a tiny universe and the tiny ant farm to be a tiny city. This is because each cell; each molecule; each atom around us is a homucular holographic reflection of the entire universe. Certainly our tendency to envision this is natural.

Another interesting thing we find with the human body is that it is host to trillions of separate living biological microorganisms. These tiny living organisms live primarily in our intestinal tract, but they also live around all of our orifices, in our bloodstream, in our urinary tracts, between our toes—and in just about any organ or tissue system. They include various yeasts, molds, and bacteria. Many of these tiny living organisms are *probiotic*, in that they contribute to the health of the body. Many however—and depending upon the state of the body—are pathogenic.

The body is the host of both these pathogenic and probiotic species. Most of these microorganisms live within colonies, in different organ systems or areas. Most live in the digestive tract. They cooperate and work as teams. Within their colonies, some perform different functions, just as people work in different occupations. Some help digestion by breaking down food to aid nutrient absorption. Many battle with competitive microorganisms. Pathogenic intestinal bacteria typically interfere in digestion and poison our system with endotoxins (their waste products). Probiotic microorganisms often secrete antibiotics to kill their enemies, which are typically also our body's enemies. Probiotics are thus considered a key element of our immune system. Many health experts have concluded that probiotics make up a good 60% of our body's immune function.

The body is homuncularly holographical with respect to the earth. The body, like the earth, is a host. The earth harbors many organisms just as the body does. Some of these organisms are helpful to the planet, while other organisms—notably much of the human race—are pathogenic to the earth. Just as pathogenic bacteria poison our bodies with endotoxin waste materials, human toxic waste is poisoning the earth.

Chapter Two

The Elemental States

For thousands of years, natural scientists and philosophers have studied nature's elements. The ancient Chinese texts of the Emperors and the Vedic texts of the Indus valley describe the universe within the context of layers of elements. The Greeks and the Egyptians also subscribed to the concept of elemental layering. We find this elemental view of the world has influenced the sciences of chemistry, biology, and physics as taught in the Europe and Mediterranean of the middle ages and through the Renaissance. Today, modern western science assumes this stratification of elements as fundamental to the understanding of chemistry, physics, astronomy, physiology, and biology.

Unmistakably the elemental layering of the elements by these ancient and modern technologies has many similarities. While the Vedic methodology discussed the gross elements as earth, water, fire air, and ether, the Chinese science discussed these similarly, as earth, water, fire, metal and wood. The Greeks embraced the Vedic version, as seemingly did the Arabs, Romans and parts of the Church. As these features matured through the European sciences of the alchemists of the middle ages and Renaissance, they gradually took on the references of solids, liquids, gases, fire and the aether.

Similar elemental derivatives have also played a key role in the Egyptian, North American Indian, Japanese, Mayan, and Polynesian cultures as well. In the North American Indian tradition, for example, the elements of nature are related as Brother Sun, Mother Earth, Grandmother Moon, the Four Brothers of the wind and the Four Directions. The Japanese *godai,* meaning "five great," also reflects five physical elements, namely *chi* (earth), *sui* (water), *kaze* (wind), *ka* (fire) and ku (sky or void). In analyzing and applying these traditions to modern science's observational schema, it seems apparent that earth relates to solid matter, water relates to liquid matter, fire relates to thermal radiation, wind or air relates to gas, and ether, space, metal, void or sky relates to the medium of the electromagnetic. As Dr. Rudolph Ballentine (1996) observed, the elements of metal and wood from Chinese tradition appear connected to the references of air/wind and ether/sky, respectively, from the Vedic tradition.

In the traditional schema, each of the elements was connected to a personification or consciousness. Many were also connected to particular body organs and their pathways. In both Ayurvedic and Chinese therapies, for example, each element moves through the body within specific channel systems. These channel systems are called *meridians* in Chinese medicine and *nadis* and *chakras* in Ayurvedic medicine.

The Layered Body

The human body is also made up of a multi-layering of elements. Each layer is made up of matter of different density and functionality. Our skin is covered with layers of structured epidermal cells enclosing the body's contents. The body is framed with a skeleton composed of a rigid matrix of osteocytes bonded with various minerals and proteins. The body's insulating systems consist of layers of adipocytes (fat cells) made of a blend of fatty acids and proteins, which keep the body warm in the winter and provide extra fuel when needed. Water is also a necessary elemental layer within the physical body. Water provides a balanced medium for the functioning of cells, blood, and lymph. The body has thermal metabolic systems (often called thermoregulators) that balance the body's temperature with heating and cooling processes such as sweating, shivering and energy production. Meanwhile an electrical nervous system delivers sensory pulses from the various parts of the body to the control mechanisms of the frontal cortex and limbic system of the brain. Here the self—the operator of the physical body—can gain feedback regarding the body's performance.

Consider how this might compare with an automobile and its components. Like the body, each of the systems of a car will be made using a unique technology and materials. While the outer shell and the framing may be made of a composite of metals, the insulation liner may be made up of some type of vinyl material. The electronic systems in turn will likely be composed of copper. The engine will have liquid cooling and lubrication systems, one using coolant and the other using oil. The engine will be made of heat-resistant metal, allowing the pistons to generate heat without cracking the engine block. The inside flooring of the car may be made of carpeting and vinyl, as will probably the dashboard. A system of air ducts and fans will supply air that can either be cooled or heated. An instrument panel will bring together wiring from various sensors around the car, which send in the electrical pulses giving the driver feedback on the car's performance.

If we consider the materials each part of either the body or the car, each component is made from a different type of material, each with a unique density and structure, differentiating that material from others. The metals, plastics, and vinyl making up the car's structure have a density radically different from the car's coolant and exhaust air, for example. These later materials may have less density and more fluidity, yet they are just as important to the mission of the car and the survival of its passengers. Each type of element provides a different operational capacity due to its specific molecular structure.

The molecular bonds of structured solids are rigid and structured, while the molecular bonds of liquids are weaker and changeable. Since molecular bonds consist of coherent standing waves of varying types, we can directly identify the difference between these elements relative to their atomic composition, density, boiling point, and so on.

The relative differences and similarities between these types of molecular structures allow us to categorize them. This categorization is not difficult, as several of these categories are obvious to the naked eye. Others, as we have illustrated, require a technical probing into the landscape of waveforms.

In order to understand the full breadth of elemental categories, we must firstly accept the reality of the limitation of our physical senses. As we've described, the "objects" we see around us are not solid. They are combinations of interfering waveform patterns which give the impression of solid-ness to our sense perception. We perceive these interference patterns as solid because the standing waveforms of molecular matter are reflecting or refracting light to render images of color and shape. Meanwhile, our brains fill in any gaps in the picture, tying together a congruent image with expectation.

The gross rhythmic physical body mirrors the physical universe "body" in its layered components of different elements. The densest elements are made of stratified, crystallized, or latticed structures which exhibit slower waveforms. These elements make up the structural components of the body including the skin, tissues, and boney network. They make up the universe's structures in the components of soil, sand, rocks and crystals. A less dense and more subtle liquid elements circulate through the cytoplasm, blood, tissues, and lymph systems of our body. The liquid elements of our universe circulate through the rivers, streams, aquifers, and oceans. The more subtle gas elements are seen most obviously in the lungs, but they also dwell in various other regions, including the inner ear and bloodstream in the form of pressurized gas elements. Our planet also circulates various gases in the form of atmosphere, which moves around and through the earth's structures. The body also circulates the less dense thermal element, consisting primarily of currents of infrared radiation—regulating and moving within the body's metabolic processes. The planet also circulates thermal energy in the form of infrared radiation and chemical combustibility. Finally, electromagnetics circulate through the body's nervous systems and more subtle energy networks. Nature also channels electromagnetic radiation through the atmosphere and body of the earth.

Each elemental state of matter provides a medium through which particular types of waveforms move. Solids provide a medium for seismic waves. Liquids provide a medium for the classic wave motion. Gases provide a medium for pressure waves. Thermals provide a medium for infrared waves. Aether/plasma, as we will describe further in detail, provides a medium for various electromagnetic waveforms.

Each elemental state has definitive characteristics that provide a synchrony for their particular waveform type. They also interact to allow information to seamlessly travel through the universe.

The Solid Elemental State

This dense elemental state of matter has been delineated by most every traditional and scientific technology. The solid elemental state is made of molecular standing waveforms in their most rigid state. Molecules in the solid state are closely inter-connected within the bonded structures of lattices and crystalline structures. These stable bonds generate rhythms with slower oscillations and longer wavelengths. The ultimate example of such a longer wavelength is the seismic wave, with a tectonic wavelength of sometimes a kilometer or more. Other rhythmic motilities among solids include the physical motions of physical movement among living organisms. The oscillation frequency and wavelength of a person running is analogous to its seismic motion.

Solids are made up of rhythmically-electrometric bonds that form rigid structures. The energies attracting molecules together into rigid states are often called *intermolecular forces*. These forces are typically explained as forces of polarity, as positives attracting negatives and vice versa. Forces of attraction between molecules that have positive and negative charges are called *dipole attractions,* while attractive forces between non-polar molecules are referred to as *dispersion forces*. These relatively simple polar forces involve magnetic fields. Solids are accompanied by a largely unseen but pervasive magnetic influence. This supports a slower, lower frequency depth of oscillation and a polarity grounding to the other elemental states.

The lower and longer waveform emissions of solids reflect the stability of its bonding patterns and electron orbit characteristics. For this reason, some researchers refer to the rigidity of the solid state as akin to frozen light. This is because research upon sub-atomic and molecular structure shows solid elements locked together into a cadence of shape and structure. Thus the solid elemental state is composed of atoms strongly bonded together in a tight, uniform manner. This state is often termed as being *morphous*. These uniform structures are much less reactive, and thus more difficult to penetrate or break apart. Solid structures are typically denser than gas or liquid substances, but this is not always the rule—as some liquids are denser than some solids.

An example of this is glass. While it is considered a solid substance to our gross senses, strictly speaking glass is a semi-liquid substance. This is because on a microscopic basis, its molecules are not arranged in crystalline or matrix format: They are positioned in more loose *amorphous* arrays. Glass is thus often referred to as an *amorphous solid*.

Crystalline structures are probably the most well known of solid structure types. Crystals such as diamonds account for some of the strongest and hardest materials known to humans. Their molecular features will also interact with and dramatically alter various types of waveforms, rendering them one of the most active mediums in the universe.

The complex orbital structural shapes seen among crystals display a variety of different geometric bonding angles. Crystalline structures are observed with cubic, tetragonal, orthorhombic, rhombohedra, monoclinic, triclinic, and hexagonal shaped molecular structures. Solid crystalline arrays will contain at least three units. These create precisely metered bonding angles and stacked arrays, which often spiral as they grow within rock formations. Crystals can be made of covalent bonds as in graphite or diamond, or ionic bonds as the various salts are composed.

Solid element compounds readily form positive ions, creating strong bonding structures. We see solids often composed of base combinations of *alkali metals*, the *alkaline earth metals*, the *halogens*, and the *carbons*, which are referred to as the IA, IIA, VIIA and IVA elements of the periodic table. As these elements combine with others, they form the basis for many of the solid element molecular structures. For example, IA elements like sodium and potassium will typically bond with VIIA elements like fluorine, chlorine, and bromine to form complex stable lattice structures. Alkali metals are typically not found in nature alone, simply because alone their rhythmic state is highly reactive and ionic, which allows them for form ion salts with halogens, for example. Alkali earth metals such as magnesium and calcium provide the foundation for many stable molecular structures within the solid elemental category.

Metals are also common among the solid layer, as they can easily exchange and share electron orbitals. This makes them prevalent as positive ions, allowing bonds with various negatively charged elements. Delocalized electron clouds have also been known to surround metal ions to form structured bonds. Metals are good conductors of electricity because they are more stable, even amidst electron exchanging. Gold, copper, silver, and palladium are all known for their good conductivity for this reason. Copper is the superior wire material for this reason because it maintains a lower temperature during electron exchanging. Metals also tend to be reflective and lustrous in color, specifically because of their dense structure. This is because unlike the crystals, they absorb far less radiation.

Solid molecular structures also include salts, phosphates, and sulfates. These dense alkalizing elements provide "grounding" platforms for many electron-rich biochemical functions of the body. Alkali mineral salts, for example, are extremely active within the body, but they will also bond together with a wide variety of molecules into rigid structures. Thus, they will often be primary components for various cell organelles and membranes. Mineral salts may form ions, which chelate and bond to help form amino acids, fatty acids, and glycosides. These basic building blocks are the foundation for the proteins, fats, and phospholipids that make up our cellular membranes. Mineral salts also rhythmically coordinate within the axons and neurotransmitter fluid to provide information channels for electromagnetic signaling.

Carbon provides a base for many solid substances. Most of our planet and body is thus composed of molecules with carbon backbones. Carbon is not metallic, nor is it halogenic. It has the ability to resonate with most elements, forming strong and stable bonds. Carbon is one of the few elements that can bond with itself. This is because carbon can share four electron waves among its valence orbitals. It can share either single electron waves among hydrogens, or coupled electron waves among carbons. This versatile bonding configuration allows carbon the unique ability to form chains and branches. Carbon atom chains can be unbranched single-chains or branched. These two possible bonding structures render a variety of possible molecular structures. When these chains incorporate oxygen, nitrogen, phosphorus and other elements, the structural possibilities are numerous.

Solid elements can easily transmute. This means they easily conveyed between elemental formats. An example of this dynamic is silica. Silica crystals bond with elements of sediment in rocks and soil. As water penetrates these sediments, silica runs off into rivers and lakes; eventually finding its way to the ocean. Here the ocean's living diatoms assimilate the silica particles, utilizing them as building blocks for their skeletal system. As their lifetimes end, the tiny diatom skeletons sink to the ocean floor. These skeletons become churned with the ocean floor's processing conveyer system, eventually converting to become part of fossil fuels, sulphites, and sulphurs, which eventually either escape or sink to become processed into the earth's crust and mantles.

Because they arise from earth's solid layering, food and its various nutrients are conveyed from the solid elemental state. Their bonding patterns retain the frozen state of energy attributed to solids. Food releases its energy when its structured lattices are broken down by the reactive thermals within the body's enzymes. This is often referenced as the reaction between earth and fire, although fluids are certainly an important part of the digestive process as well.

In traditional *Ayurveda*, the solid or "earth" elemental state is connected with the sense of smell. It is therefore linked with activities of the nose. Its action is said to be excretion and its central organ of this action is said to be the anus.

The Liquid Elemental State

Just about every ancient, traditional, and modern scientific technology recognizes the elemental layer of liquid matter. As opposed to the slower oscillations of the solid layer, the water or liquid elemental state oscillates with a greater amount of speed and variability. We also know from chemical analysis that liquids have quite different electronic characteristics than solids. Rather than providing stability, rigidity, and structure, liquids will conform to the shape of the solids they surround or are contained within. We can easily observe these effects. Liquid molecules contain weaker electronic bonds between each other due to proximity, allowing the ability to move around each other without rigid sequencing. However, this does not mean liquids are any less organized. The magnetic moments of molecules within water display a great amount of organized polarity and consistency. Hence they consistently display the same characteristics including surface tension, vapor points, and solubility. Liquids also display an amazing ability to convey and even conduct radiation and electricity.

As water interacts with the more structured solid layers, we find that some molecules dissolve while others precipitate. When a substance is dissolved into a liquid, the properties of the liquid will typically change—it may taste different, have a different boiling point, and may look quite different. This is because the bonds between liquid molecules are substantially weaker than in solids, allowing a greater level of bonding penetration and change following exposure to new substances. This is also liquid's strength. Being able to absorb exposures and adjust without substantial damage to position or motion gives water the strength of flexibility.

The type of waveforms existing within liquids relate to fluid pressure and osmotic differentials. This contrasts with the lattice-driven molecular foundation. We can see fluid's rhythmic motion in tidal waves, ocean waves, moving rivers and streams. Osmotic pressure exerting within the bodies of living organisms or between various gases within the atmosphere provide another dimension of oscillation.

Every type of fluid has a specific waveform character. Each type of fluid resonates differently with its surroundings. This is particularly seen in two liquids of different types inside the same container: They often will not mix. This is most obviously seen in the case of petroleum mixed with water. Oil will separate from water because oil is *hydrophobic*. The molecular structure of oil is made of hydrocarbons and the molecular structure of water is an oxygen-hydrogen bonding structure. Water's bonding structure creates a polar molecule. This means water molecules tend to attract each other magnetically. Oil molecules, on the other hand, do not have a distinct polarity because of the complexity of the various atoms and bonding structures. However, like water, oil molecules still will have significant surface tension. Oil's molecules are consistently attracted to each other. The two liquids are thus out of phase with each other. Oil's bonding patterns create a variance of magnetic fields, which repel water's magnetic fields. However, an *emulsifier* like liquid soap will attract both oil and water. One side of soap's molecule will attract water and the other side will attract the oil molecules. This is why dish soap is useful for cleaning up oily dishes.

The bonding interference patterns in both petroleum and water molecules have more stability and strength at their surfaces. This effect is referred to as *surface tension:* A blend of *cohesive* and *adhesive forces,* balancing each other to create a polarity barrier at the surface. This is a rather simplistic explanation, however. What we are observing with surface tension is the cohesive interaction of standing wave interference patterns. Their cohesiveness is another way of saying their molecules are interacting together *in phase* with each other.

Careful observation of large bodies of water will unveil smaller streams moving within the waters at various directions. Liquids also have the ability to rise (or fall) with pressure rhythms within channels or solids. This characteristic of the liquid elemental state is called *capillary action*. Capillary action allows living organisms to move fluids around in conveyor motion. This action allows nutrient circulation and detoxification—both processes linked to the dissolving and converting mechanisms of fluids. Capillary action is closely related to surface tension, as polar barriers are formed at the edges of the capillary stream.

We can easily observe fluids as good mediums of a variety of waveforms. While solids absorb and transmit longer and slower waveforms, their rigidity makes them less able to respond to shorter oscillations without a considerable amount of alteration. They will thus provide more negative interference for shorter waveforms with higher frequency. For this reason, we will see light transmitting through liquids while solids tend to block light. At the same time, we find liquids will often create the same kind of prism effects that crystals and glass may create as light refracts and diffracts through them. However, liquids will provide a variance of refraction or diffraction, as its molecular bonds move and readjust.

Liquids also radiate through other elemental states. As water moves through the atmosphere of air, it becomes altered by the various pressure and movements of air. Moving air—wind—deep out to sea radiates its waveforms through waves that hit the beach. Water also moves over various rocks, soil, and sediment, changing the structure and appearance of those solids. After a heavy rain, water will create channels into soils and hillsides, leaching nutrients. This effect is not unlike refraction. Just as light refracts through water as it is altered by those waveform interactions, water is channeled through rocks and soils.

Many liquids become solids with temperature changes. This conversion is precisely regulated. Every liquid has a specific freezing point, with little variance for that substance outside of other environmental factors such as atmospheric pressure. We observe this tendency between water and ice. Of course we also can observe a solid converting to liquid form at its particular melting point.

What is the molecular difference between these liquids and their respective solid versions? Not too much when considering only chemical composition. However, the substances have quite different waveform functions in the liquid state. We can skate or even drive onto a frozen lake, but we would quickly sink if we tried that during the summer. The bonding waveforms of frozen water are simply more rigid. Their molecules resonate more cohesively in a lower thermal environment.

Different liquids have different rhythmic characteristics. Each type of compound in liquid form will have a different boiling point, a different viscosity, and a different density. Each of these characteristics translates to different specific waveforms, reflected by spectroscopic analysis. The difference in density or viscosity will also translate to a different waveform interaction between each molecule. The sub-atomic waveforms within the molecule create various interference potentials with other atoms and molecules. These also create magnetic field differences within the liquid medium. Varying polarity, surface tension and cohesion are further complicated by quantum differences in spin, angular momentum, phase and so on.

These distinct features also create different interference potentials with elements in other states. This in turn determines how the liquid will interact with other elements. This is referred to as *solubility* when considered with respect to solids.

In the Ayurvedic tradition liquids are connected to the sensation of taste, and correspondingly, the sense organ of the tongue. Furthermore, the action of liquids is connected to procreation according to Ayurveda. Anatomically, the genitals are said to be the organ most connected with this elemental state according to the world's oldest recorded science.

The Gas Elemental State

The ancient philosophers of Chinese, Ayurvedic, Greek, and Arabic tradition referred to this next less dense rhythmic elemental state as *air*. The Japanese referred to this elemental layer as *wind*. As more precisely described in modern science terminology, we can safely say that air is simply a mixture of gases. Gas is simply another type of waveform interference pattern. This unique interference pattern enables gas molecules to be organized within a larger space when compared to solids and liquids. This elemental state is considered less dense as a result. Gas molecules are seemingly further away from each other, apparently floating in a medium of unique pressure. While they seemingly function outside the obvious boundaries of physical structure and shape, this does not mean gases are not organized. Molecules in gases have higher waveform energy states than the more visible liquid and solid elemental states. This translates to increased and higher frequency motion.

We have all seen gases move together within a gas cloud. It is obvious there is a certain affinity existing between gas molecules—otherwise there would be no fixed atmosphere. If oxygen and nitrogen molecules did not mix in precise proportions within air, we would all suffocate. If gases simply moved chaotically without any organization or structure, we would also be sporadically bombarded by a barrage of dangerous radiation from space. Our bodies would instantly burn up from unshielded radiation. The precise layering system within our atmosphere is primarily due to a confluence of pressure, temperature, density and temperature. These qualities are external characteristics influenced by the waveform quantum, magnetism, polarity, and affinity between elements within the air mixture.

Each type of gas has a distinct pressure level at a certain temperature. There is a proportional relationship between any particular gas temperature, pressure, and volume. This is because there is a kinetic force exerted per unit of area of any particular type of gas. The most well known kinetic force acting upon gases is atmospheric pressure. This is thought to be created by gravity, pulling the molecules toward the earth. The atmospheric pressure at sea level is about 760 mm, but this can depend upon circulating pressure differentials caused by temperature and other effects.

As we pull away from sea level, the pressure decreases. This is assumed to be because the higher altitudes have few air molecules.

Pressure is typically measured using a barometer, which simply measures the amount of pressure that the surrounding air has upon a column of mercury. As the air pressure pushes on the mercury, it forces it up a cylinder, which is graduated for measurement.

It is important to note that gas can conform within a space or container by exerting pressure upon it. Gas can typically be contained within a large space quite easily, but containing it into an enclosed tank will typically require some pressure—assuming the nozzle or entryway into the tank is narrower than the tank space.

While the molecules in gas are seemingly floating about, they cannot be easily pressed together. According to the *Ideal Gas Law*, temperature, volume, and pressure are all interrelated. It takes a change in temperature to change the pressure within an area of gas. A change in pressure or temperature within a natural environment requires another precipitating change to occur—like volume (or density). While gas theory assumes gases acting in isolation, naturally occurring gas mixtures provide substantial interactivity with the other elemental states.

An illustration of the interactive forces affecting gas is respiration. As air is drafted into the lungs through breathing, oxygen is diffused through the alveoli of the lungs. This type of respiration is the result of the pressure differential between the carbon-dioxide gas in the blood and the nitrogen-oxygen gas pressure in the lungs. This pressure differential creates an osmotic transduction of oxygen across the alveoli membrane and into the bloodstream. While oxygen is drawn into the bloodstream, carbon dioxide is drawn out of the blood and into the alveoli, where it is purged from the lungs.

A second type of living respiration takes place when the oxygen molecular bonds are broke apart in the cell. This is called cellular respiration. The orbital waveform bonds keeping the oxygen atoms to participate win the oxidation and reduction process taking place between adenosinediphosphate and adenosinetriphosphate (ADP and ATP), resulting in the production of heat and cellular energy. Here waveform energy lying between the oxygen atoms in O_2 is released.

As gas is heated, it will typically expand. This will push the gas molecules outward, theoretically increasing its volume and decreasing its density. As this takes place, molecules seemingly display increased energy and velocity. Therefore, as the famous French balloonist and chemist Jacques Charles determined in the early nineteenth century, the temperature and volume of gas is exactly proportional. In other words, as a gas it heated, its volume expands relative to its temperature. This is referred to as *Charles' law*.

Furthermore, the Italian chemist Amedeo Avogadro proposed in 1811 that this relationship is also proportionate to the number of molecules present in the gas.

Interpreting this into waveform interference language, an increase in interference is precisely inversely proportionate to volume. A larger gas volume will be present with more wave interference. This would be logical, since heat is transisted via waveforms, which would interfere with the waveforms existing between and within the gas molecules. Just as throwing a whole handful of pebbles into the pond would result in a variety of ripples interfering with each other and whatever other ripples there were, adding thermal waveforms to a volume of gas would increase wave interference patterns. This in turn would increase the motion of the gas, because after all, activity, or kinetic energy is a byproduct of waveform interference.

Chemists describe these gas characteristics using the *kinetic molecular theory*—in particle economics. The assumption here is that gases are made up of small particles floating around in constant but random motion. Secondly, collisions of gas particles are considered elastic, with no loss of total kinetic energy. Third, this theory assumes that gas particles are much smaller than the distance between them. Fourth, it assumes there are no attractive or repulsive forces between gas particles. Lastly, it assumes the average kinetic energy is proportional to the temperature of the gas.

In nature, gases flow within pressure waveforms via effusion (a pouring forth) and diffusion (a penetration or combination). We see effusion and diffusion of gas amongst other gases quite frequently in nature. This process occurs in a uniform fashion. If a chemical plant explodes, its toxic gases will predictably drift through the atmosphere, spreading in accordance with wind and weather conditions and according to the composition of the gas.

Gas 'particles' are not little balls, bouncing around chaotically like jumping beans. They, like the rest of matter, are coherently interacting waveforms. They collectively oscillate with unique waveforms. Gases occupy particular locations when their waveforms have constructive interference patterns within and with their local environments. While they interact within their own space, their waveform interactions precisely connect with the rhythms of the other elements.

Many gasses have a characteristic odor. As we review the olfactory sense structures, we understand a particular waveform will stimulate nerve endings in the olfactory tract located at the top of the sinus cavity. The tract contains a number of olfactory bulbs—afferent nerve fibers. These nerve fibers are sensitized by the particular characteristic waveform interference patterns occurring within the substances. Once a particular waveform type interfaces with these olfactory nerve receptors, it triggers a neural signaling process to register that particular waveform quality onto the prefrontal cortex, which translates that waveform information onto mapping system of the neural network for the self to perceive.

In order for each gas to display a particular waveform it must be unique. This uniqueness allows that particular gas to diffuse outward and interact with other gases in specific ways. This specificity depends upon the waveforms inherent among those gas molecules. We know each type of atom has a unique waveform frequency. We also know that each type of molecule has a composite of distinct atomic frequencies—and thus also has its own unique frequency. We know that as gases interact, their respective waveforms also interact, producing constructive or destructive coherent interference. These rhythmic results predict their relative effects within the environment.

The gaseous elemental state facilitates the expansion of specific waveforms to accommodate particular functions. These activities relate to physical movement, pressure, penetration, wind motion, travel, and communications. Ayurvedic science thus connects the gas or "air" elemental state to the sensation of touch and the sense organ of the skin. Furthermore, Ayurveda connects the gas elemental state to retentive action, cold, and dry.

The Thermal Elemental State

Another state of matter was first described as *radiant matter* by Sir William Crookes during his cathode ray experiments, and were confirmed in the late nineteenth century by Sir J.J. Thomson.

Most of today's chemists and physicists will agree with the first three elemental states of matter—solids, liquids and gases. Every basic chemistry course and most physics curricula describe these three 'states' as forming the foundation of molecular matter. However, the ancient sciences go further, describing additional elemental states. Various traditional sciences include *fire* or *heat* as a basic elemental state. While we know thermal energy can be derived from fire, fire is considered by today's science as a specific phenomena derived from the combustion of matter.

Thermal energy radiates throughout our environment, just as solids, liquids and gases do. Just as solids can be dissolved in liquids and gases can be bubbled up through liquids, thermal waveforms can radiate through the different states.

Looking at solids, liquids and gases from a wave-matter perspective, these states are *waveform structures* rather than *particle structures*. If we view solids as fixed particles, liquids as slowly moving particles and gases as quickly moving particles, we cannot perceive thermals in the same light. However, in the waveform view, these states *move through* molecular matter, rather than the opposite. For example, the H_2O molecule can be observed in liquid, solid or gas (vapor) states because these states are moving through the molecular combination as opposed to the molecules moving through the liquid or solid states. For this reason, ice and water act completely different. They might as well have completely different molecular formulae because they are completely different from each other. These elemental states are waveform types, and thus determine structure.

The same holds true for the thermal state. Thermal radiation moves through matter, warming and cooling matter and thus determining the form of temperature. This would be reciprocal to the reality that solids contain the quality of shape, liquids contain the quality of fluidity, and gases contain the quality of pressure. Likewise, thermals contain the quality of temperature.

Thermals often radiate within the infrared waveforms, although thermals also can be transferred through visible, microwave, ultraviolet and x-ray radiation as well. Even infrared waves come in a variety of waveforms. The infrared region is typically divided between several wavelength types: *Near infrared* waveforms are closer to visible light with a shorter wavelength of about .7 to one micrometers. *Short-wave infrared* waveforms range from about one to three micrometers in wavelength. *Mid-wave infrared* waves range from three to five micrometers. *Long-wave infrared* waves range from about seven to fourteen micrometers. *Far infrared* radiation ranges from about 15 up to about 1,000 micrometers in wavelength. This alone should illustrate that thermals are not waveform dependent. They have a particular waveform structure but these waveforms are not thermals.

Just as do the other elemental states, interactions can alter states. Should we apply a Bunsen burner onto a flask of chemicals, we will quickly change the state of those chemicals from one waveform type to another. Heat provides a catalyst for rhythmic change. Anytime we blend one type of elemental state with another, the same response results: A characteristic change in the rhythmic character of one or more of the substances. In the example mentioned above, bubbling a gas through a liquid will generally result in an alteration of both the liquid and the gas. The liquid will pick up various molecules available in the gas, and the gas will be affected by the liquid. In the same way, if we allow a water to encounter a solid, we will observe some of the solid dissolving into the water. The water's chemistry will change and the solid area will be eroded by the water.

The orbital oscillation structures of atomic and molecular bonds are altered by thermal waveform interference. Often this results in a jump in orbital shells. There is either a destructive or a constructive interference between the sub-atomic waveforms. This results in an altered waveform condition. This alteration may or may not be observed or it may not be observed.

The thermal system is a precise mechanism for establishing homeostasis among living organisms. The mammalian physiology produces thermal energy to keep the body's cellular, tissue and organ systems running effectively with the right balance. Furthermore, any sort of stressor or toxicity within the body will typically result in an increase in thermal output, as the body increases metabolism to purge the unwelcome visitor. After the cleansing, body temperature should fall, helping to rebalance the body's cellular homeostasis. In this way, a fever should not be seen as a disorder or ailment. Rather, it is part of the process of cleansing. Without increased thermal activity, our bodies would be overburdened with toxicity and would probably cease functioning.

Within nature's ecosystem, thermal changes result in similar mechanisms of homeostasis. As the warmer waters of the ocean are evaporated and moved through wind and storm systems, an upwelling of colder water brings nutrients to the surface. This temperature conveyor system cycles nutrients throughout the oceans and water over the land.

This mechanism is also prevalent among various closed biosystems. In the late eighteenth and early nineteenth century, Henry Le Chatelier observed and documented this effect. He observed particular chemical reactions responding to stressors. The result of the reactions reduced the influence of the stressor. This law became known as *Le Chatelier's Principle*.

The sun produces around half the thermal energy in our environment. Other sources of thermal energy are the various living organisms, the earth (another living organism), and reflective thermal energy resulting from various types of waveform interactions. The human organism produces a substantial amount of thermal radiation. The human body and related mammals apparently radiate thermal energy in the wavelength range of about ten microns.

Nerve-endings within our skin cells are sensitive to thermal energy being radiation from other elements. When these skin nerve receptors— better thought of as *antennas*—pick up the waveforms of infrared waves, they translate and send signals through the central nervous system to the imaging system of the mind. These signals allow the neural network and the self to interpret the thermal waves. Those interpreted as potentially harmful will result in the sensation of burning.

The thermal elemental state produces various coherent wave structures, which can result in a combination of conduction, convention, and/or simple radiation. These waveform effects allow thermal heat to circulate within the body through not only cellular temperature gradients, but also through subtle energy channels.

In Chinese medicine, the circulation of thermal energy plays an important role in the diagnosis of illness. Just as liquids circulate through blood vessels, the Chinese determined that thermals circulate through subtle channels called meridians. These channels have gateways which increase or reduce the flow of heat. For example, a Chinese medical doctor might determine through examination that a person has excess heat in the liver. A treatment to 'drain' or reduce this heat in the liver through acupuncture and botanicals might thus be applied.

European and western medicine has also long respected the relationship between thermal energy and health. For this reason, thermometers are used extensively to measure one's wellness. Temperature is also used to estimate ovulation timing. Depending upon the person's metabolism, during ovulation the basal temperature will typically rise a about a degree Fahrenheit. Basal temperature is also a strong indicator of the health of the thyroid gland. Low morning basal temperature is associated with suppressed thyroid gland functioning. Temperature increases at local sites are also associated with fighting infection at the site. The importance of thermal balance around the body is thus widely recognized among all medical practice.

Increases in thermal energy directly affect waveform energy levels within atoms and molecules. This is because thermal energy of a particular range tends to interfere with electron waves constructively, boosting their waveforms to heightened frequency levels. This interference can also cause a waveform emission, forcing the molecule(s) to lose sub-atomic elements. Either of these events can create the potential for a kinetic event, as the waveform energy converts to motion.

We can consider this on a practical level: When water is heated with thermal radiation, the hydrogen atoms and water molecules become more active due to their respective electron waves being boosted to new harmonic levels. This increase results in water boiling. Accompanying the boiling is its giving off of increased heat in the form of thermal radiation. Is the intensity of thermal radiation given off by the water as high as the intensity of the burner heating the water? No. The resulting thermal energy has become muted and altered as some converts to increased wave-form (electromagnetic) energy and some transmits through to kinetic and some to thermal energy. We can say the thermal energy *refracts* and even *diffracts* through the water, analogous to light diffracting and refracting through certain substances—as the light's rays are slowed, bent and/or split as they move through the substance.

With regard to the unit perception among the thermal elemental state: If we were to look at a large flame of fire burning in a fireplace, we would immediately notice there is not just one large round flame. Instead, there are thousands of little flickering flames, jumping up and down. These little flames pop up and down, continually being recycled and drawn into the larger fire, from whence new flames flare up and emerge from the fire. Upon emerging, they will merge back into the fire, and become indistinguishable from the rest of the fire. Yet these flames will still crackle noisily as they pierce the air in reach of oxygen and combustible fuel.

Are these little flames subunits of the larger fire just as electrons are considered subunits of atoms? Similar to atoms and electrons, the flames are difficult to locate, yet they encircle the core fire. Yet should we try to capture any particular flame and separate it from the fire, we will certainly become frustrated. Either we will end up with a separate fire with new flames to contend with, or we would snuff out that part of the fire. This is not much different from our inability to capture the illusive sub-atomic particles.

As we step up from solid to liquid to gas to thermal, we notice the waveforms increasing in frequency. From the slowest seismic waves of solids to the classic water waves, the pressure waves and the thermal radiation, we see waveforms increasing in speed and tangibility. The reason science tends to identify heat differently from the other states has to do with this element of tangibility. Though we can all feel heat, thermal energy is too subtle to see, smell, or hear. Although every living organism circulates thermal energy throughout its body, these waveforms are working at high rates of speed—beyond our typical sense range—other than the heat-sensitive nerve receptors located around the body.

As the ancient Vedic and Chinese sciences pointed out thousands of years ago, the thermal elemental state is related to the digestive enzymes, physical activity, metabolism, war, violence, passion, and anger. Ayurvedic tradition also says the thermal elemental state is specific states of conversion, biodegradation, energy, change, vision, destruction, and digestion. Its ability to stimulate energy conversion makes it essential for living organisms. Thermals are also considered to be cleansing. It is no coincidence that equipment is sterilized using thermal radiation. Because heat damages pathogenic microorganisms through overheating and affects molecular bonds, we can utilize its waveform interference patterns for survival. Thermal energy in smaller doses also warms, cleans, relaxes, quickens, stimulates, and moves.

The Plasma-Aether Elemental State

As rhythmic matter becomes less dense and the wavelengths become shorter, we begin to lose our ability to perceive it. This is because our senses only have the ability to perceive slower frequencies with longer wavelengths. This is why we cannot see air while we can see earth. We can see water but most fish cannot. Water is the 'air' of fish and other water organisms. Like ours, their senses can only see waveforms within a narrow range of densities and wavelengths related to the molecular (waveform) makeup of their physical sense organs.

There is no coincidence that the perception of light requires circular-shaped eyeballs for clarity. The transmission of light is also connected to the realm of the circular. This certainly begs the question: Are the light sources circular or is our perception of light circular? To this question we bring the observation that as a fish might peer through his rounded eyes and light-refracting pond to the outside "air" world, most objects would appear circular or round because of the 360 degree angular refraction of light from the fish eye perspective. Incidentally, nearly all of the instruments we use to receive and perceive the electromagnetic spectrum are circular and/or curved in shape. Telescopes, microscopes, dish receivers, oscilloscopes, mass accelerator tunnels, and even eyeglasses utilize circular optics to perceive various elemental states of the electromagnetic spectrum.

The electromagnetic spectrum ranges from the relatively longer waveforms of radiowaves to the higher frequency, shorter wavelength gamma rays. In between, we find microwaves, infrared waves, visible waves, ultraviolet waves, and x-rays. Within each of these broader categories—and most likely well beyond the gamma rays—exist waveform spectra that are significantly distinct in formation and communication. Many of these waveform types are used today to carry communications. Radiowave is the primary band used for communications, but microwaves are increasingly used, especially for high security communications.

For thousands of years, ancient scientists and philosophers assumed that electromagnetic energy moved through a medium called the *aether*. The concept of a fifth elemental state termed aether—sometimes termed *ether*—might prove a bit confusing as to the use of the term. Chemical ether usually exists in the gaseous state, with a molecular structure containing oxygen atoms bonded with alkyl groups. There are various types of chemical ethers, including dimethyl ether, diethyl ether, ethylene oxide, and others. The alkyl bonds in ethers are incapable of forming hydrogen bonds between each other. Ethers have low reactivity and lower boiling points, and thus have been used for a number of purposes over the centuries, including in the case of diethyl ether (also termed sulfuric ether), for anesthesia and hypnosis. The later use proved to be dangerous, and ether's use in medicine became controversial and eventually was discontinued in the mid-nineteenth century, after fatalities and other disasters were associated with its use.

The aether we are discussing was first introduced as an elemental state thousands of years ago by the ancient Vedic sciences. The ancient Greeks such as Aristotle and Ptolemy also professed its existence. René Descartes advanced the notion of light in the seventeenth century as a wave traveling through aether, much the same as sound waves travel through air. During the late nineteenth century and early twentieth century, physicists debated the existence of the aether—also called *luminiferous ether*—considered the medium through which all electromagnetic radiation moved. It was contended that light must be moving through a medium, just as waves move through water and sound move through air. A number of experiments were performed in this regard. One of the most famous was the nineteenth century *Michelson-Morley Aether Wind* experiment. In this experiment, light was measured using different trajectories and reflections to see if an aether wind existed which would slow down the speed of light. This experiment proved unsuccessful, and in 1881, Michelson declared that the existence of the aether was improbable.

About twenty years later Albert Einstein and Hendrik Lorentz (joint winners of the Nobel Prize) arrived at the mathematical basis for the special theory of relativity. This presented the argument that although most motion is relative to the motion and position of the observer, there was no relativity factor where the speed of light was concerned. Einstein considered the aether to be "superfluous," seemingly because he felt electromagnetic "space" was actually a vacuum. In other words, to Einstein, the medium through which light moved was empty. He also contended that light was a photon unit with wavelike properties and a fixed speed. Lorentz however, was not satisfied with this conclusion. As he noted in a 1913 lecture, he felt, *"a certain satisfaction in the older interpretation according to which the aether possesses at least some substantiality..."*

Probably the only rationale supporting the absence of aether is of light moving at a constant speed. Lorentz declared this assumption of light never moving at a speed other than the mathematical speed of light a *"daring assertion."* He further stated it a *"hypothetical restriction of what is acces-*

sible to us, a restriction which cannot be accepted without some reservation." Many other recognized physicists over the past half-century have also questioned both the premise that light's speed is constant and that space medium is a non-existent vacuum. These have included respected scientists such as Paul Dirac, Geoffrey Builder, Dayton Miller, Edward Morley, Geoffrey Builder and more recently Ole Rughede. These physicists all have contended that the existence of the aether simplifies the explanation of electromagnetic wave motion and its various dimensional affects.

We are reminded that research led by physics experts Dr. Dimitri Nanopoulos, Dr. Nikolaos Mavromatos and Dr. John Ellis documented that the speed of light varies to frequency. Professor Dr. John Moffat also illustrated that the speed of light has slowed down over time.

Another state of matter was first described as *radiant matter* by Sir William Crookes during his cathode ray experiments, and were confirmed in the late nineteenth century by Sir J.J. Thomson.

Sir William Crookes' cathode ray experiments were reviewed closely with other data, and in 1928, Irving Lanmuir named the new elemental state of matter *"plasma."* His work proposed plasma to be the *"fourth state of matter."* Plasma was described as the medium through which ions traveled, a state known for strong magnetic affects and interaction with electromagnetic radiation. Interestingly, plasma was never connected to the aether bashing, so it has retained acceptance among physicists over the years. Nonetheless, the correlation between plasma and aether is unmistakable, as they both are professed mediums for electromagnetic radiation.

Einstein's vacuum-light proposal was also rejected by most of his contemporaries—many physicists of renown. The theoretical assumption that space should be considered null and void, while aether's existence provided a clearer and more logical explanation for the movement of radiation and light was reason enough for this doubt in the space-vacuum assumption.

In order for a wave to travel there must be an original disturbance or oscillation and a medium to oscillate. Waves cannot have an existence separate from this initial disturbance and the medium: In fact they do not travel in themselves. They merely undulate a particular medium with a particular frequency. We see waves travel through water, through the atmosphere, and through the earth with precisely the same properties we have observed among electromagnetic waves. Consider that radiation such as radiowaves, ultraviolet waves, visual waves, and infrared rays traveling through our atmosphere—a proven medium. As they do, their waveforms interfere with atmospheric contents, creating a variety of results, including rainbows.

When our space ships travel into space, they are also met with these same waveforms, also with tangible results. The *NOAA Space Weather Prediction Center's* website begins with the following statement: *"Like ships at sea, satellites sail the ocean of space. And, like their terrestrial counterparts, satellites must endure severe storms in the environment in order to perform their mission."* Does this sound like a vacuum? Perhaps the space ships are moving through a type of space outside of the real vacuum Einstein spoke about? This would hardly be logical. Since 1965 *NOAA* has documented over 300 *"space anomalies."* These are described as variations in the space environment. Types of anomalies include *surface charging, deep dialectric or bulk charging, single event upsets* (such as cosmic ray or solar proton storms), *spacecraft drag, total dose effects, solar radio interference, telemetry scintillation, debris, spacecraft orientation, photonics noise, materials degradation* and *meteorite impact.*

We also find that in space there are still the effects of light and other solar radiation. Objects such as the outer skin of the spacecraft can be seen by the astronauts, just as meteors and other debris can be seen. The visible spectrum of the sun's rays are reflecting off of these objects, just as they reflect off of objects within our atmosphere.

Most of the waves of the electromagnetic spectrum including the visual spectrum, the infrared spectrum, the ultraviolet spectrum, radiowaves, and gamma rays are observed originating from and traveling from distant locations around our universe. As they pierce through our atmosphere, they may be altered as they refract and diffract through this new medium. Consider how light refracts through water. The 'atmosphere' of water is also disturbed by the radiation as it warms up, while the water itself alters the waveforms of the radiation. Light moving through water does not have quite the same effect it has as it moves through the atmosphere. For this reason, objects appear distorted under water.

The fact that they are altered by these mediums indicates they were also being conducted through a medium of some sort before impeding on the earth's atmosphere. Logically if space were a vacuum, there would be no relationship between the effects of radiation inside and outside the atmosphere.

Again waves do not travel in themselves. They undulate a particular medium; and that medium is thus disturbed. The wave itself may appear to move, but this is not the case. True wave mechanics illustrate waves are not units moving through time, but rather are informational transmissions undulating a particular medium.

As electromagnetic waves are conducted, there is a disturbance of the magnetic field, and these magnetic field disturbances have been measured moving in perpendicular directions to the undulation of the forward wave motion—the *electronic pulse*. The existence of disturbances within the magnetic field indicates electronic waves are being conducted within a medium affected by magnetism. Just as water has specifications such as density, surface tension, temperature and so on, the medium of the electromagnetic space-aether has properties of magnetism and electronic polarity. Just as we might consider the dimension of the liquid state to be dramatically different from the solid state, the electromagnetic space-aether state also has dramatically different properties. For this reason we might conclude that the elemental states might be perceived as *planes*.

Just as the medium of water is disturbed at perpendicular directions with wave motion, the medium of aether-space is disturbed with the magnetic motion of electromagnetic waves. This unseen medium of magnetism pervades the universe. Due to the magnetic fields around our planet, we can understand which direction we are facing with a compass. We can correlate magnetism to our relative position because direction is related to the flow of the magnetic field through the aethereal space we are occupying. Just as water flows with tidal currents and capillary action, magnetic fields travel with specific currents, tidal action and capillary function.

This magnetic aether has been isolated by research referring it to the *plasma* elemental state. This plasma elemental state has been postulated in research probing the electromagnetic qualities of ions as they separate from molecular hosts. For this reason, this plasma region is typically described as a gaseous elemental state within which ions travel. Yet many researchers admit that 99% of space is probably composed of plasma. As such, the plasma elemental state would not be the exclusive medium of ionic matter, since all elementals rotate between ionic and stable molecular states.

The prevailing assumption was that the plasma state and its contents were unorganized and chaotic. In 2006 and 2007 researchers from the Max-Planck Institute joined with the University of Sydney and the General Physics Institute of Russia to study the behavior of plasma "particles." Using infrared images taken by the orbiting Spitzer Space Telescope, particles of interstellar dust were observed as self-organizing and polarizing, forming helical and spiral structures. The researchers noted these observations indicated the potential for life in this dimension. The type of conscious rhythmic organization observed within DNA molecular structure was consistent with the other elemental levels of matter.

As we investigate this area of research—generally outside our realm of vision and comprehension—we are continually faced with a recurring mechanism: That of magnetism. As we investigate phenomena relating to the positioning and movement of the various bodies throughout space and time, we are faced with the interrelationship between magnetism and sub-atomic behavior. As the word "electromagnetic" implies, any electronic movement disturbs the magnetic field because the two are irreparably connected. Furthermore, when it comes to the concepts of relativity, bringing in the concepts of location, direction, and speed, we also find magnetism an active participant. For example, we know that migratory animals follow the magnetic fields with tiny polarizing cells. Many migratory species will somehow locate the precise spot they were born in order to procreate. Human travel also incorporates magnetism, as airplanes, boats and even automobiles incorporate magnetism for propulsion and navigation.

Magnetism also relates to the movements and rotations of the earth. Magnetic polarizing factors interact with tectonic movement, tidal movement, and seismic activity. The polarizing feature of individual molecules within the body enable ion channel gates to open—allowing functional cellular processes to occur. What is this influential property of magnetism pervading our universe?

In the fourteen century, the famous fourteenth century British Franciscan friar-scientist William of Ockham declared a well-respected scientific principle known today as *Occam's razor*. This principle says a theory or scientific assumption should presume the fewest amount of unobserved variables; be the most logical conclusion; and be the simplest explanation. The Latin term for this is *lex parsimoniae*, or the *principle of least action*.

Chapter Three

Living Earth

The earth moves with a rotational and magnetic harmonic balanced by layers of cycling atmospheres, circulating waters, and vibrating terrestrial layers. These move synchronically with the forces involved in solar orbiting and galactic spiraling, and the various radiation of thermal and electromagnetic origin. All the while, the earth gracefully adapts to the intruding forces of humanity. It is for this reason the ancient Vedic culture compared the earth to the cow: A gentle and calm mothering nature amidst the grasping for milk.

While covered by at least 60% water and most of the remaining topmost layers channeled with the vascular flows of aquifers, springs, streams and rivers, the earth's circulatory system is of the same quality as that of the human physiology: Liquid-dissolved nutrients are pumped throughout the system—supplying minerals, proteins, lipids and phyto-nutrients throughout the ecosystem. This circulatory system gives sustenance to every living organism traversing within or on top of the earth's surface.

Just as our bodies have an epidural layer of skin enclosing and crossing the various layers of liquid, gas, heat, air and ether, the earth has a similar enveloping system. This outer envelope or membrane is commonly referred to as the *crust*. As we probe deeper within the earth's crust, we find tremendous energy, motion, and composition among the various biological chemistries of the earth. Using observations gained through seismology and drilling, researchers theorize the earth is composed of several layers: the *crust*, the *upper mantle*, *the lower mantle* and the *outer* and *inner cores*. Within the crust layer—estimated at 1% of the earth's volume, about 30 kilometers deep including the oceans—researchers have observed a composition of various minerals and metals: We have observed calcium, magnesium, potassium, sodium, silicate, iron, aluminum, gold, silver, as primary, with numerous other secondary elements within the crust. These metals and minerals are typically found as oxides—structurally bound to oxygen. The most prevalent elements in the earth's crust are silica oxide (about 60%), aluminum oxide (about 15%), calcium oxide (about 5%), and magnesium oxide (about 4.5%). We also find about 1.5% of the earth's crust is water, circulating through the veins and arteries of aquifers and underground rivers, and of course the veins of lava.

While the earth is far from precisely round, its approximate radius has been measured at around 6,400 kilometers. The earth's mantle is estimated to be nearly 3,000 kilometers thick and is estimated to cover 70% of the earth's volume. There are believed to be at least three strata of the mantle, the first thought to be a few hundred kilometers deep, the second being thought a few hundred more (guessed at between 500 and 700) and the lower mantle theoretically ranging from about 700 to almost 3000 kilometers. It is also believed there is a thinner crust of mantle layer—curiously referred to as *D*—between the mantle and the theoretical core. It is thought that this D layer is a layer of great movement and circulation.

The mantle is primarily differentiated from the crust in temperature, movement, and composition. The mantle composition appears to have more magnesium levels for example, and less silicon than does the crust. It appears the mantle is also subject to higher temperatures and liquefaction, which is thought to be responsible for the movement and tectonics of the crust. While the mantle tends to be in motion from heat, the crust appears to transfer that heat into more structured movement. Assuming this, the upper mantle is often divided into velocity zones—lower and higher. These are related to volcanic movement, convection and seismic movement. It is thought the middle mantle layers contain complex mineral compounds not found at the surface. Researchers propose this is the result of extremely hot mantle temperatures, which are thought to be well above 1000 degrees Fahrenheit and possibly trend over five thousand degrees in the regions nearer to the core. Of course, these are fantastic temperatures, well outside of human experience. Notably, it is also thought the mantle has a circulatory system, which rotates minerals and nutrients between the outer crust, the inner core, and mantle.

Much of this information, including the following theoretical information about the core, has been deduced from the timing and movement of seismic waves and their echoes through the earth—much as sonar might be used to measure depth and fish count under water. Seismic waves are a type of wave that travels through the solid plane. Typical examples of seismic waves are earthquakes, volcanic eruptions, and reverberations from explosions.

Seismic waves are extremely long, typically about a kilometer in length. It is precisely because of this length that earthquakes and volcanoes can move so quickly through earth. While seismic waves are felt by most organisms through contact with the earth's crust, they are more precisely measured by an instrument invented in 1880 called the *seismograph*. There are a number of types of seismic waves. Two types of seismic waves have been the subject of more study, and these are used to deduce what may lie within the earth's surface: These are the *P-wave* and the *S-wave*.

P-waves are primary compression-oriented waves. They are explicitly formed because of a dynamic movement or jolt. S-waves are secondary shear waves. They typically result from the effects of P-waves. After measuring seismic waves around the world, Dr. Inge Lehmann postulated in the 1930s that the earth's core was actually not solid, but was made up of a layer of liquid surrounding a solid inner core. By measuring the effect of an earthquake felt in one location around the world, Dr. Lehmann found that the remote S-waves of the earthquake were apparently penetrating a first layer with some sort of refraction, followed by a bouncing off of some sort of inner core. Dr. Lehmann proposed in her 1936 paper that this type of bending and bouncing back (refraction and reflection) were consistent with a solid layer covered by a liquid layer. Between those two layers, Lehmann suggested, was a thin envelope: This came to be known as the *Lehmann Discontinuity*. This hypothesis was apparently confirmed in 1970 when more advanced seismographic equipment was available. Dr. Lehmann received the William Bowie medal from the American Geophysical Union in 1971 for her pioneering work.

The earth produces more than simply seismic wave motion. The earth is also a gigantic thermal waveform producer. Its core produces intense heat, as we will discuss further. The earth obviously has a liquid layer and several layers of gaseous shells over its surface. The earth produces sound waves along with various electromagnetic waves. The earth's electromagnetic field vibrates at about 7-12 hertz. It was Nikola Tesla's work on global electromagnetic resonances in the early twentieth century that first established the earth's broad network of waveforms. Then in 1952 Winfried Otto Schumann, after measuring the relationships between lightning and the earth's electromagnetic field, proposed the existence of a *waveguide governance factor* of the earth's network of waves. Dr. Schumann proposed this wave-guidance system was connected to the ionosphere: This effect was named the *Schumann resonance*. The theory was confirmed through seismic calculations a decade later. Interestingly, the frequency range of the Schumann resonance is very close to the range of human alpha brain waves. Thus we might conclude that while in the alpha brainwave state—a state of relaxation but not quite sleep or meditation—we are harmonizing with the earth's waveforms.

While still being debated, the prevailing hypothesis is the inner core is made of some combination of nickel and iron. As for the outer core, some have suggested this layer is intermixed with a number of lighter compounds or even liquefied compounds. Seismographic equipment has since become more sensitive and computerized models are increasingly utilized. Additionally, drilling projects have been able to pull deeper and deeper samples from the crust. These additional observations have indicated the crust, mantle, and core layers are likely to be much more complex than has been proposed. This became evident for Texas A&M

University's Jay Miller, head of a 2005 hole-drilling expedition. Dr. Miller stated that conventional theories of the earth's crust and composition *were "oversimplifying many of the features of the ocean's crust…. Each time we drill a hole, we learn that earth's structure is more complex. Our understanding of how the earth evolved is changing accordingly."* In addition, the project's co-chief researcher was quoted saying *"Our major result is that we have recovered the lower crust for the first time and have confirmed that the earth's crust at this locality is more complicated than we thought."*

Since seismic data has been the main tool used to assemble most of these theories regarding the layering of the earth, and some of the deeper hole-drilling has discovered even this first layer has been oversimplified, it would be logical that theories of the mantle and the core have also been oversimplified. Most geologists admit that we really do not know what is beneath the earth's crust. As our knowledge of wave movement and dynamics has expanded over the last few years, there is even some question about whether the seismic data indicating a layered mantle and core might be deceiving.

For example, the Physical Acoustics Laboratory in Colorado's School of Mines Department of Geophysics (van Wijk *et al.* 2004) determined that an acoustic wave traveling through a solid piece of aluminum bounced off of tiny holes or notches in the aluminum, rendering the resulting wave with the same data as if that aluminum were layered— which of course it was not. It was concluded the bouncing effect picked up by Dr. Lehmann and others using seismic data could have likely been caused by caverns or regions of different densities within the mantle. The concept of a round, solid core may thus be an oversimplification. This region or cavern seismic effect was termed as a *pinball effect.*

A 2008 study done at Sweden's Uppsala University indicates that the seismic data reveal an elasticity evident within the trajectories of wave conductance. This indicates an irregular-shaped core. Focused upon the paradigm of a single-body core, the researchers contend that such a core would have to have an irregular cubic structure in order to account for the elasticity of the seismic readings.

This data could also lead to another possibility. We consider first this pinball effect. Then we consider the elastic seismic readings. To this, we add observations of tectonic movement throughout the earth's crust, and the various evolving thermal and lava flows arising from the crust and upper mantle. These various symptoms render a likelihood of a *living organ system* within the body of the earth.

This is not a new concept: Athanash Kircher's *Mundus Subterraneus* published a number of charts in 1664, including one depicting earth with a molten core and various interconnected sacs quite similar to a cell's organelles interconnected with a circulatory system.

Most geophysicists agree that observations of magnetic fields, lava flows and drilling combine with theoretical data indicating a tremendous amount of heat is being generated within the surface of the earth. What is the source of this heat? We can logically assume this heat is generated from a durable generating source—which might assume a nuclear active core. We can also conclude the earth also has some sort of electromagnetic generative properties because of the strong presence of enduring magnetic pulses observed at the surface and arising with magma flows.

We also find vein networks of nutritionally relevant fluids such as mineralized waters, petroleum, natural gas and liquid magma circulating through the various layers. Among these circulating veins, we also find various veins of static strata with rich metals such as gold, silver, copper, crystals and so on. Within petroleum conversion we also find a cycle of processing which converts dying matter digested at the crust eventually accumulating into organic pools of petroleum—a digestive process creating a cache of ignitable liquid gold.

Environmental research indicates the earth is now straining under the stress of its crust being poked, punctured, polluted, and robbed of vital fluids. While there are a number of precious nutrients—such as coal, gold, silver and diamonds—the one element not only precious, but critical to the functional nature of the earth's body is petroleum. Geologists have observed that the slow process of decomposition of ocean biomass yields an intermediate substance known as *kerogen*. Through an additional process called *catagenesis,* kerogen is converted into hydrocarbons—the chemical basis for petroleum. The resulting hydrocarbons accumulate and migrate through the porous rock of the crust, accumulating into reservoirs up to six kilometers deep. Natural gas is considered a derivative of this process—the result of further conversion process called *thermal cracking.* Thermal cracking is thought to take place among some of the hotter temperatures of the earth. Meanwhile coal theoretically accumulates using a similar process of conversion, but forms the decomposition cycle for land plants as opposed to ocean biomass.

The interesting part of this process is how analogous it is to the processes of digestion in mammals and photosynthesis in plants. In all three processes, energy reserves are created using a conversion of seemingly unrelated raw materials and radiation. While digestion utilizes enzymes, microorganisms, pressure, and thermal heat; photosynthesis utilizes enzymes, pressure, and sunlight. The creation of petroleum, natural gas or coal is comparable because it also uses pressure and heat, and various microorganisms that break down organic matter into biomass. This microorganism activity is also comparable to the probiotic activity within the human and many other species' digestive tracts. As to whether enzymatic activity is involved in the process of catagenesis remains debatable. Geologist and Professor Dr. Nikolai Kudryavtsev proposed in the 1950s that hydrocarbon production by the planet was an *abiogenic process*. He suggested hydrocarbon production took place without biological activity. Of course, since no human has been able to penetrate the intensely hot mantle, there is no direct evidence of this. Certainly, we can easily see the initial process as biological, as bacteria assist in the decomposition of plant matter just as they do in our digestive process.

The obvious interpretation of these behaviors is that the earth is quite simply a living organism. The motion of the earth, the eruptions, the heat generation, and the digesting of organic matter into *highly combustible* sources of energy lends credence to this interpretation.

The Gaia Hypothesis

The word *gaia* comes from the Greek *ge* which means "land" and *aia* which means "mother" or even "grandmother." Gaia is also the goddess in Greek mythology who was born from the Supreme Being along with Eros. She married Uranus and her offspring included Pontus and Uranus among others. She is also known as *Terra* in Roman mythology.

The earth as a living being has pervaded the ancient sciences in almost every early society, including early Chinese (Taoism and Buddhism), India (Hinduism), Japanese, Egyptian, Greek, Roman, Mayan, Polynesian, and other indigenous cultures. The idea of consciousness at a larger level of existence made sense to these early cultures because they were tuned into their environment. They lived intimately with the harmonics of the earth and saw its rhythms exhibiting emotion and consciousness. They observed the limited scope of lower species and logically assumed their own limitations. Aristotle apparently said in 354: *"The earth is an organism that is born, lives and dies. Its convulsions; earthquakes and, volcanoes are bouts of fever accompanying gasping and spasms...."*

As the dark clouds of the industrial age drew over civilization, the concept of the earth's living nature moved out of vogue. While this nature continued in the many texts of the early cultures, the rush of modern science in its attempt to overcome nature blinded our view of this consciousness. This does not range too far from scientific peerage: Eighteenth century Scottish geologist James Hutton (1726-1797), considered the father of the science of geology, saw the planet earth as an ancient living creature.

The concept lost grace through the industrial era, until some three decades ago, when the living earth view was revitalized by Dr. James Lovelock in a number of well-received scientific papers and an eventual book called *Gaia: A New Look at Life on Earth*. Dr. Lovelock's published exploration of this concept—in collaboration with microbiologist Dr. Lynn Margulis—reviewed the earth's systems of oceanic circulation, soil-nitrification, plate tectonic motion and its self-regulating biosphere—which he termed *geophysiology*. These infrastructures indicated to Dr. Lovelock the evidence for a "complex entity," or living organism. Dr. Lovelock presented the earth's cybernetic homeostatic feedback system—illustrated through salinity, surface and ocean temperatures, biomass decomposition, and atmospheric processing.

Some of the evidence Dr. Lovelock pointed to include that although the sun's energy has increased by at least a quarter over billions of years, the planet's surface has maintained a relatively constant temperature. Dr. Lovelock also explained that although mineralization from rivers and other land sources should have increased the ocean's salinity, it has remained at a little over three percent for billions of years—long a mystery to ocean researchers. Another point he cites is the ability of the earth to maintain a steady mixture of gas despite various reactive elements around the planet such as methane, which should be combustible with oxygen. Dr. Lovelock extensively discusses the various regulatory loops the earth maintains in its recycling of carbon dioxide, calcium carbonate and so many other constituents—activities again illustrating the case for the earth as a living organism (Lovelock, 1988).

While we spend billions of dollars endeavoring to other planets in search of life, we have overlooked the most obvious living organisms—the planets themselves. Showing every sign of a life form including growth, digestion, reaction, survival response, and response-stimulus, there is every reason to believe the earth is a living organism, only living on a much larger scale than we do. Perhaps the situation is not much different than that of a flea living on the body of a dog. The flea simply does not have the scope to understand the dog's existence as a living organism. The flea's sense perception cannot perceive in that field or realm. Its eyes are simply too small to perceive that level of largeness. Proud of our tiny brains and "advanced" culture, humans may well be missing the bigger picture much like the flea does. While we observe thousands of species of simpler living organisms not having the scope to perceive our existence let alone the environment range we see; why should we as scientists assume that only *our* scope and perception captures the whole picture?

Assuming petroleum is an important part of the earth's nutritional and metabolic processes—functioning much like our blood—this would make the drilling for oil akin to the female mosquito's harvesting of blood from the skin of a human or mammal. Ironically (and holographically), the traditional oil-drilling rig remarkably resembles the proboscis and body of the female mosquito.

The Human Crust

Our skin has a number of layers just as the earth does. The outer crust is the *epidermis*. This is lined by the *stratum cells—corneum, spinosum and basale*. The epidermis is translucent—allowing light to pass through with some restriction. Under the epidermis is a *basement membrane*, attaching the epidermis to the next layers. The dermis layer is deeper than the epidermis, and contains various vessels and organs, including nerves, blood vessels, nerve cells, sebaceous glands (sweat glands) and hair follicles. Under this layer lies another layer called the *hypodermis*. The hypodermis also contains blood vessels, nerve cells, and adipose (fat) cells.

There are a number of different types of nerve cells within these skin layers. These include *thermoreceptors*, which respond to changes in temperature. Their range of sensitivity lies between 29 degrees and 43 degrees Celsius. There are also several types of *mechanoreceptor cells*, which include *Pacinian corpuscles*, located in the hypodermis layer. These cells are sensitive to skin compression and friction with oscillations in the range of 50-1000 hertz. *Meissner corpuscles* lie in the dermal layer. They pick up lower frequency rhythms. They can detect subtle shapes and surfaces with oscillations at 10-60 hertz. *Ruffini ending* cells are also found in the dermis and the hypodermis. They tend to respond to a stretching of the skin and directional forces with a frequency range of .4 to 100 hertz. *Merkel disks* are found in the epidermis. They respond to compression at low frequencies of 5 to 15 hertz. With this variety of receptors, our bodies can gauge the temperature, shape, roughness, sharpness and so many other characteristics of the world around us. The location of these nerves also allows different parts of the body to sense different things slightly differently. For example, each finger can sense pressure at different degrees, allowing our bodies to gauge subtle pressure or force differences.

It is the rhythmic oscillation of the outer world that our nerve receptors are designed to respond to. As the surfaces of other bodies encounter our skin, they emit or cause an oscillation. As this oscillation vibrates through the cells, it signals the appropriate receptor. Depending upon the range and type of rhythm, these receptors transmit special information signals through the nervous system into our brain mapping system for resonation upon the brain and mindscreen.

Just as the dumping of so many synthetic toxins by our industrial manufacturing-based society is poisoning the earth's crust, our skin has become a dumping ground for so many toxic synthetics. The various skin creams and cosmetics most women apply to their skin is one example. Research has indicated women who use cosmetics daily may be absorbing nearly five pounds of chemicals. In a 2004 study of more than ten thousand personal care products used by 2,300 people, the *Environmental Working Group* determined the average adult uses nine personal care products containing 126 different chemicals. Of the 28 products tested by the study, one-third contained at least one chemical classified as carcinogenic. Nearly 70% of the products had ingredients potentially tainted with chemicals linked in one respect or another to various complications such as hormone imbalances, fatigue, skin irritation, or cancer.

Some of the more typical chemicals or chemical types contained in cosmetics include parabens such as ethyl paraben, propyl paraben, butyl paraben, and isobutyl paraben. Parabens have been considered carcinogens since their components were found in breast tumor tissues (Soni *et al.* 2005). Others include:

Polyvinlpyrrolidone is derived from petroleum and is also called povidone-iodine. Research has suggested it interacts with the thyroid gland, the kidneys, and creates cellular systemic toxicity (Zamora 1986). It is also toxic to synovial cells and articular cartilage cells (Kataoka *et al.* 2006).

Diazolidinyl urea's prevalence in cosmetics is widespread and second only to parabens. This chemical is a preservative, releasing formaldehyde when in contact with the skin. For this reason, it is known to cause contact dermatitis (Hectorne and Fransway 1994). Other effects such as headaches, fatigue and depression are also suspected from formaldehyde absorption.

Stearalkonium Chloride has been in use for many years by the fabric and paper industries. Its softener and anti-static properties are well known. However, this chemical is also a known skin irritant and allergen.

Propylene glycol is the chemical that killed hundreds of people recently in South America as an ingredient in toothpaste. It is also an ingredient in certain cosmetics. Propylene glycol is a solvent: a primary ingredient in antifreeze and brake fluid. OSHA requirements for working around propylene glycol require protective gear be used to prevent skin contact. Exposure has been known to cause headaches, nausea, eye irritation, and skin irritation (LaKind *et al.* 1999).

Diethanoloamine and Triethanolamine or DEA and TEA respectively, are lubricants and surfactants, which help to wet the skin and spread ingredients. Both can form carcinogenic nitrosamines when combined with other ingredients. Dermal application on mice damaged liver cells and altered fetal brain development (Niculescu *et al.* 2007). Cell death in the hippocampus was also shown in numerous other animal studies (Craciunescu *et al.* 2006).

These are just a few of the hundreds of potentially debilitating chemicals used in cosmetics. Many cosmetics and personal care items are also made with a host of synthetic colors, synthetic fragrances, and petroleum derivatives. These can block pores; disrupt hormone receptors; and cause cancer-causing oxidized radicals, allergies, contact dermatitis; and a number of other side effects with daily use. Some of these ingredients can take hundreds of years to biodegrade.

The interesting holographic irony in this affair is many if not most of these dangerous synthetic chemicals are petroleum-based. This means our own 'crust' of skin and 'mantle' of body is suffering from the very substances we are damaging the earth's crust and mantle to obtain. Indeed, the same type of damage we are inflicting upon the earth's crust and water layers is also damaging our skin and tissue layers.

The damage to our earth's crust is not only in the form of drawing out its precious commodities. As we produce these chemicals, byproduct waste streams of chemical effluent flow from the manufacturers' factories. Our massive dumping of chemicals upon the soils and waters of the earth is depleting the natural balance of nutrients and bacteria that digest matter before its journey through the earth's assimilation mechanisms. The massive dumping of pesticides and herbicides by growers and homeowners, along with the thrash-and-burn mentality of our treatment of forests, rural land and fallow zones is dramatically affecting the earth's ability to sustain life in Toto.

We might also consider that cotton—one of our key crops used to cover the human body 'crust'—is also one of the most chemically treated and most polluting crops. Five of the top nine pesticides—naled, propargite, triflualin, cyanide and dicofol—are either Category I or II (Environmental Protection-rated) chemicals, which are the most toxic classifications. There are estimates that about one-third of a pound of chemicals goes into each cotton T-shirt. Just in the San Joaquin Valley California alone, it is estimated that 18 million pounds are used annually on one million acres of cotton. Furthermore, it has been estimated that no more than quarter of crop duster-emitted pesticides actually reach the crop. The rest drifts for several miles, through neighboring lands and waters, including residential areas. In 1995, endosulfan (a common cotton pesticide) leached into an Alabama creek from nearby cotton fields. An estimated 245,000 fish were killed over a range of sixteen miles.

Cleaning agents are also particularly harmful for both crusts—the skin and the earth. According to the U.S. Poison Control Centers, about ten percent of all toxic exposure is caused by cleaning products, with almost two-thirds involving children under six years old. While we might be shocked to find a child toying with a bottle of drain cleaner containing sulfuric acid, hydrochloric acid and lye, we do not think twice about feeding this same product to our mother earth. While we wear gloves to protect our skin from the harmful affects of ammonia and bleach while we do our cleaning, we assume the earth's skin is impervious and unconditional.

In a 2002 U.S. Geological Survey report on stream water contaminants, 69% of stream samples revealed non-biodegradable detergents, and 66% of the samples contained disinfectant chemicals. Phosphates—central ingredients in many commercial laundry soaps—have been banned for dumping in over eleven states in the U.S. because of their dangerous effects upon the environment. Yet many people still use these soaps without any consideration of their effects.

We might consider that our individual disconnect with the planet does little harm. However, when we consider there are about 300 million individuals in the U.S. alone, the toxic burden becomes cumulative.

Planet Biowaves

The oscillating rhythms of the earth and the exterior physical body resonate with the same seismic waveforms. The seismic wave has a very long wavelength. The long wavelength allows it to be felt by touch. Seismic rhythms are felt through the nerve sensors in the same way that movement is felt. Although we might sometimes be able to see the earth move during an earthquake, or even hear the cracking of the earth during one, this is only a result of the movement creating further waveforms, which the eyes or ears can sense. Otherwise, seismic waves and other grounded waves are only felt but not seen, heard, tasted or smelled. This is because seismic waves occur at the gross physical *solid layer* of matter.

Researchers are puzzled about the source of seismic waves. While geologists define the earthquake as a release of stored energy, there is not a known source of that energy. We are thus left with the experience of an incredible source of shaking and trembling without a known source. Here again the solution to this mystery is provided by an understanding of the earth as a living organism. This is because we know living organisms—each due to the occupancy of an intentional inner self or personality—produce biological mechanisms resulting in heat, movement, circulation, digestion, and so on. Earthquakes account for only a fraction—some say about 10%—of the total energy expended by the earth. Other than life, where might this energy come from?

By watching our news media, one might think earthquakes are random and chaotic. However this is far from the truth. The fact is, earthquakes occur in a rhythmic fashion, occurring with periodicity and often in clusters. There are generally considered two types of earthquakes. One is called a *tectonic earthquake,* while the other is called a *deep focus earthquake.* While the tectonic earthquake theoretically originates at the upper crust within a hundred kilometers of the surface, a deep focus earthquake apparently originates through a process called *subinduction*—theoretically forming up to seven hundred kilometers deep, within areas of volcanic activity. These zones were developed in the 1920s and 1930s through the independent research by two professors—Dr. Hugo Benioff of California and Kiyoo Wadati of Japan; and are thus named Wadati-Benioff zones.

Earthquakes are related to a combination of rhythmic movements of the crust's tectonic plates and the circulation of lava through the mantle and crust. Their interaction creates a harmonic of confluence. Seismologists and geologists have studied both of these effects in detail, and have extensively charted out the earth's various tectonic plates. They have also attempted to map the movement of volcanic lava. Clustering earthquakes have been tracked and recorded over the last two hundred years with increasing detail. Distinct and periodic plate movement has been isolated in many parts of the world, with increased stress at the plate connections in California, Japan, New Zealand, Alaska, Portugal, British Columbia, and elsewhere around the world. Nearly ninety percent of all earthquakes and about eighty percent of the largest earthquakes occur in the Pacific belt called the *Pacific Ring of Fire.* This ring surrounds the pacific plate, which is for some reason the most active plate in the world. As researchers have looked at a map of directional earthquakes, they have noticed plate movement tends to move toward the Ring of Fire from most plates outside the Pacific plate, while the Pacific plate tends to expand outward from the middle during its activity.

Some half a million earthquakes occur each year. Of these, 100,000 might be felt by humans and only 100 might cause damage. Since 1900 there have been an average of one 8.0+ earthquake per year and eighteen 7.0-7.9 earthquakes per annum. This average, surprisingly, is fairly stable.

As the tectonic plates have been cross-referenced with volcanic activity, a rhythmicity or type of pulse appears to be evident.

Certainly, the mountains and mountain ranges are obvious exhibits of the earth's pulse. These huge growths of thermal energy appear around the earth with a synchronicity of their own. As we observe the mountains and crevasses of mother earth, we notice they present—if viewed cross-sectionally—an undulating wave motion of mountains, valleys, and deep-water ravines becomes apparent. As we investigate the formation of mountains more deeply, we notice underneath these dramatic rock growths are veins of thermal and lava flows, pulsing through the earth's crust with a paced synchronicity resembling the pumping of blood or lymph through our circulatory system.

While still somewhat controversial, there is building evidence linking the rhythmic lunar cycle with the earth's tectonic movement. Correlations between the moon's cycles and earthquakes have revealed more specifically a link between eclipses and large earthquake activity.

In order for a solar eclipse to take place, the new moon and the sun have to be within one degree of the same declination. Due to the massive difference in the two bodies, this also means the moon must be positioned within the earth's orbit in such a way that blocks the sun completely. A lunar eclipse, in contrast, takes place during a full moon, which moves into the earth's shadow. This requires both the sun and moon to be exactly opposite each other with the earth in the middle.

The solar eclipse tends to affect the side of the earth where it shows itself more fully, simply because the two bodies are lined up with that side opposing the sun. This theoretically pulls the waters of the earth towards the moon. This affect appears evident in the tidal rhythms. However, along with tidal changes come other rhythmic changes. These take place because of the combined effect of blocking the sun's various waveform energies, complicated by with the moon's own rhythmic force fields.

For thousands of years, the period around the eclipse of the new moon has been associated with earthquakes. The Greeks documented this cycle. Early Greek writers Thucydides and Phlegon documented this link between earthquakes and new moon eclipses. Prior to that, Egyptian and Vedic cultures noted this trend. Over the past two decades, a number of seismology researchers have also proposed this link. A rash of recent quakes has also followed this trend. The San Francisco quake of 1989, the 1999 Turkish quake, the Sumatra earthquake and the tsunami of 2004 all occurred shortly before or after a solar eclipse. The Italian Physical Society published a study (Palumbo 1989) linking various other earthquakes to lunar cycles as well.

While this theory remains unproven, earthquakes appear to be related to the lunar cycles somehow. In 2004, UCLA's Department of Earth and Space Sciences published a report in _Science_ (Cochran _et al._), which correlated earthquake prevalence with high tides—both water tides and solid land tides. Others have confirmed this correlation: Notably Columbia

University's Maya Tolstoy who in 2002 revealed tidal pressure correlation with deep ocean continental plate earthquakes (Handwerk 2004).

Earthquakes and magnetic fields are related. While researchers initially thought of the earth's magnetic fields as rotating outside the earth's crust, anisotropic scaling has modeled and measured magnetic fields flowing both vertically and horizontally, even subsurface to the outside crust. These changing magnetic fields predispose the movement of magma together with tectonic movement.

The amount of the earth's magnetic field ranges from 45,000 to 60,000 nT (nanoTesla) over the U.S. This huge variance is thought to be created by the existence of "buried magnetic bodies" under the earth's surface. This reality has been confirmed by the fact that magnetic field variances have been found in areas where there is an underground tank or other fixture buried. This especially has been the case for steel under-ground tanks, which have resulted in variances of thousands of nanoTeslas of magnetic fields. This fact, when combined with anomalies indicated above, points to the probable existence of natural geophysical pockets beneath the earth's surface as discussed above, which might also be construed as organs or active cells of some sort.

The rhythmic relationships between earthquakes and animals are well known yet puzzling. There have been many stories of animals acting dis-turbed well in advance of an earthquake. Well before the earthquake can be felt or heard physically—at least to our senses and equipment—animals have evacuated nests, begun running wildly and in general acted erratically. There have been many anecdotal observations of dogs, cats and other animals acting wildly even days before an earthquake. In 1975, the Chinese government evacuated Haicheng based upon the observation of erratic activity of animals. Much of the city was evacuated, and as a result, only minor human loss and injury occurred from the 7.3-magnitude quake that followed days later.

In 2003, The Quantum Geophysics Laboratory at Osaka University (Yokoi et al.) reported that mouse circadian diagrams showed unusual circadian rhythmic activity just prior to the 1995 Kobe earthquake. The researchers reported drastic increases in locomotive activity, changes in sleep, and other effects on the mice before the quake.

As we have mentioned, most animals have an incredible sense of di-rection and destination. This is witnessed during the incredible migrations of tens of thousands of kilometers by caribou, lobsters, whales, turtles, birds, and so many other organisms. As researchers have looked for clues as to where this sense of direction comes from, it was found that most of these animals have a sort of magnetic compass within certain cells of the brain. Just as a compass needle points north, these animals are able to sense the magnetic fields of the earth using them to create geomagnetic maps of sorts. Birds are apparently able to organize a grid-like mapping of the earth, enabling them to steer back on course around obstacles or weather fronts. Studies of turtles, lobsters, and mole rats have altered magnetic fields around the animal, resulting in their reorientation towards the appropriate direction. Mole rats for example, burrow tunnels—sometimes 200 meters long or more—towards the south and place their nests at the end of those burrows. If the magnetic field changes, their burrows will redirect to the southern-most direction of the magnetic field. Loggerhead turtle hatchlings immediately have this sense of direction upon birth. From the beginning of their lives they are able to sense where to migrate (Trivedi 2001; Handwerk 2003).

Furthermore, the research of Heyers *et al.* (2007) has unveiled recently the likelihood that birds actually see the geomagnetic field. Neuronal tracing within the eye and forebrain of migratory birds indicate highly active rhythmic impulses consistent with visual signal pathways during magnetic compass bearing. *Cryoptochromes* contained in the neurons in the eye and in the forebrain appear to confirm this hypothesis.

While many still insist the universe works through chaos, there is little evidence of this as we observe living organisms traveling the planet. While we might have assumed living creatures accidentally wander throughout the earth, their journeys are paced and directed through the earth's geomagnetic fields. These geomagnetic fields are themselves timed with the pulses of the planet—synchronized with the sun, the moon, and the various planetary bodies. This synchronization of body waves indicates intention and consciousness.

Kinesiological Waves

Just as the earth has a network of rhythmic information flowing through its core and mantle, the human body has a network of orchestrated waveforms moving through it. As we have discussed, the type of waveforms very by their frequency, amplitude and shape. The two types of waveforms discussed at this elemental level pertain to the slow-moving physical-seismic waveforms and the slightly higher thermal radiation. At the same time, however, we know that the information complex within the human body does not isolate waveforms. These two waveforms interrelate with nervous energy waveforms, chemical-bonding nuclear force waveforms, sensory electromagnetic waveforms, and of course, beta, alpha, theta, and delta brain waveforms, among others. The slower-moving waveforms observed within the earth's solid matter are referred to as seismic or tectonic. In the body the network of slow-moving grounded waves are more closely related to physical and neuromuscular movement as well. Here these will be referred to collectively as *kinesiological waves.*

Kinesiology is typically defined as the science of physiology, encompassing the relationship between the body's anatomy and its practical use, or mechanics. *Kinesis* is derived from the Greek word for *"motion."* The types of motion observed among the various muscles, ligaments, bones, and related tissues in general kinesiology include *flexion* and *extension; abduction* and *adduction; circumduction; pronation and supination;* and *rotation.*

Over the past three decades, focus upon the fundamentals of kinesiology have been practiced by numerous chiropractors, osteopaths, and physical therapists. Through clinical work and research among their peers, these professionals have uncovered some interesting mechanical relationships between the body's various muscle groups and related tissues. Certainly physical therapists and massage therapists have practical experience and application of how the body can more easily heal through therapeutic exercise and manipulation. Furthermore, their experiences have also documented that the body responds quicker when physical therapy includes heat and water. As to why the body responds so positively to these waveforms, the research has not offered clear explanations.

Chiropractors and osteopaths have found other interesting relationships between the parts of the body. Osteopaths have discovered that the body has a remarkable ability to reprogram its bone, ligament and muscles alignment if guided with pressure in certain locations, especially along the spine, pelvis, skull, and cerebrospinal channels. Chiropractors, though often given to the forced adjustments called *subluxation,* have also determined that muscle imbalances often are the cause of pain and injury. In the 1960s, George Goodheart Jr. D.C. found that weak muscles could be strengthened, and over-stimulated muscles could be relaxed using techniques borrowed from ancient Chinese healing modalities—focused on manipulating different channels of waveform flow. Dr. Goodheart continued the work of Terrence Bennett, D.C., who decades earlier observed a connection between the varying pulses among the central nervous system and the blood vessels that feed these areas. Dr. Bennett determined a specific pulse called the *neuro-vascular reflex.* When certain points are stimulated, Dr. Bennett observed, organs around the body were stimulated, which increased healing times. The relationship between the blood flow, these neurovascular reflexes, and the cerebrospinal fluid increasingly became evident in clinical application.

As Dr. Goodheart began integrating the therapies of the eastern modalities such as Chinese medicine, he discovered that there was a precise inter-relationship between the body's nervous system and its muscle and bone strength. In 1964 Dr. Goodheart coined the word *applied kinesiology* to describe techniques utilizing these points therapeutically. AK now refers to the practice of stimulating the release and/or targeted strengthening of particular muscle groups by manipulating either key meridian points, massaging muscle insertion or attachment locations, or applying pressure to *neuro-vascular points* and *neuro-lymphatic points*. Through the 1960s and 1970s, Dr. Goodheart and his students wrote a number of scientific papers and studied these techniques in clinical settings among a variety of injuries. Eventually, methods to teach AK and certify therapists were developed. One of the better known methods was developed by John Thie, D.C., which he described as *Touch for Health* (1973). Dr. Thie's AK incorporated Chinese meridian philosophy, acupressure, shiatsu, massage, physical therapy, osteopathic and chiropractic techniques, providing gentle tools to re-strengthen weaknesses found in the body and encourage its inner healing abilities.

The techniques used in *Touch for Health* are seismic in their frequency and structure and electromagnetic in their connectivity. Prior to the application of a particular technique, the TFH therapist will perform a series of muscle strength tests to determine which muscle groups are weak and which might be over-stimulated. The theory is that many muscular aches, injuries, and pain evolve from an imbalance of muscle groups. No muscle works in isolation. Every muscle is supported and balanced by other muscles to provide the maximum in flexibility and strength. If one muscle(s) is unbalanced—being too weak or over-stimulated, that muscle or group can be gently manipulated to encourage the body to become balanced. The techniques used to accomplish this include applying gentle pressure, circular and pulling massage motions, acupressure holding points, and neuro-point pressure application. These combined might weaken over-stimulated muscles or strengthen weaker ones.

Out of the AK movement, an awareness of more subtle information flows within the body's muscular and nervous system became apparent. Doctors observed that the body's muscles became weakened simply by touching a substance that when eaten or otherwise used, might deter the body's metabolism in some way. AK practitioners found that holding a substance that injurious to the health of the body up to the solar plexus would weaken the brachioradial muscles if the practitioner were to press downward on an outstretched arm as the patient pulled upward. Testing the muscle group's strength with the patient holding healthy substances to the solar plexus, or without any substance provided a standardization or placebo for this test. This technique is called *manual muscle testing*, or MMT. MMT is used primarily to determine where there is a muscle weakness, though this substance testing has increasingly become one of its interesting applications.

Though controversial, standardized studies—using randomization and double-blinded protocols—have been performed using MMT and AK. While some results of been equivocal, many have illustrated these techniques have validity. In 2007, Cuthbert and Goodheart reviewed over one hundred studies that employed either AK or MMT, including twelve randomized controlled trials. The review concluded that both MMT and AK were reliable methods with clinical usefulness and applicable when studied without research bias. It should be mentioned this review was recently critiqued by Haas *et al.* (2007), who disagreed with the results of the review. Yet clinical results and evidence of AK's and MMT's effectiveness continue to confront the medical profession. A study from Graz' Interuniversity College for Health and Development (Waxenegger *et al.* 2007) examined muscle testing against the effects of a cholesterol prescription with eleven patients with elevated cholesterol. A "significant correlation" was made between the brachioradial muscle test given prior to the treatment, and the success of the cholesterol treatment using standardized testing procedures.

One can judge this test quite easily. Double-blindedness can easily be accomplished by putting the food or medicine within identical closed containers, so that neither the tester nor the person being tested knows what is being tested. One container can contain an unhealthy food and the other a healthy food. Both can be tested against the arm's strength with the subject holding the container next to the abdomen or other sensitive area. It is for this ease of double-blinded control that this test has continued to be convincing to practitioners. (We would comment here that should some subtle bias be involved in the patient's weakening of muscles despite the blindedness, there is still a flow of information taking place between the muscles and the object or the practitioner. If the practitioner's knowledge of the substance being tested weakens the muscle then somehow that information is being passed to the patient's muscles on a subtle basis.)

Using the double-blind techniques, it is apparent that the electromagnetic properties of the substance produce the effect. We might compare this simply to heat. When heat is applied to a muscle, it responds in kind. If the heat is therapeutic—as in the case of an overstressed muscle, the heat will allow the muscle to be stronger. If the muscle is not in need of heat, the muscle will likely weaken with heat, as it becomes more relaxed. In this case, a slightly colder application would result in the muscle responding with more strength as its contraction rate increases. Here the muscles are responding to thermal waveforms. Through spectrometry we know that all substances produce unique waveforms, depending upon their molecular makeup.

Grounded Thermals

Accompanying the seismic rhythms of the planet is a wealth of thermal energy. This is also referred to as *geothermal.* The infrared radiation produced by the earth has been speculated upon for centuries. It is estimated that temperatures increase from 1,000 degrees Celsius for molten volcanic depths of fifty miles down to exponential temperatures as the center is approached. Current theories guess that the primary cause for this tremendous heat is from the radioactive decay of isotopes deep within the earth. Other possible heat sources include heat created from the magnetism, heavy metal sink, and rotational distortion. Each of these theories is compounded with the general notion that the core of the earth is some kind of nuclear dynamo. Currently there is no acceptable explanation for the sustainability and consistency of these thermal waveforms.

The earth as a gigantic, ancient organism could logically explain such enduring heat productivity. Living organisms produce heat through metabolism. Living organisms are driven to convert one element to another in order to obtain and transfer energy. Heat is not only a byproduct of metabolism, but it is required for healthy sustenance. Living organisms by nature must maintain a particular core temperature balance.

We know our own bodies need to maintain constant heat to stay healthy. Should our core temperatures drop to below about 95 degrees F we will be faced with the prospect of a slowed and dysfunctional metabolism. Because most of the body's enzymatic and catabolic function is heat-sensitive, a degree or two drop will deter certain reactions or conjugations. While the earth's geothermal heat production is only a small fraction of the earth's thermal production, it is just as vital and relatively dynamic. For this reason, most mammals, including humans, were designed to remain connected to the ground.

The ground not only provides a rhythmic component that harmonizes the earth's seismic waves with the body's own kinesiological waves, but the earth's thermal radiation is designed to balance the body's heat. Indigenous peoples throughout history have and still sleep in caves, on the ground in huts or otherwise on grounded flooring. Natural earthen structures provide shelter and natural insulation. In most ancient cultures, those who could not find caves built huts or houses of earthen or adobe bricks. Within the rigid resonating structures of clay, we find great thermal insulation from either hot or cold temperatures. They are also natural conductors for the geothermals resonating from the earth. They also provide a natural thermoregulation system by staying cooler during hot weather. One of the coolest places to be on a scorching hot day and one of warmest places to be during a chilly night is on the earthen floor of a cave or adobe house.

The unique thermoregulative abilities of the earth are illustrated by the traditional Eskimo *igloo*. While seemingly made of the same ice and snow covering the freezing environment outside, the inside of an igloo will often be from fifty to seventy-five degrees warmer on the inside, heated only by the human body and the earth's geothermals.

Feng Shui and Vastushastra

The science of *Feng Shui* is the positioning of the natural elements within a spatial design conducive to effective energy flow. *Feng Shui* teachers describe this energy flow as the *Qi*, or life energy force of the universe (as opposed to *qi*, the life force of the body). *Feng Shui* literally translates to "wind and water." The term appears to have its origin in a quote from the third century BC <u>Book of Burial</u> translated as *"The Qi that rides the wind stops at the boundary of water."*

The science and application of *Feng Shui* can be traced back to at least four thousand years ago, although some claim evidence of its use dates back a couple more thousand years. The *Yanshao* and *Hongshan* cultures appeared to use *Yingshi*, meaning to "Lay out the Hall." Its use was evident during ceremonies surrounding the sun's post-winter solstice. Part of its role appeared to be to maximize the solar thermal input into a building, increasing its warmth and light.

The science was apparently quite a bit more complicated than maximizing thermal heat and light, however. Astrological positioning was used to chart out the correct orientation and location for building temples and governing houses. This appeared to be associated with both spiritual practice (orienting for proper worship) and successful endeavoring. Astrological compasses with the positioning of significant star groups oriented to the earth's poles were used to establish the correct spatial positioning. The science was used to create both powerful and learned institutions. It was extremely practical in application, yet significantly complex in its considerations. *Feng Shui* also appears to be related to an older Vedic science of architecture called *Vastushastra*, which relates the positioning and location of buildings and environmental conditions to conduct beneficial energies.

The *Ba Gua* is a significant feature and tool used in today's *Feng Shui* practice. This is an octagon with each side corresponding to a particular aspect of life. The eight sides are marriage; fame; wealth; family; knowledge; career; benefactors and children. Each side also corresponds to a particular color, with marriage being red-yellow; fame yellow-red-green; wealth red-green; family black-red-green; knowledge green-black; career black-green-white; benefactors black-white; and children white. These colors trend towards the colors of the earth.

Traditionally the temple was the primary focus for the application of this art. Today *Feng Shui* emphasizes homes and businesses because this is where there is the greatest commercial application along with the greatest level of interest. To provide an arrangement in ones house or business in such a way that attracts success or a happier family life is certainly seductive. Whether the correct positioning of objects around ones house actually creates success in these areas is hard to say. Little in the way of objective research has been able to confirm the science. As a result, we are left with observation and anecdotal information.

Nonetheless, *Feng Shui* is well respected, and there are many reports of success through its methods. Today banking institutions and various other commercial concerns consult with *Feng Shui* professionals while they design and lay out their offices and stores. Many of the effects of *Feng Shui* have been said to have been experienced rather quickly, and are also easy to apply in ones home.

Here are a few of the elements a *Feng Shui* expert may apply in their consultation:

Because the *Qi* and various other waveforms travel through doorways, the positioning of furniture and décor with respect to the doorway is critical. The entryway to a building is called *ch'i kau,* meaning "the mouth of *Qi*" The character and flow of energy through this entryway is considered critical to the harmony existing within the room or building. Thus, the entryway is recommended to be clear, uncongested, simple, and inviting. Using water, flowers, or other ornamental natural elements appears to be useful to enhance this flow.

Feng Shui advises not sitting with ones back to windows, doorways or entryways, but rather facing these. It is said this invites in the energy rather than halts it for the person. Ones seat should also be seated out of line with the doorway as well. Having a solid wall behind the desk and behind the chair apparently lends support and strength. This calms the environment for the sitter.

Bringing the elements of nature into the inside environment is considered important for the flow of *Qi.* This means plants, especially in empty corners, natural wood furniture, natural fibers, and other elements of nature. Water, sun, and fresh air are also important for the clear passage of energy according to *Feng Shui.*

According to these principles, our sleeping place is of great importance, because we want to sleep soundly and have pleasant dreams. Often this is where *Feng Shui* is immediately helpful. *Feng Shui* consultants typically advise that the bedroom be peaceful and uncluttered, with no signs of the "outside" world. It should be a sanctuary of sorts, without computers, televisions, books, and other distractions. The bed is recommended to either be lifted above the ground without boxes or other clutter under the bed, or be the ground itself with natural matting or wood. The bed should be positioned with the headboard side against a windowless wall for support and strength. Again, the bed should be out of the direct line of the entryway. This is said to allow the circulation of *Qi* through the room.

Feng Shui experts suggest the kitchen be clutter-free and thoughtfully adorned with natural elements. The kitchen should be clean, bright and uplifting, as this energy will be reflected upon ones meals and appetite.

Although there seem to be no hard and fast rules, most *Feng Shui* consultants seem to agree that ones living space should be simple and clear of clutter and knick-knacks. Each item we display in our living environments has a unique energy according to *Feng Shui*. Therefore, each item will generate its own waveform and energy. As various waveform energies become combined, they interfere and can create different, sometimes disturbing effects. Simple arrangements with a single motif will provide calm and encourage thoughtfulness according to *Feng Shui*. Many *Feng Shui* experts emphasize that ones personal space is a reflection of the state of the person. It appears this works both ways: Our surroundings reflect us and we in turn reflect our surroundings. We would add this is also the case with regard to the condition of our physical body as well as the rest of the environment within our immediate control.

Some *Feng Shui* consultants advise one to locate the *power spot* in ones environment. This should be a place where a natural element is placed, which theoretically spreads the *Qi* through the room. Many *Feng Shui* consultants will suggest particular orientation to north, east, west, or south by certain elements in a house. This appears to also be related to the location of the house or building in general.

There are also various other specific recommendations *Feng Shui* consultants may suggest for specific locations or habitats.

The ancient architectural science of *Vastushastra* (*Vastu* meaning "building" and *shastra* meaning "teaching") may well be the predecessor to *Feng Shui,* as it encompasses the design and construction parameters, as well as the lay out of a building. Because of the age of this ancient science, there is a potential that may also have influenced many traditional architectural standards still used around the world today.

Construction parameters of *Vastushastra* are oriented around the positioning of the building with respect to the magnetic poles. Ultimately, a building should be squarely faced with one side facing the northern direction. Furthermore, *Vastushastra* suggests building on slopes elevated towards the southwestern sides and lower towards the northeastern sides. This may be related to the origin of this science in the northern equator. The shape of the building is a consideration in *Vastushastra*. While square and rectangular-shaped buildings are connected to growth, hexagonal-shaped and circular buildings are linked with prosperity. Other shapes, such as triangular shapes and ovals are not advised. Square buildings are especially suggested for businesses.

Entrances to the building are considered important by *Vastushastra*. A building is divided into 32 sections on the outside walls. Each section is said to be linked to a particular affect because of its positioning with respect to the rest of the house. Certain sections of the house (called "gateways") are tied to particularly important effects. Entrances on the third and fourth gateways support prosperity. Gateways twenty and twenty-one on the west side support growth, for example. Gateways twenty-seven and twenty-nine on the northern side support general happiness and various opportunities, respectively. Other gateways may not be so hospitable. Gateway eleven on the southern side supports discontent, for example.

According to *Vastushastra* the center of the home—called the *Brahmasthan*—should be clear of barriers like walls, beams, pillars or other fixtures. Through this section flows the vital currents helping to keep the mind clear of clutter. Stairs are suggested to be on the west or north face of the building. Any turning or spiraling of the staircase should be clockwise during the walk up. Again this may be related specifically to a northern latitude position.

Interior planning inside the home, business or temple (actually, it was advised that every home be a temple) has some similarities with the *Feng Shui* model. However, more detail about the layout is defined spatially. Beds are suggested to lie in the plane of east and west, while boys' rooms are recommended as positioned on the northeastern side. The girls' rooms are suggested as positioned on the northwestern side. The dining room is suggested to be on the west side. The bathroom is suggested to be on the east side. Kitchens are said to be better on the southeastern part of the house. The master bedroom is considered best positioned on the southwest part of the building.

In an office building, *Vatsushastra* suggests that management should have offices on the northeastern part of the building with toilets on the opposite sides.

Other principles include entryways being clear from any kind of obstruction, with doors opening to the inside to allow guests open access.

Many other principles accompany this science. We can see that a number of these ancient principles of *Vatsushastra* have been in use in many cultures and are still basic fundamentals in some of today's construction and design principles.

One of the fundamental teachings of both *Feng Shui* and *Vastushastra* is that nature itself is naturally arranged in such a way that harmonizes the flows of nature. If we could take a forest and put it under a roof and put a protective floor and walls around it, we would probably accomplish the best living environment as far as energy flow goes. The outside environment offers the greatest flows of the *qi* and *Qi* forces as each living organism in nature harmonizes with the Greater Living Nature.

The lessons of both of these ancient sciences have been confirmed by many studies over the past few decades. Horticultural therapy was confirmed as a healing therapy by Azar and Conroy (1992). Camping had a therapeutic effect upon subjects attempting addition recovery in Bennett *et al.* (1998). Bishop and Rohrmann (2003) illustrated how real environments had more positive subjective responses than simulated environments. Davis-Berman and Berman (1989) illustrated how adolescents were positively affected by wilderness settings. Freeman and Stansfield (1998) showed how urban environments create increased stress and anxiety. Gesler (1992) illustrated how natural geography affects medical disorders therapeutically. Hammit (2000) illustrated that even an urban forest environment resulted in the reduction of stress and anxiety. As for the placement of windows and entrances, Heewagen (1990) established that the size, design and placement of windows in a house had significant psychological effects upon those living in the house. Honeyman (1992) established that increased vegetation in the surrounding area significantly lowered stress levels. Kaplan (1983; 1992; 1992) also established that being surrounded by nature had positive psychological effects, increased well-being and significant positive behavioral responses. Pacione (2003) established that natural landscapes even in an urban environment significantly affects well-being. Ulrich (1979; 1981; 1983; 1984; 1992; 2002) showed in over two decades of research that nature's landscapes, natural scenes, natural environments, and being surrounded by plants have a significant therapeutic effect.

This concept of getting outside and harmonizing with the earth's rhythms could well explain why many northern climate cultures do not experience more *seasonal affective disorder* (SAD) than their lower latitude peers. While research has pointed to a lack of sunlight as the main factor contributing to this debilitating disorder causing depression, anxiety, fatigue, and lethargy, there are curious exceptions. For example, according to a study from Iceland's National University Hospital (Maqnusson and Stefansson 1993), Icelanders experience lower SAD prevalence than do people on the east coast of the United States, using the same testing methodology. In another study from Turku University's Central Hospital (Saarijarvi *et al.* 1999), the prevalence of SAD was higher among women and younger ages, but notably also more prevalent among those with a higher body mass index. These characteristics of course are also consistent with people who are less likely to go outdoors. There are many other studies indicating lower rates of SAD among those who exercise outdoors. Those who get outside and connect with the earth, air and sun more directly are less likely to experience depression, fatigue and other symptoms such as SAD. As to whether SAD can be separated from the effects of sunlight; the Icelandic study above provides some assurance that even in darkness, the natural rhythmic effects of connecting with the planet outdoors have positive effects.

Grounding Nutrition

While most of us have experienced "eating dirt" at one time or another during our early years, most of modern society now detests this practice as unhealthy. To the contrary, soils, rocks and the waters that have filtered through them are full of various trace and macro minerals. The body requires about 80 known macro and trace minerals. While the macro minerals have been the subject of most mineral research, we know each of the trace minerals are also important because enzymatic reactions require them.

Some of the more prevalent macro minerals include, among others, calcium, potassium, magnesium, sodium and phosphorus. These are called macro because they exist in larger quantities in the body, thus they are considered more important. However, trace elements are possibly even more critical to health, because the balance between both macro and trace elements is one of the governing factors in overall physiological homeostasis (Wilson, 1998).

Important trace elements include fluorine, zinc, cadmium, and chlorine—although there are about eighty in all. While in larger quantities these elements are toxic, they are nonetheless essential for the healthy existence of the body. We see these trace elements involved in just about every enzyme process, without which the body's operations become imbalanced. All of the body's cellular and hormonal functions can abruptly change without a balance of both macro and trace minerals (Wilson, 1998).

Zinc, for example, has only recently become recognized by modern research as a critical trace element. Now it is identified as a key element bringing together various enzymatic and immune system processes. Zinc has been found to be important to cell differentiation and detoxification through its interaction with superoxide dismutase (Mocchegiani *et al.* 2007).

Because minerals become ionic, they supply unique waveforms critical to particular biochemical processes. They provide the ionic bridge in various catalytic processes throughout the body. Ions also balance the body's conducting mechanisms, and provide the cells and intercellular tissues of the body with the means to a pH balance.

pH is the measure of the rhythmic-electrical or ionic quality of a medium or substance. This is reflected by the acidic or basic nature of a substance. In traditional chemistry, acids and bases are defined in hydrogen atoms or proton proportions. An acid is typically described as a substance with an excess of hydrogen ions (H+), which acts as a net proton donor. A base, on the other hand, is considered to either be a hydroxide (OH-) donor or a proton acceptor—a substance with excessive electrons. Net charge is also used to describe these solutions. An acid solution is one with positive charge, while a base solution has a negative charge.

An acid is often referred to as *catonic* (with cations) because it has a positive net charge, while a base will have a negative charge and thus be referred to as an *anion*.

The term pH is abbreviated from the French word for hydrogen power, or *pouvoir hydrogene*. pH is measured as an inverse log base-10 scale, measuring the proton-donor level by comparing it to a theoretical quantity of hydrogen ions (H+) in a solution. Thus a pH of 5 would be an equivalency of 10^{-5} H+ *moles* worth of cations in the solution (A *mole* is a quantity of substance compared to 12 grams of the six neutron carbon isotope.) Put another way, a pCl (chlorine) concentration would be the negative log of chlorine ion concentration in a solution, and pK would be the negative log of potassium ions in a solution.

The pH scale is 0 to 14, for 10^{-1} (1) to 10^{-14} (.00000000000001) range. The scale has been set up around the fact that the measurement of pure water's pH is log-7 or simply a pH of 7. Because pure water forms the basis for so many of life's activities, and because water neutralizes and dilutes so many reactions, water was established as the standard reference point and considered the neutral point between an acid and a base. In other words, a substance having more hydrogen ion-like properties than water will be considered a base, while a substance containing less H+ properties is considered an acid.

Using this scale any substance measuring a pH of 7 would be considered a neutral substance, though it still has a significant number of H+ ions. The pH of healthy tissues is very similar to the pH of healthy rock and soil—both rich with minerals to balance the anions and cations for maximum metabolic efficiency. Most crops grow best with a pH of 6-7, while some will grow better slightly above a pH of 7. The human body has a similar pH, but this also varies to the part of the body. Blood, saliva, organ tissue, and urine all have a typically different pH levels from each other, with optimal range from 6 to slightly over 7. In humans, a level in the range of pH 6.4 is considered a healthy state because this state is slightly more acidic than water, enabling an ionic current to flow through the body. Better put, a 6.4 pH offers the appropriate *currency* of energy flow because there are enough electron waves present for ionic channel passage. Probably the easiest manner to home-test the body's pH is to measure both the urine and the saliva and average the two.

We can thus better describe acids and bases in the more suitable context of waveforms. The amount of electronic waveform frequency deficiency within a substance typically is considered its level of alkalinity, while the amount of electronic frequency abundance among an ion is considered its level of acidity. The quantifiable need for the ion's waveform frequencies to return to their harmonic levels would then be considered the substance's level of acidity or alkalinity.

As minerals ionize they exhibit strong magnetic fields and polarity abilities. Because our bodies utilize mineral ions for most enzyme catalytic metabolic activities, our bodies both utilize and generate magnetic fields. Researchers have known magnetism increases biological reactions for many years. We mentioned Dr. Grad's research at the McGill University in the late 1950s with regard to the growth rates of barley sprouts being affected by consciousness. The central subject of these investigations was a Hungarian refugee named Oskar Estebany, who purported to be able to exert a healing effect with his hands.

While it was only suspected that magnetism was involved in Mr. Estebany's ability to make an effect, other trials confirmed these effects. In one, Dr. Justa Smith at the Rosary Hill College (1973) compared Mr. Estebany's ability to increase enzyme reaction rates with magnetic fields. After Mr. Estebany affected an increased reactivity effect among enzyme reaction rates, Dr. Smith applied the magnetic field and compared the rates. It turned out that the increased effect Mr. Estebany had upon the enzyme rates accompanied a 13,000 gauss magnetic field.

Dr. Smith had spent a number of years studying these effects prior to her tests with Mr. Estebany. She authored a book on the topic, published in 1969 called *Effect of Magnetic Fields on Enzyme Reactivity*. While this research was considered radical at the time, it didn't take long for mainstream science to confirm her findings. In the early-to-mid 1990s a flurry of research was published from around the world showing magnetic fields in the 2,500-10,000 gauss range affecting reaction rates of ethanolamine ammonia lyase, coenzyme cobalamin-dependent enzymes, horseradish peroxidase, and various others.

By 1996, more than fifty different enzyme reactions had been found to be influenced by magnetic fields. In two linked studies by University of Utah's Charles Grissom in 1993 and 1996, single-beam UV-to-visible spectrum and rapid-scanning spectrophotometers with electromagnets built in were applied to two different cobalamine (B12) enzymes. One enzyme (ethanolamine ammonia lyase) had significantly different reaction rates in response to magnetic fields, while the other enzyme (methylmalonyl CoA mutase) did not appear to have any responsiveness to the reactions. It could be concluded that some biochemical processes are more sensitive to magnetic field influence than others. The effect is still mysterious because of the limitations of our senses and instruments, but it is evident that the body houses a flow of magnetic fields.

There have been a number of controlled studies showing the body responds physically to magnetic stimulation. Amassian *et al.* (1989) stimulated the motor cortex with focal magnetic coil giving a sense of movement to paralyzed appendages. In Maccabee *et al.* (1991) stimulated almost the entire nervous system with a magnetic coil. This stimulation instigated responses among the distal peripheral nerve, the nerve root, the cranial nerve, the motor cortex, the premotor cortex, and the frontal motor areas related to speech along with other nerve plexes.

The body also utilizes mineral ions to create sophisticated entry gateways throughout the intracellular systems of the body. These mineral ion gateways are called *ion channels*. These are tiny *gates* lying within cell membranes, typically consisting of protein and minerals. Through these ion gateways, voltage potentials are negotiated. As these gates are stimulated with particular ionic charges, they open up, allowing into the cell key ions and nutrients. In some cases these gateways conduct informational signaling through *receptors* into the cell from *ligands* (special signaling biochemicals) traveling outside of the cell. These ion-protein channels and receptors exist within the phospholipid cell membranes, providing the primary passageways through which the cell's ionic balance and messaging is conducted.

Many of these ion channel systems utilize the earth's minerals as signaling mechanisms for switching on and off metabolic processes. The earth with its metal and crystalline structures provide the basis for these necessary minerals, which are prevalent within natural food diets, mineralized water and soils and rocks themselves. Without them, our bodies will quite simply switch off.

Therapeutic Rocks

The earth's rocks and stones are used therapeutically in a number of applications. Many practitioners now advise patients to walk and lie upon natural rocks and stones for their healing effects. The art of hot stone massage—also called *sacred stone massage* in Ayurvedic medicine, and *Lomi Lomi* in Hawaiian medicine—uses basalt rocks, volcanic rocks or other thermally-enhanced rocks to provide a massage therapy well known for its success in relieving tension, stress, muscle fatigue, and nervousness.

Crystals have been used for thousands of years in a number of healing modalities as well. The crystal is a special geometric formation of mineral composite: A formation of mother earth's geometric potencies. The special crystal formation captures the various electromagnetic fields pulsed through the bonding lattice of that particular element. Each type of crystal has captured a unique spectrum of waveform bonding, utilizing this to create a specific orbital formation. As a result, when other electromagnetic energies pass through the crystal, they become modulated in a unique way, just as spectrometry readings change with different elements. This is also the premise behind the prism's ability to break light radiation into individual spectra.

Crystals can perform this type of waveform modulation variously—depending upon the type of crystal. Thermal waves, sound waves, microwave waves and various other electromagnetic waveforms are transmuted by crystals as they interfere with crystals' highly structured bonding patterns. This of course is how the crystal radio operates: As discussed previously, the crystal is able to transmute radiowaves into pulses of electricity, producing the movement of a magnetic speaker cone. Another application of the ability of crystals to modify electromagnetic fields is their *piexoelectric* effect. Discovered first by Pierre and Jacques Curie in 1880, certain crystals will display electrical charge after undergoing pressurization. This ability went unused until the 1950s, when crystals became commonly used in electrical testing devices and record player needles.

Ancient Ayurveda records have documented crystal use quite extensively. From these, we can mention some of the different crystal patterns considered resonating with the body's different energy centers or chakras. While the first 'foundation' chakra is said to resonate with the red trigonal crystals such as bloodstones, carnelians, agates, and amethysts; the second 'sacral' chakra is considered to resonate with the blue cubic-shaped crystalline structure of diamonds, fluorites and garnets. The third 'solar plexus' chakra connects with the green hexagonal structures of emeralds, apatites, and aquamarines; while the fourth 'heart' chakra is thought to oscillate with the tetragonal crystallines of zircon, wulfenite, and chalcopyrite. Meanwhile, the throat chakra is said to resonate with the orange orthorhombic crystals of peridot, topaz, and alexandrite; while the sixth 'brow' or 'third eye' chakra is thought to resonate with the monoclinic crystals of jade, malachite, and azurite. Finally, the seventh 'crown' chakra is considered to resonate with the yellow triclinic-shaped crystalline structures of turquoise and rhodonite.

Today the most valuable commodities known to humans include diamonds, gold, rubies, sapphires and other "precious" crystalline stones and metals. Humans utilize these most of these latter stones not for practical purposes, but as prized possessions and symbols of wealth, attraction or love. Thus we can see our intimate connection between with our conscious earth mother.

Harmonization: Becoming Grounded

As we seek to find harmony and rhythm within our own lives and bodies, it is imperative that we synchronize our activities with our natural environment. This means harmonizing with the processes and functions of the planet earth in a conscious manner.

The first step in this process is awareness. We might consider becoming aware of the living nature of the kind, gentle nature of the earth. We might consider becoming aware of its ability to react and respond, just as any other living organism does. We might consider the possibility of actually caring for this large, gentle creature. If we assume the earth is a conscious living organism, then it would also be logical to assume that her consciousness is on a significantly higher level than human consciousness. This points to a larger realm of understanding and a subtler means of communication.

Our becoming aware of the earth as living might be analogous to an insect becoming conscious of a human. If an insect suddenly became conscious of our existence as it approached, we would probably realize it immediately because we could watch its actions. Its actions—the way it circled us, or got out of our way when we didn't want it around, and say kept other insects from landing on us by buzzing them—would certainly make us realize at some point that the insect was aware of us. After a few buzzes, we would probably figure the insect was not one to swat away.

Of course, we see no senses evident among the earth. However, neither would an insect be able to see our senses. Furthermore, many creatures, such as bacteria, jellyfish, and plants have no evident senses either. Four decades ago researchers discovered that plants appear to be aware of humans, responding to their actions and words without any obvious sense organs (Tompkins and Bird, 1973).

Awareness should parallel activity. As we become increasingly aware of the earth's consciousness, we can make immediate changes:

❖ We can use natural cleaning agents like baking soda, vinegar, lemon, clay and oil combinations to accomplish our household cleaning tasks.

❖ Natural oil- and glycerin-based soaps and deodorants without fragrances and heavy metals like aluminum or titanium are now available.

❖ We can use natural pesticides and fungicides made from mint-oils, clove-oils, sulphur, copper and other natural elements. Borax is a natural crystal that accumulates when lake beds evaporate, leaving boron-based crystals, and effective at removing many pests (read labels).

❖ We can use plant-based lubricants such as aloe, avocado, cucumber, olive oil and other natural seed oils to moisturize our skin.

❖ Botanicals such as witch hazel, comfrey, and calendula can reduce pimples and blackheads.

❖ Minerals are critical to our body's health. Various clays such as *bentonite clay* and *Montmorillonite clay* are edible as multi-trace mineral supplements. While straight clay can also be eaten; only clean, virgin clay from tested areas should be consumed. Clay is also therapeutic and can be used in a bath, on bee stings and for skin inflammation.

❖ Wheat grasses are also good mineral sources, as they root deep into the soils for minerals. Unprocessed mined rock salts are good trace mineral sources and good for cooking. Seaweeds are chock full of minerals, drawing the earth's minerals from seabeds and sea waters.

❖ Soil-based probiotic organisms are necessary for intestinal health. We typically obtain these as when we played in the dirt as children. Our sterile, pharmaceutical-antibiotic-world has probably killed these off for many of us. *SBO* probiotic supplements are now available.

❖ Walking barefoot on rock, grass, sand, or even dirt is calming and tension-relieving, connecting us with the earth's various waveforms.

❖ We can connect with the earth by simply sitting or lying on the ground. This can be upon a rock, a field of grass, or up against a tree trunk. For this reason, camping can be an exhilarating experience.

❖ We can choose to live in homes made of natural materials. Natural brick, straw-bale or mud-stucco houses are naturally cooler in the summer and warmer in the winter. Wood or stone floors can keep us in touch with natural's rhythmic elements.

❖ The tactile sense—touching—is critical to our state of harmony. A lack of contact with others creates irritability, depression, and anxiety., Dr. Harlow's caged monkey experiments in the 1960s proved. We can seek opportunities to appropriately touch the people around us. A caring handshake, a gentle touch or a hug is healthy and will help build honest relationships.

❖ We can touch and feel the natural objects and life forms around us. This gains a tactile sense of the life's rhythms revolving around our lives. Wood, rocks, grass, plants, dirt and natural fibers are examples.

❖ We can choose to living in an outdoor, rural environment. An out-door environment radiates the earth's natural pulses within and around our bodies. Otherwise we can just get outside to the park or other natural setting frequently, or at least keep the windows open.

❖ We can also connect with the earth by looking at nature. The beauty of a mountain stream or ocean view can bring about a feeling of re-laxation and tranquility, evidenced by the many therapeutic retreats in mountainous and ocean-view settings.

❖ We can also bring the earth indoors. This means bringing stones, wood and plants into our households; using natural materials for rugs, seats and massage tools.

❖ We can wear natural materials such as organic cotton and wool. These materials provide the ultimate in thermal conduction. Natural fibers are cooler for summer wear and warmer for winter wear.

❖ We can choose natural crystalline stones, and natural metals such as gold, copper and silver for jewelry. These metals provide beneficial magnetic properties, along with some detoxifying properties—as they will attract and pull toxins from skin while providing ions important for many metabolic and enzymatic processes.

❖ We can choose cooking and eating tools made from natural metals. Copper, stainless steel and glass are excellent for cooking utensils. Glass in particular is extremely stable to cook with. Copper can also be used for drinking glasses and water pots, as it provides beneficial copper ions, useful for detoxification and general mineral nutrition. Using grounded elements will increase vitality and wellness for both our bodies and our living planet.

Chapter Four

Biological Water

All of us are attracted to the beauty and rhythms of water. Think about it. Higher-priced houses are typically near or on ocean fronts, lake fronts or river sides. Among those houses near the water, those houses with water views are the most expensive. For those houses not near the water, we often add artificial bodies of water such as ponds, fountains, and pools.

We yearn to connect with the rhythm of water. When we go on vacation we typically travel to water. We go to the lake, the ocean, the river, or the waterfalls. Again, we seek to connect with this flowing medium. Why?

The dimension of water is one of motion, flexibility, sound, and diffusion. Water stimulates relaxation, letting our body energies harmonize with the flow and rhythms of its movement. Once within water, we can flow with its tides and currents; surf its waves and eddies; and synchronize with its motion. Into water we can dive, piercing the surface like a knife; sliding into the depths. In water our bodies can soak, feeling the water pressure embracing our skin and filling our pores.

Water is the most abundant molecule on the planet. As most of us know, water is theoretically composed of two hydrogen ions and one oxygen atom, bound together with covalent orbital bonding. This means the oxygen atom theoretically shares an electron wave with each hydrogen ion. Through covalence, each hydrogen shares part of its energy orbit potential with part of the oxygen's orbit potential.

The oxygen atom has eight electron-waves. After the first completed orbital valence shell of two is filled out, oxygen has six electron-waves settled within an orbit shell capable of eight to fill out that shell's potential. The electron-waves (one each) from the two hydrogens each share their electron-wave potentials with oxygen, filling out oxygen's second valence shell. Since the oxygen is sharing an electron-wave with each hydrogen atom, each hydrogen atomic orbital shell of two electron-waves— also becomes virtually complete. The final water molecule of H_2O is now extremely stable and cohesive.

Water's molecular arrangement of valence sharing also creates two covalent hydrogen bonds on each side of the oxygen atom. This bonding situation creates a unique bond angle. While one might mathematically calculate a geometric 109-degree tetrahedral structure, water's hydrogens sit at 104.5-degree waveform angles, as observed using nuclear magnetic resonance and infrared absorption technologies. Chemists theorize this special bond angle is due to subtle repulsion-attraction forces between the two hydrogen electron-waves bonding with the oxygen.

While we might think all water is alike, there are a variety of different types of water, depending upon where the water has been and how it is molecularly structured. Which compounds it attaches to and what compounds dissolve in it depends upon its molecular characteristics. This relates to its quantum spin of its electron waves; its particular magnetic polarity; its ionic properties; its acidity; its mineral content and hardness; and its dissolved solids and pollutants.

While water's total dissolved solid level is typically considered its mineral content and even hardness level, its dissolved solids can also include toxic and industrial compounds. Furthermore, some consider 'hardness' to be specific to water's calcium levels, while others consider iron and other minerals adding to water's 'hardness.'

The molecular structure of water will vary with the environment. Environmental conditions can change the polarity and spin of water, which may in turn affect the water's specific gravity and surface tension. Water's polarity, spin, and surface tension directly affect its ability to resonate and adhere to cell membranes and other body fluids. This adhesion ability in turn affects water's ability to hydrate the body.

The earth's fluids and the body's fluids are holographic. Three-quarters of the living earth is covered with water. Water circulates through all living organisms as a vital fluid. In humans, 90% of our blood is water and about 75% of our muscle tissue is water, depending upon our age and health. At the cellular level, some 60% of the body's water is ideally intracellular and about 40% is extracellular.

Geologists estimate that about 96.6% of the earth's water lies on the earth's surface in the form of oceans, rivers, and lakes. Another 1.75% is estimated to be in the form of ice, and about 1.7% of the earth's waters are thought to be underground in the form of aquifers and underground rivers. Water is also held in soil moisture, in the atmosphere, and within living organisms. Still, these combined are minute compared with the other waters of the planet. Some estimates are that there are about 332,500,000 cubic miles of water on the planet (Gleick 1996).

Over two-thirds (about 71%) of the earth's surface is water. Over 96% of the earth's water is salt water. Of the remaining fresh water, more than 68% lies within glaciers and ice. Thirty percent of fresh water is under ground, leaving just under 1% of fresh water (which is 3% of the earth's total water) on the surface. Rivers only account for about 2% of this surface water, or 1/700th of 1% of the total water. Thus, useable fresh surface water is only about 3/10th of 1% of the total water on earth. This means that over 99% of the earth's waters are not readily useable by humans. Ground water pumping increases that only slightly, as does the current capacity to convert salt water to fresh water with desalination.

Very little of the water in our environment is drinkable. Surface water supplies about 90% of the water moving into vapor, and plant transpiration provides the other 10%. The sun's radiation breaks the water's inter-bonding, leading to vaporization. This process will occur until the air is saturated with vapor, leading to a relative humidity of 100%. Most of this water vapor stays over the oceans, while about 10% moves over land where it can fall into fresh water reserves. Water vapor moves towards colder regions. As the saturated air is cooled, condensation occurs. In this form water vapor changes into liquid water, retained within cloud formations. While water droplets are not exclusively held in clouds (fog and air also contains water droplets), clouds form the majority of traveling water.

Because the atmospheric pressure is less at greater altitudes, the air temperature is also cooler. Cooler air allows the water droplets to be retained within the clouds as they move about the atmosphere. This cooler air (below dew point) accompanies condensation levels higher than evaporation levels, causing clouds to form. Because air rises, and the air below a cloud is denser than the cloud itself, clouds float.

Cloud water constantly evaporates and condenses as clouds gyrate and temperatures change. There are two general theories of what causes rain. Some believe that precipitation requires the water droplets to condense upon particles of dust or salt. An accumulation of water droplets colliding with these particulates are thought to produce precipitation. Others believe—referred to as the *Bergeron-Findeisen process*—ice crystals form in clouds as temperatures cool. This causes the ice to fall out due to their increased weight. If the temperature near land is higher, the ice crystals melt. If the ground temperature is less, they drop as snow or ice rain.

In the 1980s, David Sands, PhD, a professor of plant sciences at the University of Montana and a team of researchers found that certain bacteria seemed to be present in most precipitation and ice events, leading them to propose an effect called *bio-precipitation* (Caple *et al.* 1986). In 2008, further evidence has uncovered that these bacteria may play a critical role in forming the ice which eventually falls to the earth as either rain or snow.

Once rain or snow falls, the surface water freezes or flows through rivers, some of which filter into the ground and become part of underwater rivers. These rivers form pockets of water storage. They also flow out to the oceans—just as surface rivers do. This outflow of fresh waters through surface and underground rivers complete the water cycle. The circulation process around the earth is very precise. For this reason, we can almost predict within a few inches how much annual rainfall each particular region around the planet will receive.

As water moves, it has considerably more energy than when it is stagnant. While moving within a living system, water brings together various minerals and energetic vibrations, sparking enzymatic reactions that create a host of byproducts. Interestingly, the composition of ocean water and the composition of human blood are very similar. Both contain similar levels of electrolytes, including magnesium (3.7% by weight in ocean, 4.8% by weight in blood); sodium (30.6% vs. 34.8% respectively); potassium (1.9% vs. 1.1%); calcium (2.1% vs. 1.2%); sulfide (7.7% vs. 10.9%) and chloride (55.2% vs. 40.1%). We can thus say with certainty that not only are the earth and the human organism both made up of about 70% water, but the content of most of that water is also extremely similar. We note that other living organisms vary slightly in their water content from humans as well.

Within living organisms water travels in channels of current. On the earth, we see this same motion, in the form of channels of rivers, streams, and aquifers. Within the oceans, we also see channels of tides, eddies, currents and streams of differentiated water moving. We also see these same circulatory effects as we follow fluids circulating through the blood vessels, capillaries and the lymphatic system. Within the basal membranes, we also see extracellular water moving in channels, and we see streams of water flowing in and out of the cell through ion channels embedded within cell membranes.

Because water belongs to the chemical family of hydrides, its melting and boiling points should correspond with the rest of the family— regulated theoretically by its molecular weight. However, this is not the case with water. If water were aligned with the characteristics of other compounds in the hydride family it would not be in liquid form at the temperatures of living systems—notably on the earth and in the body. It would evaporate and exist as vapor at these temperatures. Water is a very unique substance.

The extraordinary stability of the rhythmic bonding between two hydrogen waveforms and an oxygen waveform combo renders water the unique ability to have a higher boiling point than expected and a lower melting point than expected, as well as an abnormal amount of *surface tension*. Surface tension is created by the weak hydrogen bonds that occur between separate water molecules. These weak bonds pull together, creating a surface area and adhesion. Water's surface tension allows it to retain its shape and structure throughout its travels. Without surface tension, water would simply seep into its surrounding environment and disappear.

The reality of liquids as waveform enablers was illustrated in 2007 when amateur inventor John Kansas developed a radio wave frequency generator able to ignite salt water. This effect, corroborated by a bevy of university physicists and chemists, astounded the scientific world. As seemingly harmless radiowaves collide with saltwater, it ignites a flame of over 1500 degrees F. How could saltwater become so flammable? It is a testament to the strength of the bonding orbitals within water. As water's rhythmic bonds interact with radiowave rhythms, their interference pattern translates to thermal waves.

Water's surface tension also translates and refracts other types of waveforms. As the visible light spectrum collides with water, part of the light will refract through the water and part of the light will reflect off the water's surface. Its refractive strength is also evident when we see an upside down reflection of a scene on the other side of a lake. We also notice this effect when we hear and see sonic waves transmitting through the medium of water. Many species of life, such as whales and dolphins, utilize sonic waves to communicate.

Water's surface tension renders it the ability to form a sort of thin membrane enabling it to resist entry to various substances and objects. This aspect gives water and its surface tension the ability to support and protect living organisms.

Water's inward surface tension combined with the angular orbital bonding renders raindrops, dewdrops, and bubbles almost perfectly spherical in shape. With this convex inward surface tension we also find, as D'Arcy Wentworth Thompson pointed out almost a century ago, that the surface tension of water is at least partially responsible for the appearance of rainbows through water vapor.

Water's surface tension and unique structured fluidity is due to the existence of *weak hydrogen bonds* existing between different water molecules. These weak electromagnetic forces contain a dipolar element that draws the water molecules together into structured formations. When water freezes, we observe rigid formations as crystal lattices. While in the liquid form, this rigidity is lost. Still, water's intermolecular weak hydrogen attraction creates a unique activity known as *water clustering*.

The first observations of water clustering were presented to scientific community in the 1950s. X-ray diffraction techniques illustrated water molecules clumping into groupings, sometimes only for short periods. The clusters seemed at first to be flickering and random. Over the years, observations with diffraction as well as nuclear magnetic resonance have established that these weak dipole interactions between hydrogen bonds do certainly create periodic clustering of groups of water molecules in a variety of situations. Clusters from 25 to 90 water molecules have been observed, forming a number of distinct geometric structures.

One of the more recognized researchers in this area of water clustering was Dr. Mu Shik Jhon, who from the 1950s until his death in 2004 investigated and participated in more than 280 research papers over a distinguished career in science. As a former president of the Association of Academies of Sciences in Asia and president of the Korean Academy of Science and Technology, and an assistant professor at the University of Virginia, Dr. Jhon's research into the molecular structure and properties of water are well established and documented. Dr. Jhon's research indicated that although cluster sizes can vary greatly, there appears to be more frequent structuring of water within five or six molecule groups. Seemingly because resonating electron-wave orbits are positioned at particular angles, the resulting clusters were observed in his research to have either pentagonal (five-sided) or hexagonal (six-sided) structures. Dr. Jhon's research also indicated that hexagonal water clusters appeared to have more stability in the presence of calcium and sodium, key macro-minerals that regulate most living organisms.

More complex and larger water cluster structures have also been observed. Research by Professor Martin Chaplin of London South Bank University reported that water would form icosahedral clusters quite frequently as well. This cluster is formed with twenty 14-molecule tetrahedral units, totaling 280 H_2O molecules.

These types of clusters are believed to form and disappear within short windows of time and thus have been described as fleeting. While these fleeting clusters of H_2O molecules certainly are not well understood, their existence nevertheless stands up to scientific scrutiny. The debate today regarding water clusters appears to be what actually causes these clusters to appear, and why they dissolve so quickly.

Physicists Giuliano Preparata and Emilio Del Giudice have mathematically confirmed this nature of water in a 1991 *Physical Review Letters* paper, calling the clustering of molecules into groups *"coherent domains."* This paper followed an experiment where the radiation fields of water were tested before and after additives were added. They reported the electric dipoles in water molecules appear to interact with coherence. They described a type of *"collective dynamic"* existed in the water, because the resulting polarity fields of the water were consistent with the polarization of the additive.

At 32 degrees Fahrenheit or 0 degrees Celsius, water will crystallize rapidly into lattice form. This provides another type of surface area, which retains the dipolar and electromagnetic resonance into a very solid lattice structure. During the process of forming this lattice, crystallization occurs. Partially frozen water will form into beautiful crystal shapes, many resembling delicate snowflakes. As a result, we see a wide variety of crystal shapes form as water becomes frozen. As researchers have observed, these crystal structures can vary greatly in size and dimension.

Water Memory

In the 1990s, Masaru Emoto and an assistant began taking photographs illustrating water crystal formation under different circumstances and influences. He first published these findings in 1999. Emoto's photographic images implied that water crystal formation varied not only to water sources, but also to interactions with music, spoken and even written words. Water exposed to different types of music theoretically formed different crystals: classical music created full symmetrical crystal shapes while hard rock created unsymmetrical and disoriented shapes. Water exposed to different types of words or phrases theoretically formed different crystals: uplifting words theoretically created full, symmetrical crystals while words of hatred or anger theoretically created disoriented shapes (Emoto 2004).

This research became well publicized yet controversial among the scientific community. Some researchers have decried Emoto's reports as lacking the rigor acceptable for peer-review. While the photos themselves create little doubt regarding the variability of ice crystal formation, the question of his research boils down to the extent controls were applied to the process of taking and choosing photographs for publication. Because we now understand that the same water can form a variety of ice crystals shapes, Emoto chose one photo to publish to represent each scenario. Was there any bias in the selection of crystals to publish? This concern has yet to be adequately resolved. Therefore, although Emoto's crystal photograph research is intriguing, and may indicate water's ability to reflect consciousness, Masaru Emoto's research falls short as quotable evidence linking consciousness with crystallization. (The author contacted Mr. Emoto with these questions, and was referred to someone else who did not respond.)

Still there are a number of undeniable characteristics within water that confirm a sort of 'memory' or reflection of consciousness. Water's memory capacity is easily observed should a rock be dropped into a pond. For minutes even hours afterward, the resulting rhythmic waves traveling away from the entrance point of the rock specifically reflect the size of the pebble, its velocity, and even the shape of the rock to some degree. A small rock will create a different rippling than a large rock might. A flat rock will create a different waveform than a round rock might. The size and shape of the ripples will also reflect the velocity of the toss into the water. A harder toss will be reflected quite differently than a light lob might. These ripples will reflect the information about the rock and the throw for some time, affecting various other events occurring within the pond.

We all also know that when water is heated and cooled, it will retain a 'recollection' of that temperature input for a period of time. The hotter the initial flame, the longer the water will remain hot. An electric stove will heat water at a different degree than a gas stove might. Thermal heat from the sun creates still another temperature range. As water is cooled, again it reflects the cooling source. Water cooled with ice will cool differently than water cooled in a refrigerator or even freezer. Though we might expect the result to be proportionate to the temperature the water is exposed to, in reality, different sources have different effects. Those varying effects will also dramatically change the water's characteristics for hours or even days following the changes. Although this 'thermal memory' appears resident among most substances, the presence of water hastens the thermal conduction and retention process, which makes water an efficient thermal conductor. For this reason, food is most often cooked in water as opposed to any other substance. Water is also used in various other thermal conducting mechanisms such as heated floors, baths, and so on. When in need of an instant cooling mechanism, water is also sought after for its ability to immediately conduct cool temperatures. Water quite easily 'remembers' the temperature of its surroundings, easily transporting that memory to other substances.

Water's memory is also observed with regard to water's solubility and surface tension characteristics. When we dissolve a substance into water, the water's specific gravity and surface tension will change, reflecting the properties of the added substance. This solvent will also change the water's boiling and melting points. As the solute is precipitated out of the water, the water will often retain a variance in these characteristics as compared to the original water source. This variance may be caused by the presence of additional hydrogen ions in the water, or by the existence of additional water clustering. However, water's character is altered establishes a sort of memory capacity. While other substances can be used as a solute, water is an efficient solute because of its molecular properties.

As we have touched upon earlier, water's ability to retain memory has been clinically examined through the medical science of homeopathy, beginning with Dr. Samuel Hahnemann's original research with dilutions two centuries ago. While the *like cures like* portion of homeopathic therapy is well-accepted by modern medicine (the basis for vaccination among other therapies) the notion that water will retain a distant memory of a substance diluted to the point of theoretically diluting every molecule of the substance away is not acceptable to much of mainstream science today.

As we discussed earlier, research on water memory was advanced greatly by Dr. Jacques Benveniste. Dr. Benveniste, a successful French medical doctor, discovered accidentally in 1984 that white blood cells responded to an allergen in a solution despite there theoretically being no remaining antibody molecules in the solution. This led to hundreds of studies among Dr. Benveniste's team and other research labs, which mostly replicated these results. These results confirm that a water-based solution somehow retains memory after full dilution. Over 300 trials were performed confirming these results.

In 1991, Benveniste developed a system of amplifying molecular signals through sensitive electromagnetic microcoils and transducers. After a few years of application, the process was refined to the point that his research team was able to record molecular emissions into digital form. The molecular signal associated with the digital recordings indicated frequencies in the 0-22 Khz range. Incredibly, the digital recordings could be played back through an amplified transducer in the presence of a particular reactive organ, such as cholinergic activity among pig hearts. His digital playback resulted in the same physiological result the biochemical hormone might have—without the physical biochemical substance present. Dr. Benveniste demonstrated this effect emphatically when he was able to send disks or email recordings to labs in remote locations. In these cases, the playback of the recordings would have the same effect (Benveniste 1997).

As Dr. Benveniste continued his research, he discovered that the effective transmission of the signal had some dependency on the mixing system employed. Without the proper mixing process, the diluted mixture's ability to affect the same result was substantially decreased. On the other hand, proper mixing resulted in a significantly greater effect (22.6 vs 3.2 coronary flow changes in an acetylcholine dilute, for example) when compared to the solution *prior to mixing* (Benveniste 1999).

The ability of water molecules to retain particular electromagnetic waveform interference patterns is illustrated using these data. As we established earlier, a waveform basis for matter is supported by a century's worth of physics research. We might compare this with voice vibration. In order for voice vibration to be instructive, there must be a precise manipulation of waveforms striking the eardrum, which contain the message of the speaker.

The transmission of sound and visual signals via radiowaves parallels this fundamental process. It also provides an illustration for the ability all waveforms have in carrying specific informational signals.

As we consider a particular area of ocean and its net motion at any particular place and time, we could certainly connect a myriad of atmospheric and tidal events to the current motion and condition of the water. There may have been a large windstorm thousands of miles away that drove some of the waves. There may have been a hurricane storm front, which drove another set of waves from another direction. There may have been a large tidal change due to the moon's orbit, which also influenced the tidal pulses within the ocean's motion.

This holds true for sonar wave composition. There should also be a net effect of the various intelligent transmissions of dolphins and whales as they communicate amongst their own schools. If we take this net effect of all the movement within the ocean, we have the basis for a medium of memory for the reflected rhythms of all these conscious inputs. If we were to have an instrument sensitive enough to pick up all these subtle waves moving through the medium of the ocean from these various sonar waves, we have a confluence not unlike the random number generator and the global consciousness project results from the research of Dr. Schmidt, Dr. Jahn and Princeton's Dr. Nelson as mentioned earlier.

We might also compare this process to the laying down of music tracks at a recording studio. In order to provide a platform for each instrument to be recorded onto one master recording, there must be a process of combining separately recorded music sound tracks. Each track is recorded separately and then mixed together onto one master track. When this master track is played, all instruments are played together. By listening to that one master blended recording, an intelligent studio listener would probably be able to discern the different tracks and instruments recorded separately. This method of mixing reflects a memory device able to blend a number of inputs.

A number of researchers have confirmed water's ability to retain or reflect the touch of a therapeutic practitioner. Dr. Edward Brame, Dr. Douglas Dean, Dr. Bernard Grad, and others have either led or co-authored studies—some confirmed by infrared spectroscopy analysis—that healer-treated water maintained molecular changes reflecting therapeutic touch. In more than one of these, the molecular bond angles had slightly shifted. In others, decreased surface tension of the water confirmed a subtle molecular change. In still others, the rate of growth was effects (Dean 1983; Dean *et al.* 1974; Grad 1964, Grad *et al.* 1984; Schwartz *et al.* 1987).

Emission bombardment effects have been well established by physicists in mass spectrometry research. Because each electron orbit has particular waveform or quantum characteristics, bombarding an atom with radiation of the right frequency will boost certain electrons into lower or higher valence shells, affect the orbit's spin or angular momentum, and/or eject an electron/wave out of the nucleus "orbit" altogether. This effect has been the subject of numerous mass particle accelerator studies, diffraction results, and spectrometry analyses. Brainwaves and body waves are certainly types of radiation as well. There is currently no documented evidence known to this author that would eliminate the ability of these "biological" waveforms to influence electron/wave orbits of molecules touched or otherwise interfered with.

Water Physiology

The makeup of our cells is primarily water. Each cell is also bathed in water. The basal fluid provides an ionic-balance between the water inside the cell and outside the cell membrane. The ionic nature of the water inside the cell membrane is typically charged with negatively charged potassium ions, while the water inside the cell membrane is charged primarily with positively charged sodium ions. This creates what is often called the *sodium/potassium pump*. The sodium/potassium pump provides a dual-balanced mechanism of attraction between the sodium and potassium ions.

This ionic charge difference creates an electromagnetic attraction between the inside and outside surfaces of the cell membrane. This gives the membrane a stable partition. Yet while providing a partition, the ionic gateway channeling contrasting electromagnetic moments provides a vehicle for the exchange of important fluids through the cell membrane. Through tiny channels in the membrane created by spaces between stacked phospholipid molecules come nutrients and water into the cell, escorted by sodium ions. Just within the surface of the membrane is a tiny rhythmic pump network called the *protein pump*. These pumps push the escorting sodium ions that slip through the cell membrane back outside, carrying with them the waste generated by the cell. This entire process is often called *cellular diffusion*. Diffusion allows each cell to bring in nutrition and pass out toxins through the cell membrane, utilizing this ionic attraction and pump process.

As we have discussed previously, there are different types of ion channels throughout the body and within each cell membrane—each of which brings different types of nutrients into the cell. For example, there are specific channels that bring glucose into the cell, often called *glucose receptors*. These channels are connected to specialized gates which are turned on and off by receptors which sit on the surface of the cell membrane. Receptors are switches turned on by special messenger cells like hormones, equipped with ligands that electromagnetically signal the receptor switch. While most ligands are thought to require a touching of biomolecular particles, research is increasingly indicating many ligands communicate from a distance, using rhythmic waveforms.

In the case of channeling glucose into the cell, glucose receptors are switched on by the signaling hormone *insulin*. In the same way, other channels and pumps provide for the flow of the fluids and nutrients that hydrate and nourish our cells. Many researchers theorize molecules line up and move through these conduits one by one when the gateway is open. While this is certainly logical given the size of these tiny channels, this theory is limited by the unproven atomic particle theory. Assuming matter is composed of waveforms, the process of entry would rely upon the specific molecular signals being emitted by the ligands. Input into the cell membrane through receptors and channels appear be similar to light entering into the eyes, or sound entering through the ears: In the case of light, inbound radiation is received and processed through receptors on the retina cells of the eye. In the case of sound, radiation is channeled into the ear canal, stimulating the eardrum *receptor* and signaling through the inner ear transmission system. In each case, waveform emission stimulates the activity, rather than biochemical particles.

A system of sodium and potassium pumps also occurs within nerve cells. In addition to providing channels for the exchange of nutrients and toxins, the nerve cell sodium/potassium pump provides a mechanism of the exchange of information. As electromagnetic pulses traverse the length of the nerve cell, sodium ions are pulled into the nerve cell, while potassium ions are pushed out. Each ion movement instigates the movement of the next ion wave, much like a row of dominos laying down one after another. The electrical conduction through sodium and potassium ions provides the pathway for rhythmic pulses. This mechanism appears to function somewhat similar to the 1s and 0s of computer machine language code. As most of us know, this rather simple computer machine code system of 1s and 0s allows for dramatically complex instructions. The key to the sodium/potassium mechanism is the rhythmic balance established between the potassium and sodium ions. It may go without saying that maintaining this delicate balance between the potassium and calcium ions is the key to healthier cellular function.

pH is the measure of the ionic capabilities of a particular solution. We can thus utilize pH as a tool to determine the ability of our body waters' ability to channel waveform instructions efficiently through the membrane. The body's fluid pH is critical to metabolism because it is through ionic conducting mechanisms that information is transmitted through the cells, organs and other tissue systems. The ionic nature of the body's waters provides the vehicle for the rhythmic flow of consciousness throughout body. Should the ionic balance of the body's fluids not be properly maintained, we will soon discover a number of key metabolic processes being blocked or distorted. Without the proper flow of information conducting through the ion channels of the cell membrane, the cells cannot act harmonically.

We find this condition evident when a person has consumed alcohol or psychotropic drugs. The body's ionic fluids are altered, most notably the neurotransmitter fluid, which provides the medium for the pulses traveling through nerve cell ion channels. As the neurotransmitter fluid chemistry is altered, the information pulses traveling from nerve to nerve are altered, affecting coordination, physical response, perception, mood, and other functions.

To some degree, we might compare this mechanism with looking through a clear water stream followed by looking through the stream after it became darkened with mud. Through the clear water, the rocks below are easily visible. The muddy stream on the other hand, either may block our vision of the rocks below altogether or may alter their appearance. They might appear smaller or deeper than they are. Because the fluid is clouded by mud, the ability of light to travel through the water is affected.

This is illustrated by the same general effects of intoxication. A heavy drinker will usually have cloudy or bloodshot sclera. The pupils will not readily be responsive to light, and thus may appear overly dilated or constricted for that environment. They may speak with lethargic or slurred speech. They may find difficulty forming words from their thoughts, or may find it difficult to control their language. They might also be easily angered or upset, as their perception may be clouded. They may misinterpret an events and or context. They may also have a lack of control over their emotions or appropriateness. Alcohol intoxication in particular will often also result in a decreased sense of balance and coordination with a heightened sense of confidence. These are all symptoms of an altered ion channel motion. The ionic—thus magnetic—imbalances within the body's fluids created by the intoxicant disrupt the normal ionic flow of minerals, nutrients and neurotransmitters through the various ion channels.

Much of the body's biochemical processes are *hydrolytic*. In other words, most of the body's biochemical metabolic reactions are water-dependent, requiring hydrolysis. The hydroelectric energy and ion transport through the cell membrane allows many of these processes—such as the ATP energy manufacturing cycle—requiring water as the essential ion buffering and transport foundation. Thus water is one of the main substrates within a vast range of metabolic biochemical processes within the body.

When the body becomes dehydrated, a number of physiological responses take place. Initially, when the cells among the least vital areas of the body become exhausted of water, various ionic channels will close to preserve fluid for vital tissues. As this happens, the movement of ions, nutrition, and fluids in and out of those cells will slow down or even shut down. Typically the body prioritizes and rations its availability of water very pragmatically and intelligently. The areas where cell membrane channels close first will be the less vital regions of joints, tendons and other less active areas—away from the vital circulatory-rich organs. Various messenger cells such as histamine, vasopressin, rennin-angiotensin, along with supportive prostaglandin and kinin activity support and communicate the body's water level adaptation processes. After the body orchestrates its prioritization routines, should enough water still be unavailable, channels of the surrounding tissues also begin to close down.

As a result, we find that a number of disorders are related to chronic dehydration. For example, many arthritic and joint issues appear be directly related to dehydration. In the case of joint issues, water is reprioritized to other parts of the body. As this effect matures, a lack of hydration can either directly cause or contribute to the deterioration of joints and their supporting structures (Batmanghelidj 1992).

Negative effects upon cognition and psychomotor performance has been observed when at least 2% of the water by body weight is lost due to dehydration according to Grandjean and Grandean (2007). This has been established through the use of various neurological testing and other neurophysiology tests.

Other pathologies appear to be related to even mild dehydration as illustrated by Manz (2007). These include cystic fibrosis, renal toxicity, urinary tract infections, constipation, hypertension, various coronary and artery disorders, and glaucoma. Evidence from the CF/Pulmonary Research and Treatment Center of the University of North Carolina (Boucher 2007) has revealed the possible link between hydration and cystic fibrosis as "low airway surface liquid volume." This is another way of saying that the mucosal membranes lining the airways require good body hydration.

We also can link a lack of water consumption with various digestive disorders such as GERD (gastroesophageal reflux disease), ulcers, and irritable bowel syndrome (Batmanghelidj 1992). In a state of dehydration, the pH chemistry of the stomach radically changes. Again, the mucous membranes of the esophagus and stomach wall become thinner and less protective. This is because water is the central component of mucous and gastrin, which in turn buffers the HCL component of stomach secretions. Water is also a major element in the sloughing and division of the cells of the walls of the stomach and intestines. Intestinal cells are some of the shortest living cells of the body. Cells of a healthy digestive tract will divide and slough off within days, leaving the body with virtually a new stomach mucosal lining within days. Should these dividing cells become dehydrated, genetic mutation may also take place during division. This mutation may allow the cell to function with less water, but often at a price. It may also result in other adjustments, which may decrease the healthy functioning of the stomach and its lining. Once mutation occurs, the cell may assume foreign attributes. As the immune system is exposed to these foreign attributes, a condition referred to as *autoimmunity* may arise.

These are only a few of the potential effects of chronic dehydration in the human physiology. Others may include headaches, heart problems, hypertension, allergies, asthma, hiatus hernia, low back pain, diabetes, and so many other ailments—many of which have interestingly been described as "autoimmune diseases."

Hydration

There are two central issues to consider as we determine appropriate water consumption. The first is water loss and the second is water absorption, which relates directly to quality and volume. Obviously, the two issues are inter-related. As Dr. Jethro Kloss pointed out decades ago (1939), the average person loses about 550 cubic centimeters of water through the skin, 440 cc through the lungs, 1550 through the urine, and another 150 cc through the stool. This adds up to 2650 cc, equivalent to a little over two and a half quarts (about 85 fluid ounces).

Most health professionals suggest optimal water volume around eight 8-oz glasses per day. In 2004, the National Academy of Sciences released a study indicating that women typically meet their hydration needs with approximately 91 ounces of water per day, while men meet their needs with about 125 ounces per day. This study also indicated that approximately 80% of water intake comes from water and beverages and 20% comes from food. Therefore, we can assume a minimum of 73 ounces of fresh water for the average adult woman and 100 ounces of fresh water for the average adult man should cover our minimum needs. That is significantly more water than the standard eight glasses per day—especially for men. It is not surprising that some health professionals suggest that

50-75% of Americans have chronic dehydration. Fereydoon Batmang-helidj, M.D., probably the world's foremost researcher on water, suggests 1/2 ounce of water per pound of body weight (1992). Others, like Anna Maria Sapugay, M.D. (2007) recommend 64 ounces of water per day with an additional 32 ounces for each forty-five minutes to an hour of strenu-ous activity. She suggests consuming 16 ounces of water before and 16 ounces after exercising, in addition to a few sips during exercise. Extremes in temperature and elevation increase our water requirements. Additional water is also required in the case of fever or increased sweating.

Most experts agree that as soon as we feel thirsty, our bodies are al-ready experiencing dehydration. Becoming consciously thirsty is the point where tissue and cell damage takes place. Water is critical for the smooth running of all of our cells. Areas likely to suffer during dehydration in-clude the stomach and digestive tract; the joints; the eyes and the liver. Some health experts have estimated as little as a 5% loss of body water will decrease physical performance by up to 30%. Dr. Sapugay suggests watching our urine to make sure we are getting enough water. Our urine color should range from light yellow to clear. Darker urine indicates dehy-dration. A bright yellow urine color typically indicates the supple-mentation of riboflavin (B2)—present in many multivitamins.

Pregnant or nursing women may want to increase their intake by two to three glasses per day. Most beverages can be included in that total. The Academy study mentioned above includes caffeinated beverages hydrate at a one-to-one ratio. However, we should note that caffeine is a mild diu-retic agent, so if caffeinated beverages are being consumed at above-average levels, more fluids are needed to accommodate for the increased fluid loss. An additional four ounces should be added for every eight ounces of coffee or caffeine soda included in water intake according to some experts.

We might also consider that if a dominant portion of water intake is sugary sodas or diet sodas with chemical sweetener-replacements, addi-tional water is required to provide additional detoxification.

Sodas with sugar and caffeine are thought to have a negative effect upon hydration levels. Sodas are also known to have refined corn syrup, fructose, and/or sucrose as well as preservatives like sodium benzoate. According to 1999 research led by Dr. Peter Piper from the University College in London (Kren et al 2003), sodium benzoate appears to harm the mitochondria of cells, causing genetic damage and starving those cells of their energy production capacity. Sodium benzoate is also a common ingredient in pickles, some Chinese sauces, jams and some fruit juices.

Our bodies utilize water during sleep. Therefore, it is important to hydrate immediately upon waking. A large glass of warm or room temperature water first thing in the morning will practically be absorbed immediately, replenishing much of the water loss of the night before. According to Ayurvedic principles, water taken prior to brushing our teeth will also deliver to our stomach much needed mucous from our mouth and esophagus, aiding the stomach's gastric lining.

Mineral Waters

The necessity of drinking water everyday leads to a discussion of the purity and ionic nature of our water. Rainwater typically has a pH in the region of about 5.6 to 5.8 due to carbonic acid. Levels below this (less than 5.6 pH) is considered *acid rain*. When carbon dioxide dissolves in the presence of water, it forms carbonic acid: The more carbon dioxide presence in the air, the greater the carbonic acid formation. Thus, most of our rain is primarily acidic in nature, and is not exactly the best source of hydration. Certainly we can assume that rainwater was an excellent source of hydration before humankind began burning fossil fuels for energy. Because carbon dioxide has been present in our atmosphere for millions of years, we can assume that rainwater has always had some acidity. However only extreme situations such as an intense volcanic eruption could approach the acidity created by the massive carbon monoxide and sulphur levels currently existing due to humankind's endeavors over the past century.

Water that has traveled from snow pack or rain-drenched field through rock is typically alkaline, however. As water travels through a creek or underground aquifer, it picks up the minerals and the ionic qualities of the surrounding soils and rocks. The most prevalent mineral gleaned during water's travel is calcium carbonate, which is present in most rocks and limestone. Magnesium is another major mineral present in rocks—also prevalent in most spring or well water. Magnesium is a great alkalizer and a necessary mineral for muscle and cardiovascular efficiency. A number of other minerals—both macro and trace—are also picked up from rock as water flows through, depending upon location.

As water flows through rock, minerals are pulled away from their current associations as ions, and associate with water. Common minerals can include selenium, silica, iron, copper, potassium, sodium, and many others. Mineral profiles of rock deposits show rocks can contain up to 80 trace minerals. 'Trace' usually means content in the parts per million or parts per billion. While passing waters may not pick up every mineral, the further a water has transversed the rocks of the planet, the more nutrients it probably contains, and the cleaner it will probably be. This is because minerals draw toxins through chelation and covalent boding. Mineral ions thus pull various contaminants, bacteria and pollutants out of water.

Research is still investigating the human need for many minerals. Mineral water is known as a great source of calcium, and calcium is one of the better-researched minerals. A number of studies have shown that absorption of calcium from water rivals calcium absorbed from milk (Couzy 1995; Heaney 1994). These studies have also showed no significant excretion difference, illustrating that calcium from water is utilized well by the body.

Furthermore natural mineralized water or *hard water* has been linked with improved cardiovascular health (Tubek 2006). What elements in hard water make it beneficial? Suspected are calcium, magnesium, cobalt, lithium, vanadium, silicon, manganese, and copper. Each of these minerals has been shown to be beneficial to the body in one respect or another. Perhaps a macro view is essential for understanding the true benefit of the various minerals obtained through natural hard water. Just about every enzymatic and metabolic process in the body utilizes minerals. Also, behind each required macro mineral lies supporting minerals that balance their association, as they exist within an ionic balance. In its proper view, the combined synergy of various naturally occurring trace and macro minerals in ionic association gives natural mineral water (and natural foods for that matter) its therapeutic value.

There is considerable evidence indicating that cancer risk is lowered and possibly prevented through the ingestion of a balance of trace minerals (Navarro *et al.* 2007). Selenium has been linked with lower prostate and lung cancer risk. Zinc has been linked with lower breast cancer risk. Arsenic has been associated with lower lung and bladder cancer risk. Cadmium has been associated with lower lung cancer risk. While research has primarily focused on individual minerals, we can logically assume that since a deficiency in each causes one or another issue, their combined presence is required.

Lawrence Wilson, M.D., one of the leading experts on mineral balancing, documented many years of clinical experience in analyzing subjects' hair and blood mineral testing and treatment. Dr. Wilson (1992) prescribed various mineral supplement strategies to rebalance deficiencies in one mineral or other. Dr. Wilson also established various mineral balancing ratios between macrominerals such as calcium, sodium, potassium, zinc, lead, copper and others. For example, Dr. Wilson established using hair analysis, that sodium-to-potassium is at an ideal balance at around 2.5:1. Levels below this ratio illustrate issues related to sodium pump mechanism problems, which also accompanies adrenal stress and possible exhaustion. Dr. Wilson also determined the optimal sodium-to-magnesium ratio was 4.17:1. Ratios below this, yet above 1:1, were linked with adrenal over-activity, leading to high blood pressure, high blood sugar, and excess gastric acidity. Dr. Wilson's work also indicated that the optimal calcium-to-potassium ratio was around 4:1. A higher ratio indicated sluggish cellular thyroid activity, while lower ratios indicated overactive cellular thyroid activity. Dr. Wilson also documented various other mineral relationships with various pathologies.

Dr. Wilson also found that certain minerals become metabolic "analogues" or substitutes for deficient minerals. In a zinc deficiency, for example, copper and cadmium will often substitute zinc in enzymatic and metabolic reactions. These can often create anxiety and nervousness. In situations of calcium deficiency and lead overexposure, lead can substitute in metabolic functions typical for calcium. Dr. Wilson observed that this substitution created a propensity for seizures. Using precise mineral rebalancing therapies, Dr. Wilson has contributed to the recovery of thousands of patients over the years.

To date, research has yet to determine the minimum allowances for most of the trace elements, let alone many of the macro minerals. For example, various studies have confirmed that certain trace minerals such as zinc, selenium, copper, arsenic, cadmium, and silica are necessary for health (Bourre 2006). However, their precise deficiency levels have proved difficult to establish, as protocols for determining cellular mineral deficiencies and their effects remain elusive. While blood tests have been predominantly been used to gauge mineral levels, the blood is constantly shifting minerals from different tissue locations around the body as needed. Blood levels are thus difficult to determine deficiencies among tissues or cells. We might assume that high levels in the blood indicate high levels in the tissues. This assumption conflicts with other clinical research. For example, we know that blood calcium levels can be quite high (hypercalcemia), yet bone calcium levels can be low, and bone loss is occurring concurrently. The leaching of calcium from the bones to maintain blood levels of calcium has been established, and this has been elucidated over the past few decades by Dr. James McDougal in his lectures and books. It is for this reason that Dr. Wilson's work relied more heavily on hair analysis, which indicates tissue levels of mineral content, rather than blood levels. Others have criticized hair analysis, pointing to accuracy issues. Dr. Wilson answers that concern with protocols that minimize inaccuracies of sampling and testing.

Water softeners pull out important minerals like magnesium and calcium out of water using a combination of sodium and zeolite minerals.

Zeolites are ion-exchanging minerals that attract mineral ions, drawing them out water's association. As the zeolite beads become saturated with minerals, the recharging system on the water softener passes the beads through salt crystals, replacing the mineral ions with sodium ions. This process leaves a tremendous amount of sodium in the water. In one study (Yarows *et al.* 1997), softened well water varied from 46 to 1219 milligrams per liter with an average of 278 mg/L, *outside of the sodium already contained in the water.* 17% of the households tested above 400 mg/L. The level of sodium will be proportionate to the amount of minerals in the original water—in other words, how hard the water is. This study also measured various municipal waters, which had an average of 110 mg/L, with the highest registering at 253 mg/L. This means that if a person is on a restricted sodium diet of say 1500 milligrams a day, just two liters of household water with a 400 mg/L sodium content would make up more than half the day's water needs.

A number of studies have suggested that hard waters with magnesium and calcium had a protective effect for cardiovascular disease and heart disease. More than eighty studies have documented research in this area and many of these were reviewed and summarized by the *World Health Organization,* in a 2004 report called *Nutrient minerals in drinking-water and the potential health consequences of long-term consumption of demineralized, remineralized, and altered mineral content drinking-waters.* This report suggested a number of negative health consequences from drinking demineralized water. As reported in the 2004 draft to the WHO, by F. Kozisek, *Health Risks from Drinking Demineralised Water,* research has suggested that distilled or otherwise low mineral waters is less thirst quenching and potentially has negative effects upon the linings of the digestive system.

While the effects are not clear, it appears that intestinal epithelial cells undergo alteration without mineral ions present in our drinking water. The above report also underscored research showing consuming low mineral content water also has negative effects upon the body's homeostasis mechanisms, thereby altering the body's enzyme and hormone mechanisms. Increased dieresis (urine output); increases excretion of cellular ions; decreases secretions of thyroid hormones and aldosterone (ADH); increases cortisol levels; alters kidney function; lowers red blood cell volumes; and reduces bone ossification—increasing bone mass loss. This report cited further research showing that distilled and demineralized waters causes a repositioning of water around the body, decreasing intracellular fluid, increasing extracellular fluid and increasing plasma fluids. This activates further increases in urination and mineral loss.

Epidemiological studies over the past ten years have confirmed that soft water drinking increases the risk of mortality from cardiovascular disease. Soft water has also been associated with higher risks of bone fracture; increased risk of certain cancers; heightened risk of motor nerve damage; increased risk of sudden death; and higher levels of pregnancy disorder (Monarca *et al.* 2003; Nardi *et al.* 2003; Sauvant and Pepin 2002).

The WHO's reports have suggested that concentration ranges of 10-30 mg/L of magnesium and 10-50 mg/L of calcium were enough to cause a beneficial effect upon cardiovascular disease rates. While questions have been raised on whether mineral waters have adverse effects upon kidney stones or gallstones, there are a myriad of suspected causes for these stones, predominantly related to diet. Ramello *et al.* (2000) confirmed no correlation between water hardness and kidney stone formation.

To the contrary, both calcium and magnesium—like many other minerals—are extremely important to the overall health and metabolism of the body. While calcium is involved in muscle contraction, nerve conductance, bone structure, heart contractibility, blood vessel wall and lung flexibility; magnesium is involved with at least 300 critical enzymatic functions. Notably it increases flexibility among muscles and artery walls, while stimulating nerve response. Calcium and magnesium play many other important roles in the body. Excluding them from water not only creates an absence of these important rhythmic elements, but causes further imbalances upon the various other trace elements of the body.

Research has also confirmed that naturally mineralized water increases water absorption. This occurs on a rhythmic level. The ionic structure of mineralized water resonates better with the rhythmic nature of the human cells. This is due to the similarity between the pH range of mineralized water and the pH of the fluids of the body. The body's fluids are slightly alkaline, and so are the earth's fluids due to their interaction with minerals. The earth's water is composed of many minerals such as calcium, magnesium, and others due to its interaction with rocks, sand, and soils. The mineral content of water orients the water with respect to its pH and ionic polarity. These influences orient the water's spin and surface tension. While the ions within natural mineral water cleave onto the molecular structures of the body, its water orients with the right surface tension to allow the correct adhesion.

Polluted Waters

Most tap water originates from local reservoirs fed from runoff from a variety of sources. Many are pristine sources such as mountain snowmelt or underground springs. Others, however, may be fed from rivers and streams carrying waste streams from industrial operations.

One of the biggest problems occurring with creek water or river water is the runoff of toxic waste and agricultural chemicals. Studies in the San Francisco Bay, for example, revealed that chemical fertilizer and pesticide runoff from farms was one of the greatest sources of the water pollution in the bay. Drinking water reservoirs are also often fed by streams that connect to commercial agriculture. Other sources of toxic chemicals in drinking reservoirs include industrial chemicals from all sorts of manufacturing waste streams; airborne chemical pollutants such as sulphur and lead falling out with rainfall; groundwater seepage from septic tanks, underground gasoline tanks, and fertilizers applied to lawns and farms; and the various abuses of the marine environment such as commercial shipping discharge, marine oil leakage, antifouling paint, and marine head outfall.

A study released in 2005 by the *Environmental Working Group* found 260 contaminants in drinking water supplies around the country. Of these 260 chemicals, the U.S. *Environmental Protection Agency* regulates 114 to some extent; and another five have been assigned secondary standards— non-enforceable minimum contamination goals. That leaves an astonishing 141 chemicals with no standards, in the drinking supplies of over 195 million people. Not all 141 chemicals are being served to all 195 million people at once. The report specified that at least 40 of these 141 chemical toxins were being serviced to at least one million people, while another 19 were being supplied to at least 10,000 people. In some cases, single water supplies had up to an additional twenty unregulated contaminants. In other words, some water supplies are much worse than others, depending upon their proximity to industrial waste, commercial farm waste, chemical fertilizer use, pesticide use, herbicide use, sewage dumping, and the existence of underground storage tanks.

The report also broke down the unregulated contaminants into several categories; including methyl tertiary butyl ether or MTBE from gasoline, perchlorate from rocket fuel, fifteen water disinfection byproducts, four plasticizers, seventy-eight consumer and industrial production chemicals, and twenty gas, coal and other fuel combustion chemicals. Fifty-two of these unregulated chemicals have been linked to cancer. Seventy-seven have been linked to reproductive or growth problems. Sixteen more have been connected with compromising the immune system. By far the biggest source of both regulated and unregulated chemicals in this study was industrial waste with 166 chemicals, followed by agricultural pollutants with eighty-three chemicals. Household and urban pollution came in third with twenty-nine chemicals and water treatment plants contributed a jaw-dropping forty-four pollutants to tap water systems.

Agriculture spreads 110 billion pounds of chemical fertilizer over almost 250 million acres according to a 2002 USDA report. This accounts for about one-eighth of the entire United States. Meanwhile about one-tenth of the U.S. is being drenched with herbicides and pesticides. Meat production is another dramatic contributor. There are about 248,000 feed-lots in the U.S., primarily housing cattle or pigs. These feedlots produce about 500 million tons of manure annually, laden with hormones and antibiotics. This means our thirst for commercially farmed meat is limiting our future ability to consume our body's most precious requirement: water. As Fawell and Nieuwenhuijsen (2003) document in their review of various water contaminants throughout the world, water contamination has been linked to a variety of disease pathologies, especially in areas with increased industrial and commercial agricultural economies.

Water Bottles

For many consumers the solution to water contamination comes in the form of bottled water. As a result, today billions of people drink their daily water from personal size plastic bottles. Millions of households and businesses drink from family size water 'fountain size' bottles. Fifty-gallon plastic jars are delivered to the door by billion-dollar water packaging companies. Retail stores stock refrigerators full of various sodas, waters and juices in plastic containers.

Plastic can leach a number of chemicals into the water in the presence of light and heat. The extent of leaching can vary, depending upon the grade of the plastic, the extent of exposure, and the length of exposure. There are a number of grades and types of plastic now being used to store and ship water and other drinks. The grade and type of plastic bottle is stamped onto the bottom of most plastic bottles. This number is called a resin ID code. Often it will appear within a small three arrow recycling triangle symbol. The number code indicates the grade and type of plastic.

There are seven basic codes in current use. There is polyethylene terephthalate (PET) usually marked with the number 1; polyvinyl chloride (PVC) (#3); polypropylene (PP) (#5); polystyrene (PS) (#6); polycarbonate (PC) (#7); and two types of polyethylene—high density (HDPE) (#2) and low-density (HDPE) (#4). #7 may also be nylon, acrylic, acrylonitrile butadiene styrene or a number of similar composite types.

Polyethylene terephthalate or PET is now probably one of the most distributed water container types, especially for smaller sizes. Most sodas and soft drinks, along with various waters are now stored in these bottles. PET is a thermoplastic polymer resin, part of the polyester family. PET bottles may be clear or opaque. Most PET is made with a trans-esterification reaction between ethylene glycol (yes, the green stuff put in radiators) and dimethyl terephthalate. The ethylene glycol is the typical byproduct of polymerization of PET. While resonated PET is very strong, unmodified PET's melting point is below the boiling point. Modified PET, used for most food applications, is more stable and can be heated to 180 degrees Celsius for 30 minutes without melting. Most PET will have a *glass temperature* (the level where its molecules have more mobility) of about 75 Celsius, which is about 167 degrees Fahrenheit. In other words, this is the point where molecules of the plastic may leach indiscriminately into the liquid or food present. However, this is a sliding scale and minute leaching begins at much lower temperatures. Research has confirmed that heat leaching is higher when PET is exposed to direct sunlight. At temperatures above room temperature or even at room temperature for extended periods, significant leaching of toxins from PET has been observed. Up to 19 different migrating chemicals have been observed from commercial amber PET bottles. Various fatty acids, plasticizers, and acetaldehyde have been discovered in PET research. Furthermore, foods microwaved in PET have revealed significant levels of cyclic oligomers. It is also thought that cyclic oligomers from PET bottles may also leach due to sun exposure.

That said, PET has undergone extensive toxicology and risk assessment studies. *In vivo* and epidemiological studies, (many of which were extensively documented by a 2000 report by the International Life Sciences Institute), have indicated that toxicity due to PET leaching is below levels thought to pose an immediate threat to the health of most of the population. Certainly, the body has tremendous abilities of detoxification and adaptation. We must also consider their affect upon those whose immune systems are burdened. Whether PET leaching toxins are significant toxins or merely additional toxins adding to the total stress burden on the body—it is still a concern. While small amounts might not pose a concern for the body's detoxification systems, continuous exposure is likely to cause a host of possible disorders over time. We have yet to understand the long term risks of PET.

Polyvinyl chloride or PVC plastic will typically be coated with a plasticizing resin called Di-2-ethylhexyl-phthalate, usually referred to as DEHP. Along with several types of plastic bottles, DEHP is a popular coating on rubber hoses. Many hiking and camping bottles are made with PVC coated with DEHP. Many of us have probably drunk out of a PVC bottle on a bike or hike. After a few hours in the sun, the water will have a strong plastic odor and taste. The *Environmental Protection Agency* has determined that the MCLG or *Maximum Contaminant Level Goal* for DEHP is set at zero, while the regulated MCL requirement is set at six parts per billion in public drinking water. This means the aim is zero, but the reality is much higher. According to the EPA, nausea, vertigo and gastrointestinal pains are associated with short-term exposure (drinking); while liver damage, genital and reproductive damage, and cancer are noted as potential effects of long-term use of DEHP. PVC is also a popular material used in various toys. These toys should be eliminated from kids' toy rooms, as they will release DEHP when they are played with. Medical grade silicone nipples and teethers are probably better choices than PVC toys. Even better are natural materials.

Polycarbonate bottles should be avoided especially if exposed to heat. Baby bottles and various types of water bottles are often made of PC. PC is also used in medical devices, dentistry, lenses, and clear storage containers. PCs are made using Bisphenol-A (or BPA). As a result, PCs are known to leach BPA into bottle contents as heat is applied. BPA is a monomer and considered an endocrine disruptor. BPA was first synthesized in the 1930's for estrogen replacement. Research indicates that BPA appears to have negative effects upon physiological development, growth, memory, behavior, skeletal formation, prostate size, sperm count, and may be implicated in the Alzheimer epidemic. BPA is weakly estrogenic, and more than one study has linked it with dramatically increasing estrogen levels because of its affect upon estrogen receptors. This makes PC a potential contributing factor in the dramatic rise in breast cancer. Breast cancer does seem to occur more prominently in cultures with prevalent plastics exposure.

Plastic bottles with both low-density and high-density polyethylenes (LDPE and HDPE, respectively) have been known to leach polyalkylated phenols such as BHT and Irganox 1640 into their contents, whether liquids or solids. Interestingly, both of these chemicals are also commonly used as chemical preservatives in food production—their toxicity is not well established. The stabilizers used in polypropylene production have been observed as bioactive, indicating they may also be leaching into contents. Some of suspected these may inhibit nerve conductance. Heat and light are the two causes observed most. The higher density polypropylenes appear to leach less.

While the cumulative effect of these compounds entering our bodies for many years has yet to be determined, there is some evidence that most plasticizers are estrogenic, which means that they can attach to estrogen receptors, and thus confuse the body's stimulation of different mechanisms such as growth and inflammation. Estrogen ligand and receptor balance is also critical to our flow of various other hormones and neuro-transmitters such as serotonin, dopamine, progesterone and many others. We should also note that rates of menopausal and chronic fatigue disorders and breast cancers, which are all directly related to the balance of hormones, neurotransmitters and estrogen balance, have been rising dramatically over the past 20 years. There may indeed be a link to plasticizers.

Bottled water can also undergo significant environmental exposure during transportation and storage. Once the water is extracted from its well, aquifer, or municipal reservoir—and many bottled waters are simply filtered municipal tap waters—it is usually filtered by distillation, reverse osmosis, or a combination of the two. Some might refer to reverse osmosis as triple filtration because it travels through three filtration membranes. Sometimes selected minerals are added back for taste. (Because our bodies naturally prefer mineralized water.) Once filled into plastic or glass bottles, the water is then shipped around the world.

During this distillation and shipping process, different environmental exposures may affect the purity and stability of the water. Certainly we can agree that this water is far removed from the natural ionic waters nature produces. Shipping though modern transportation will expose the water to various electromagnetic waveforms borne by mechanical equipment. These include various conveyer belts and automatic palletizers before they leave the line. Various fans and other motors like forklifts come into close contact with the water. Many of these motors are built with a variety of magnets and coils, inflicting a combined electromagnetic effect upon both the water and the plastic bottles. After warehousing, the water will be loaded onto trucks and/or shipping containers, where it may travel thousands of miles to its delivery destination.

Overseas water may be loaded onto a container ship where it undergoes a lengthy ocean voyage in the hold or on deck. Either location opens the possibility of temperature variability. Once arriving at the port, the water may undergo irradiation. From the dock, the water will be trucked and possibly juggled between storage warehouses on its way to the store. By the time this water ends up in our mouth it may have undergone months of transportation and warehousing—most of it within a plastic container. Each of these environments will induce various levels of heat, light and pressure variance. The unfortunate result may be various plasticizers and other chemicals leached from the plastic container.

In a 1997-1999 multi-year study of bottled waters by the *Natural Resources Defense Council,* over 1,000 bottles of water from 103 brands were tested for contaminants after reaching the shelves. Over one-third of the bottled water samples turned out to have contaminant levels greater or equal to levels typical of municipal tap water supplies. A third of 103 different bottled waters tested violated state limits or guidelines, with 22% violating enforceable limits, and 17% violating government water guidelines for municipal water treatment.

This does not include whatever ionic and clustering changes may have occurred within the water through the various electromagnetic exposure.

In addition to the purity issues, and the potential damage transportation and leaching effects, we should consider the environmental impact of carbon-intensive transportation. For a commodity most Americans have access from local reservoirs, wells, and aquifers, it appears financially and economically idiotic to waste our resources on shipping water around the world. Yet according to the *International Bottled Water Association,* bottled water sales have grown from almost nil in the 1980s to over twenty-seven gallons per person in 2006. In just under twenty years, total gallons of bottled water sold burgeoned from less than 300,000 gallons in 1976 to more than 3,000,000 gallons per year in 1997. Imagine the waste in resources over this twenty-year growth period. These shipments could have just as well been grain shipped to the hungry. Today as we peruse our shelves at the store, we can see the distance some of our commercial water travels: Fiji, Italy, Switzerland, and Hawaii. Often a single water shipment will take up and entire ship or at least the cargo hold of a ship.

While the water itself in a bottle of water might cost only a few cents, more than 90% of its total value comes from the activities of forming and packing the plastic bottle, shipping it, and paying the margins of distributors and shippers. Some have estimated that the cost of a bottle of water is 240 to 10,000 times the cost of filtering and purifying tap water.

Glass containers are by far the most natural containers—made primarily of sand. Glass is extremely stable when exposed to light, pressure, and heat as a result of its elemental stability.

Tap Waters

Certainly, municipal tap water undergoes harsh treatment. Chlorine and chloramines are standard treatment additives in most all U.S. municipal water treatment plants.

There is good reason for chlorine purification, however. We must point out that chlorine has been used for over a century in water treatment and because of its use, public water-borne bacterial diseases have dramatically decreased. Chlorine itself is a natural element, known primarily to be the ionic element of sodium chloride—the salt of our oceans. It is also a necessary trace mineral, used for various enzymatic and metabolic functions in the body. In heavy doses, however, chlorine can burden the body's chemical load. Though the body can readily detoxify or utilize chlorine, chlorine breaks down into other harmful chemicals when added to water. While EPA guidelines only require about .2 parts per million for the municipality to purify water, levels are often increased to handle higher bacteria levels, which can occur during seasonal low water levels—further increasing chlorine levels.

Chlorine readily breaks down in sunlight and during transport through pipes. The byproducts of this breakdown are trihalomethanes (THM) and haloacetic acids (TAA5). In the United States, the Environmental Protection Agency regulates these two byproducts in municipal drinking water. Over the past few years most municipalities have reduced total trihalomethane (TTHM) levels to 80 ug/L (equivalent to 80 parts per billion) and 60 ug/L (60 ppb) respectively. Meanwhile, many states have mandated tougher standards, some to the tune of 10 ug/L or 10 parts per billion for municipalities within their jurisdiction.

The FDA's mandated level for TTHMs in bottled water is 100 ug/L. While we might assume bottled waters would not contain trihalomethanes, the NRDC's 1997-1999 tests concluded that many bottled waters had substantially higher TTHM levels than the state mandated levels. Many had levels ranging from 20 to 90 ppb of TTHM. While levels below the EPA's and FDA's mandates have not been shown to pose a significant health risk, the bioaccumulation effect is uncertain. TTHM's are metabolized through the liver and liver enzymes, so they will certainly add to the toxic burden on the body.

As for haloacetic acids, there are five known acids that can result during in municipal water system disinfection. These are monochloroacetic acid, dichloroacetic acid, trichloroacetic acid, monobromoacetic acid, and dibromoacetic acid. Levels of these byproducts are also regulated by the EPA for public water systems at 60 parts per billion. A number of studies on rats have indicated the TAA5s are readily metabolized through the liver or through glutathione conjugation. Evidence of bioaccumulation appeared absent in these studies. Human studies have not shown toxicity at lower dosages. Toxicity at higher doses is not well established.

As to whether THMs or TAA5s are more toxic than some of the deadly bacteria the chlorine removes; well, most of us would probably prefer the chlorine to the bacteria. Many water-borne bacteria are highly and immediately toxic, causing dysentery and even death if untreated. At the end of the day, the residual risk chlorine byproducts from standard chlorinated municipal water treatments pose is significantly lower than the risk of bacterial infection and likely even plasticizers from plastic bottles. The latter two present a greater tendency for bioaccumulation and metabolism disruption.

The municipal water additive to be more concerned about is the newest version of chlorine treatment: chloramine. Chloramines like monochloramine are produced by combining chlorine with ammonia. This creates a more hardy molecule, one that does not break down as readily as chlorine does in water. For this reason, most EPA researchers believe that because chloramines last longer, less can be used. They also believe that since they do not break down as readily, there will be less THH and THM content in the water supply. In addition, chloramines treatment affects the odor and taste of water much less than chlorine treatment.

Because chloramines do not break down as readily, they provide more potential accumulation risk and thus may be more toxic. Aquarium fish will typically die in chloramine water. Chloramine stresses the gills of fish. Chloramine content above .3 ppm will kill most fish. Chloramines also react more with certain rubber materials, which can damage hoses and gaskets in hot water and washing machines. Noting its affects upon fish, chloramine is likely toxic to the human body, though this has not been established with certainty.

Ammonia is a harsh toxin. It will typically break down into nitrates, which at higher levels has been shown to disrupt hemoglobin's ability to bind well with oxygen. This can cause brain damage and various other metabolic disorders. Nitrate levels in drinking water are therefore tightly regulated. Levels over 50 parts per million are considered toxic and therefore not allowed in municipal water supplies and shared wells. Ammonia-based fertilizer, our genius invention to fertilize nitrogen-fixing plants, has backfired on us. Nitrate byproducts now slowly seep through commercially farmed soils down to the once pure deep-water aquifers that supply much of our drinking water. Though toxic levels of nitrate are monitored, lower nitrate levels also have inherent risks. In one study done at the Riley Hospital for Children in Indianapolis (Mattix *et al.* 2007), regions with higher nitrate levels in drinking water experience significantly higher rates of congenital abdominal wall abnormalities.

While bottled water is regulated in the United States by the *Food and Drug Administration,* municipal tap water is regulated by the *Environmental Protection Agency.* While the FDA rarely if ever tests bottled waters—unless there is a mass complaint or scare—the EPA requires multiple daily tests are done on municipal supplies. Municipalities are also mandated to make all test results available to the public. While the FDA mandates weekly testing by commercial bottled water companies, those records are rarely if ever inspected, and they are not available to the public or any public agency. Therefore, municipal water at least presents a known quantity. Tap water with a good filtration system provides a viable alternative to plasticizers and other unknown contaminants possible in bottled water.

Many municipalities are moving increasing amounts of their purification to ozonation, a viable alternative to chlorine, although most ozone systems still require chlorine albeit in lower concentrations.

Water Purity

The waveform nature of water is tied to its ionic and electromagnetic properties. This typically parallels water's natural mineral content and where it has flowed. We say 'natural' here because we are not convinced that mechanical distillation followed by adding minerals provides the same benefits. The ionic and electromagnetic properties of water are formed as the water flows through rocks and crevasses of the planet. During this flow, water picks up the byproducts of the living earth. As a result, water from fresh sources, even following carbon or ceramic filtration, will retain these effects. This is because, as we will discuss, both the elements of carbon and ceramic are extensions of the earth's own purification systems. The earth's purification systems remove bacteria and toxins because the rocks of the earth magnetically attract and remove impurities. For this reason, most virgin well water is quite pure and bacteria-free.

While quantum and molecular analysis might tell us that waters with different pH, polarity, spin, and surface tension will be problematic for hydration purposes, the simple solution is to draw water from the natural springs and wells around the planet. This assures us that the water has been cycled through the earth's purification and filtering structures, adding the appropriate mineral content. The process of cycling that natural water travels includes evaporation, pressurization, rainfall and/or snowfall, and then filtering through the earth's rocks and sediment of the earth. This cycle orients the water to the rhythms of the earth, lending to a specific molecular structure, polarity, pH, spin, and surface tension—the precise specifications our bodies happen to require. By nurturing and protecting the elements of nature that provide these mechanisms, we assure ourselves of the consistency of supply.

The better water's ionic quality, the healthier the water will be. The more effective it will be in hydrating our body. In other words, water that resonates with our body's liquids will be better absorbed and utilized. It will create less of a burden on our livers, kidneys and other vital organs. Our physical body is primarily liquid in composition. Theoretically, the water that resonates the greatest with body fluids would obviously be water that has been filtered through our own body's metabolism. For this reason, there is a long tradition of drinking urine for health benefits in Ayurvedic medicine. The filtering and mineralizing systems of water's travels through our own bodies is mirrored holographically by the movement of water through the earth's body.

In short, water naturally flowing through the contours of the living planet will have the magnetic polarity and electromagnetic characteristics resonating most closely with the living human body. This statement is supported not simply by a study of one hundred clinical subjects, or even one million subjects. It is supported by billions of subjects over a span of millions of years of drinking the earth's waters. This scientific conclusion comes from the most available and extensive scientific testing imaginable—one where every living organism has taken a part. This thousand-year epidemiological study been pervasive; it has significant controls and placebos. Those who have died or become sickened by clear drinking water contaminated with various chemicals provide us with circumstantial evidence that prior to the industrial age, the forces of the elements provided adequate hydration means. Those who have become diseased with dysentery and parasites after drinking seemingly good-tasting water but sitting without movement also provide evidence regarding water's need to be moving. The survival of cultures that developed nearby mountain and valley rivers and streams speak clearly to which waters are healthy.

Thus we must consider not only the source of the water, but its movement. Water sitting for long periods without movement will typically become a host for microorganisms unhealthy to the body. On the other hand, water moving over rocks, streams, and aquifers that channel water over naturally electrostatic mineral rocks of various compositions will typically be more suitable to drink both from a nutritional and sanitary perspective. This 'science' of knowing which water to drink was a practical matter of life or death prior to humankind's increasing detachment from nature's forces.

Water from a virgin stream flowing gently through the rocks fed from an underground spring in a pristine environment would likely have an extraordinary mineral content along with a minimal level of microorganisms (with many of those remaining being probiotic soil-based organisms). It is reasonable to also assume we will find this forms water clusters quite consistently and has a suitable polarity for consumption and assimilation. This water would probably be most refreshing and would be naturally oxygenated. It would probably be absorbed and utilized easily by the body.

Today's rivers and streams are not so virgin. They are likely to host various types of *E. coli* and other bacteria; organophosphates from fertilizers and pesticides; and various other toxins from our industrialized environment. Today, unless the water source is a deep underground aquifer in excess of 500 feet, we will most likely filtration, and possibly even chlorination or ozonation for purification.

While it is probably results in some of the highest water purity, *distilled water* from reverse osmosis filtration is not that great a way to filter water. There are a couple of several reasons. The pressurized vacuum-filtration method of distillation screens out most of the mineral and ionic elements that give the water its nutritional benefit. Another negative about reverse osmosis is the pressurized triple filtering can result in much of water's oxygen content being lost. At the same time, if our water is full of nitrates or other difficult toxins to filter, reverse-osmosis may be our only option, as other filters simply are not as thorough as reverse osmosis.

As for distillation, we find that the water undergoes an intense heating and evaporation process. This process effectively filters out most of the impurities. Again it also filters out practically all the mineral content. While it provides a vehicle for oxygenation, assuming it is an open system (as opposed to a vacuum system). Distillation then can provide a good basis for hydration, but will not provide the mineral and ionic nutrients needed for the body to properly drive and orient that water for use in the body. Unless the water is given a chance to filter through the natural rocks of the earth, distilled water is for most part, dead.

There are many environments where distillation is necessary. In situations where there are poisonous particles in the water or mineral-related toxins in the body, distilled water may be one of the few ways to clear those out. Should we have little or no water and we are surrounded by the salt waters of the ocean, we may need to distill the water to desalinate it. For a thirsty population, this may be a matter of life and death. The obvious solution for a large desalination process would be to allow the desalinated water to flow through an area with natural rocks to allow it to interact with the planet before it is consumed. While desalination is not necessarily an unnatural process (as the planet earth also does this), it is unnatural for us to drink it before it comes into contact with the mineralized earth.

Manipulated Waters

Over the past few years, the concept of *hexagonal water* and *alkaline-ionized water* has been gaining attention. The research is compelling. As noted earlier, the evidence of water clustering is grounded in science. Through x-ray diffusion, observations of water molecules coming together into group formation bound with weak hydrogen bonds has been clearly documented. However, many prominent scientists have pegged these waters as pseudoscience.

The debate seems to revolve around the importance of this clustering, and whether the clusters provide any specific benefit to the body. While many conservative researchers point out that the clusters are transitory and elusive, some point out that particular cluster types have extraordinary significance over others. Dr. Mu Shik Jhon's research has confirmed a link between calcium availability in water and the number of hexagonal clusters available in the water. Now whether additional hexagonal water encourages additional calcium absorption or vice versa, we do not know. Dr. Jhon's position is that hexagonal water allows for better calcium solubility. *In vitro*, it appears from Dr. Jhon's research that water with increased hexagonal structures interferes with cancer cell growth rates. Curiously, it seems that this effect is also associated somehow with calcium's being dissolved in those waters. In an attempt to sort this out, Jhon measured 3T331 tumor cell growth in three types of solutions: One with unprocessed water; one with calcium chloride ions deposited into the water ($CaCl_2$); and one with calcium chloride ions plus mechanically ionized "structured" hexagonal water. By far the fastest tumor growth occurred with the unprocessed water. The two $CaCl_2$ waters had decreased tumor growth. The $CaCl_2$ water alone had a faster immediate reduction of tumor cell growth. However the hexagonal- $CaCl_2$ water began slower but after two days its reduced rate of growth exceeded the $CaCl_2$ alone (Jhon 2004).

While hexagonal water is portrayed as significantly better water, it also seems credible that that the calcium in the water may have provided the alkalinity to reverse the tumor growth. This would be consistent with the other research involving mineralized water and calcium in general.

However, it should be noted that many, though not all, of the hexagonal water proponents also sell expensive ionizing machines. Thus their science is likely biased by the potential of financial reward. Many of these machines are simple electrolysis machines—comparable to a car battery but without the sulfuric acid. The point of electrolysis is to separate ions of a substance by passing electric current through the substance. This creates a movement of electrons to positive poles and positive ions (cations) to negatively charged poles. The question is, does electrolysis significantly change the nature of water? Pure water is balanced at a 7.0 pH (though mineralized water will be slightly alkaline), and electrolysis will simply create hydrogen gas and oxygen, which would theoretically leave the resulting water again at 7.0 pH as it disassociates. Now if the water has salt in it, then yes, electrolysis might separate the sodium and the chloride ions, which may cleave to separated hydrogen and hydroxide. But this will create nothing more than a weak sodium hypochlorite solution—commonly known as bleach.

Certainly if the incoming water is tainted with toxic chemicals, these ionizers may be able to create slightly alkaline or acidic water, depending upon the polarity of the electrolysis—which might force the precipitation of certain chemical toxins. This might leave the water a bit cleaner—especially assuming there is a post-filtration included. It has been claimed that the acidic waters created by these machines make good antibacterial agents within the body, while the alkaline water portion carries significant health benefits. Fact or fiction?

The body's pH is certainly critical to health. Should the body's pH turn acidic enough, the body will stop functioning. Most health experts agree that a pH around 6.4 is healthy, and a significant rise or drop will significantly reduce health. We can ascribe to this, as studies of healthy and sick people illustrate a relationship between blood and urine pH. However, the question becomes; will drinking a more acidic or alkaline water significantly affect one's total body pH?

We have to be clear again that we cannot actually make pure water alkaline or acidic. Pure water has a pH of 7.0 by definition. pH is based upon water as the neutral point. Therefore, an alkaline or acidic water cannot exist in pure state. There must be some mineral or ion element to make the water more acidic or alkaline. Should we drink water with minerals in it we will be drinking alkaline water because of the positive ions (cations) that minerals produce. We can also drink water with lemon juice in it, which will make the solution acidic due to the addition of citric acid from the lemon.

Certain electrolysis ionizers claim to decrease cluster size. Whether these smaller clusters can indeed increase water absorption and utilization is not proven, however. There has been some research indicating the possibility that electrolysis and other ionic treatment may decrease cluster size. There is also some research indicating that possibly the size of hydrogen ions may be changed by ionic treatment. However, it still should be understood that observations of cluster sizes and hydrogen ion superstructures have been fleeting, thus their life expectancy is unknown. It is uncertain if these micro-clusters or superstructure ions are a result of ionization. The other central question is whether they will last long enough to after absorption into the body. Furthermore, the evidence that treated ionic waters truly have health benefits is simply not available.

With respect to water and hexagonal cluster structures, it is true that while ice is forming, water molecules will align more favorably into lattice structures, with a majority taking on hexagonal structure—a typical four hydrogen bond tetrahedral. However, these clusters are related to temperature, and it is doubtful the body's temperature of almost 100 degrees Fahrenheit will provide enough cooling effect to create such lattice structures. Upon heating, water's structure has been observed using neutron and x-ray diffraction data to resume the double hydrogen bond structure typical of water. Cold hard water apparently provides an increased number of hexagonal clusters. Again as the water heats, the number of hexagonal clusters will be reduced by the increased thermodynamic.

Much of our water is absorbed through either the stomach or the small intestine, where it almost immediately affects the hydration levels of the blood and the liver. During this process, water must travel through the intestinal membrane, through the gastric pits of the gastric cells or through the microvilli of the intestinal tract. Following this absorption process, the body's liver interacts with water, separating it from any dissolved ions and toxins. Dr. Jhon proposes it is the memory function of water that allows a reformation of hexagonal clusters after assimilation. While Dr. Jhon and others have firmly established the existence of hexagonal water clustering, this portion of the theory remains debatable.

The evidence to firmly establish hexagonal or ionic water may not be convincing to everyone, but there is some anecdotal discussion of its validity. Could this be a placebo effect? Truly, any chemist will accept that while water travels through volcanic rock, there is an ionic exchange taking place, leaving the water mineralized and thus "harder." To ignore the possibility that this mineralized is ionized may also prove to be short-sighted. Certainly evidence and experience clearly points to the reality that the earth's circulatory system is suited to render the best form of water for human consumption. Why not simply accept the naturally ionized waters already available from our local natural springs and wells?

Typically, a water molecule has a tetrahedral angle. The oxygen ionic element is flanked by two hydrogen ions. Between each water molecule flank is a weak ionic bridge. This weak bridge allows water to cluster somewhat, giving water its surface tension. Through the work of Nobel Price winner Dr. Carlo Rubbia, it has been determined that these water structural bonds are formed through an interaction with light, aligning each water molecule in a slightly different way. This results in each molecule of water having a unique electromagnetic frequency depending upon its unique orientation with light. These are subtle bonding effects, synchronizing on a harmonic, resonating basis.

Again, what the human body needs is natural living water. Living water is water that has been circulating through the living planet earth. As it has been circulating, it has become ionized, tuned perfectly with the rhythmic magnetic movements of the planet. The minerals like magnesium and calcium gathered from natural rock are merely symbols of the water's natural movements. This is the visible portion. Behind these minerals are the water's waveforms—some might also call these weak bonds—generated partially by the movement of the water through the rocks. The dissolved ions existing in the water are the simply one result of those bonding waveforms. The appropriate rhythmic characteristics of the water translate to higher absorption by the body and higher absorption into cells. Just as the physical body was designed to digest natural foods and produce vitamin D from natural sunlight, the body was designed to drink natural waters of rhythm.

The simple reason our bodies will maintain survival (perhaps accompanying decreased wellness and vitality) through chronic dehydration or toxic water intake is due to the nature of water combined with the body's ability to preserve, accumulate, and reuse wastewater when dehydrated. When it comes to toxic water, a strong body will work to purify and molecularly orient most any source of water for use in the body. The body will strip water of many chemicals through its passage through the stomach and liver. Before water is assimilated through the walls of the stomach and intestines, the stomach's acids attempt to sterilize it. These acids are also primarily composed of water. Gastric acids from a healthy stomach wall lining will destroy most microbial content. After this sterilization phase, the liver and bloodstream will soak up the water and strip away much of the chemical toxins present. The liver performs this task by acting both as a filtration system and a scavenger system, as it secretes aminotransferase enzymes like AST, ALT, and substances like glutathione to isolate, break down and escort toxins out of the body. Again, all of the pathways for these processes take place using water as a substrate.

Without a strong stomach wall lining and healthy gastric cells, many microbes may get through the intestinal wall and possibly into the bloodstream. Here they may become dangerous. If the gastric pit cells are not producing large enough quantities of hydrochloric acid and peptic acid; and the colonies of probiotics are not strong enough, the intestines may allow entry to these invaders—through *intestinal permeability*—sometimes referred to as *leaky gut syndrome* along with larger food molecules such as peptides. This is not to say bacterial colonies cannot make it through a healthy stomach—they can if the colonies are large enough or resilient enough. This is a relative situation. The healthier the stomach lining and the better its secretions; the less likely bacteria will pass through the stomach to the intestines. Should enough bacteria pass through, dehydration becomes a new concern, as diarrhea can cause a large fluid loss. Dehydration due to diarrhea is one of the main causes of death in dysentery and choleric situations.

In general, the process of water absorption through a healthy human body means that our bodies can handle and use large quantities of chemical-laden or bacteria-laden water and remain alive. However, the collateral damage must be considered. While hydrated, our concern with toxic waters becomes the burden on the liver and the various detoxification systems we use to cleanse toxicity. When we consider the many toxins the body must deal with coming in from our air, food and water, in addition to the other hundreds of other tasks the liver must accomplish, burdening the liver is of critical importance to the long-term survival of the body. The liver itself performs so many tasks, from secreting bile and digestive enzymes; to filtering and recycling blood and removing toxins from the body. To burden the liver with more toxicity will eventually overload it, reduces its life expectancy and performance, as each toxin takes its toll. This can lead to cirrhosis and other diseases of the liver. In other words, the liver becomes compromised by the addition of a greater chemical load, whether from polluted food or chemically laden water.

As to the choices for treating natural living waters poisoned by humanity's industrial environment, filtration is strongly suggested in any municipal or urban water sources. As mentioned, reverse-osmosis systems are probably not the best choice unless there is a chemical toxicity not removable through common filtration, such as nitrates or high levels of arsenic. There are a variety of filtration systems now available which filter in the range of 99% known water bacteria, chlorine and chlorine byproducts, lead and various other heavy metals, without exhausting the water of its mineral qualities. Carbon-based and ceramic-based filtration systems provide some of the highest rates of protection from contaminants among filtration choices. Of these two types, ceramic filters might perform better on for several reasons: Ceramic filters appear to use rhythmic ionic forces of attraction and filtration similar to the earth's process of purification through rocks. Ceramic filters are made predominantly from stone or earth-based materials, so they are probably more environmentally sustainable than carbon-based filters. Ceramic filters also last longer—they treat more gallons. It also appears from manufacturer specifications that carbon filters may capture more healthy minerals from the water as compared with ceramic filters. A consideration to make regarding the ceramic filters is that algae may grow on them, requiring periodic washings.

Toxic Oceans

Illustrating this, among the beaches of California, long known for their pristine blue ocean waters, there were a record 25,000 beach closings in 2006. These closings were the result of polluted ocean waters—so polluted swimming was considered toxic.

As we look to the ocean increasingly for nutrition, we face concerns not only of profitability and population size, but also of sustainability and stewardship. Over the past few years two important studies have been released—the _Pew Oceans Commission Final Report_ and the _United States Congress' Oceans Commissions Report._ These two reports send an urgent message to our marine harvesting and coastal development industries, summed up by Dr. Wallace Nichols, Research Associate at the _California Academy of Sciences_: _"Too much is being dumped into the oceans, too much is being taken out, and the ocean's coastline habitats are quickly being destroyed."_

According to Communications Officer Jeanette Lam of Canada's _Fisheries and Oceans, "several factors are affecting marine populations: rising water temperatures, illegal and over-fishing, bycatch, pollution, and food chain imbalances."_

These combined factors are increasingly problematic to a human population seeking to sustain life on the planet. Research by Slovenian chemists Tatjana Tisler and Jana Zagorc-Koncan (2003) has shown that we are drastically underestimating the toxic pollution effects of industrial waste. Our typical method for toxicity research has been to study each individual chemical and its possible toxicity. What we are missing with this type of research is the combined effects of the thousands of chemicals we are putting into our waters. As these chemicals mix, they create a toxic soup of new chemical combinations. Some of these are combinations are exponentially more toxic than the individual chemicals.

The issue of soup toxicity is tantamount for the health of our oceans. Marine life studies are showing higher levels of mercury, lead, arsenic, cadmium, and other heavy metals in combination within fish, bottom dwellers, and marine mammals. Polychlorinated biphenyls (PCBs), petroleum, _E. coli,_ and other bacteria from waste run-off are bioaccumulating up the food chain along with these heavy metals.

A newer toxin to our oceans is plastic—bioaccumulating at levels surprising many ocean experts. As plastics break down into smaller particles, they are absorbed by filtering marine plants and aquatics and passed up the food chain. Research led by Captain Charles Moore of the Algalita Marine Research Foundation (2001; 2002; 2008) found an astounding six-to-one ratio of plastic particles-to-plankton in some areas. This means that for every pound of algae—the key nutrient for nearly all marine life—there are six pounds of plastic in the oceans. This also means our marine life are eating plastic particles with their meals.

Requiring some 500 years to breakdown, plastics are known to disrupt hormones and accumulate hydrocarbons. It is estimated that about twenty percent of the plastic polluting the oceans comes from discarded plastic pellets used to make plastic by manufacturers. These pellets are being swept or blown into the water from careless manufacturers and transport companies. The other eighty percent of the ocean's plastics is estimated to come from daily consumer use and the careless littering of oceans and runoffs.

Dying Seas

Our oceans, lakes and seaways are dying. This is due to a combination of toxic chemical industrial waste, toxic personal waste from households, and the widespread use of chemical fertilizers.

Nitrogen-rich fertilizers seem through our soils, choking rivers and oceans with extra nitrogen, causing abnormal blooms of algae. These algae blooms cut off the supply of oxygen for other fish, suffocating fish and other marine species. The massive marine life die-offs and over-whelming algae blooms together are gradually transforming the entire marine environment into lifeless regions now referred to *dead ocean zones* or *hypoxia regions*. These dead ocean zones are increasing throughout the planet. They have been growing dramatically especially in regions known for heavy chemical fertilizer use and commercial farming. A huge dead zone is now growing in the Gulf of Mexico, for example. In 2007, the size of this one dead zone was estimated at 6,000 square miles and growing.

These dead zones are reversible. Dead zones in the Black Sea and the North Sea have been reversed through decreases in fertilizer use and industrial chemical dumping in those regions (Mee 2006).

In addition, many major fisheries (regions with large fish populations) around the world are in trouble and many fish species are in decline or close to extinction. While fish farms seemed an environmentally viable alternative a few years ago, ocean experts have now established that contaminant levels of toxins like PCPs, viruses, and genetic manipulation are slowly finding their way into the wild marine environment, damaging natural fish stocks.

The bycatch problem has also become a critical issue for researchers. Sharks, porpoises, turtles and other beautiful ocean species are being endangered through accidental netting. *"The problem is worse than is being reported,"* said Dr. Nichols, who is also the director of Ocean Revolution. *"Many fishing regions around the world have little or no system for reporting their catches."*

In 2004, U.S. fish oil sales grew over 54% in the natural foods market and over 31% in the conventional market over the previous year. The thirst for fish oils has been fueled by studies showing *docosahexaenoic acid's*

(DHA) various health benefits. Meanwhile micro-encapsulation techniques have attracted new entries from various supplement and nutraceutical manufacturers. The pressure on the raw fish oil and fish supply is increasing with this growth. Independent projections indicate a critical world-wide fish oil shortage between 2010 and 2015. If consumption levels continue to increase, this shortage may arrive sooner, and various species may become extinct.

Surprisingly, fish do not produce much DHA. Fish primarily obtain their DHA by eating algae or eating marine life that fed on algae. Today some of these DHA-producing microalgae are being cultivated to produce high quality DHA oil. Two of highest DHA-producing microorganisms are *Crypthecodinium cohnii*, and *Schizochytrium spp.* Oil from these species have had increasing use in both the supplement and food formulation markets. They provide a pure form of healthy DHA. Perhaps we can save some fish species from extinction now.

The twenty-two-carbon chain, six-double-bond arrangement of DHA is either stored or converted to twenty-carbon, five-double-bond *eicosapentaenoic acid* (EPA) in the body of both humans and fish. Fish oil has been touted as cardiovascular-protective. This has been shown in many studies. Because fish oil has both EPA and DHA, the assumption has been that both EPA and DHA were necessary to gain these benefits. Newer research indicates otherwise. Typical crude fish oil will supply approximately 18% EPA and 12% DHA, subject to seasonal and species differences; while algal DHA will supply a standardized 35% DHA. EPA is short-lived within the body while DHA remains longer. DHA is either stored or used by cells, and easily converts to EPA as needed.

The same or more heart-healthy benefits available in fish oil are available in DHA-algae. In a randomized placebo-controlled 2006 study published in the *Journal of the American College of Nutrition,* 116 coronary artery disease patients took either 1000mg algal-DHA alone or fish oil with 1252mg DHA+EPA for eight weeks. The DHA group experienced average triglyceride reductions of 21.8%, while the DHA+EPA group experienced 18.3% average triglyceride reduction (Schwellenbach *et al.* 2006).

Sustainability does not seem to be an issue for algal DHA production. While fish stocks are being depleted throughout the world because of our unsustainable need for fish and fish oils, DHA-algae is grown in tanks, and capacity can be added with demand. Algal DHA is one way to reduce our stress upon the oceans for those requiring DHA supplementation. Note that ALA from flax, canola, walnuts, pumpkin seeds, and chia seeds readily converts to DHA in healthy people. Those with compromised immune systems and those on western meat-rich diets have been shown to have reduced conversion rates. These folks require additional DHA supplementation.

Shellfish are also an environmental concern these days as well. As Chris LaRock, Emergency Response Officer from *Environment Canada* put it, *"Chronic non-point source pollution from urbanization is increasingly endangering marine stocks, as bottom-feeders accumulate these toxins."* Filtering eaters such as shrimp, scallops, clams, oysters, and mussels have become increasingly toxic as a result. In addition, many shellfish species are now restricted for harvest because of low populations and bycatch problems. According to Dr. Nichols, most of us are not aware of shellfish trawling techniques: *"These shellfish are being caught by literally scraping huge areas of the ocean floor, damaging or killing coral, sea turtles, sponges, rays, and other sea life. 50-90% of the shrimp hauls are accidental bycatch. It is like clear-cutting a forest to gather a few mushrooms."*

Shellfish aquaculture is another environmental concern, according to researchers and government officials like Dr. Nichols and Mr. LaRock. Traditional shellfish aquaculture is installed in coastal areas by clearing the seafloor of important plants like mangroves and fish habitats to isolate the shellfish. According to ocean experts, this practice is severely endangering precious coastal habitats.

Meanwhile, *glucosamine sulfate* derived from shellfish is partially responsible for mass shellfish extinction—along with shellfish eating of course. Anyway, marine-obtained glucosamine sulfate's absorption and efficacy is debatable. A sustainable alternative for shellfish-derived glucosamine is *glucosamine hydrochloride* derived from the fungi *Aspergillus niger.* Since it is not bound to potassium, the HCL version apparently has 83% active glucosamine versus 50.7% for sulfate. Meanwhile, research has shown glucosamine HCL to be readily assimilated. This more sustainable HCL supply may help decrease the ocean's burden.

Sea mussels have been increasingly threatened from over-harvesting, prompting some regulatory bodies to restrict both wildcraft mussel harvesting and aquaculture—which draws seeds from wild mussel stocks. New Maine state regulations, for example, limit blue mussel seed removal to four *seed mussel conservation areas,* because of depletion fears. Aquacultured mussels, typically grown in shoreline pools, have been plagued with various difficulties over the years. Seasonal die-offs due to water constriction and contamination multiplied by proper rock-attachment issues have threatened these unnatural aquaculture mussel populations. Meanwhile wild mussel populations are in danger from dredging, drag-netting, water pollution and the red tide toxin PSP, which has resulted in a good number of reported illnesses.

As mentioned above, worldwide shark populations are increasingly under pressure. Luckily, shark cartilage sales continue to lose ground as recent research is starting to illustrate the lack of significant benefit to cancer patients. The title to a recent *Health News* article sums it up: *"Shark cartilage cancer treatments are pseudoscience."* While shark cartilage's *chondroitin sulfate* has shown limited usefulness in osteoarthritis research, its large molecular structure limits its absorption across the intestinal wall. With absorption rates at less than 15% and only partial absorption of that amount into joints, alternative versions of lower-molecular weight chondroitin appear to be the future for this controversial nutraceutical. Perhaps the lack of confirmed shark cartilage benefit for cancer and low chondroitin absorption may help keep some shark species from going extinct.

Water Botanicals

Nutrients from the ocean's plant kingdom are typically not as sensitive to over-harvesting and bycatch. Most marine botanicals are either cultivated or *sustainably wildcrafted*—with stationary seasonal blooms forcing self-regulation. Because they rely upon photosynthesis rather than feeding, they grow prolifically. They can also easily be grown in controlled environments such as human-made lakes and tanks. Because they do little if any water filtering as shellfish do, algae cause little toxicity concern.

There are some 70,000 known algae, and they are typically divided into three general types: *Chlorophyta* or green algae, *Phaeophyta* or brown algae and *Rhodophyta*, the red marine algae. These range from single-celled microalgae to giant broad-leafed kelps.

In terms of environmental food economics, sea vegetables trump all other food sources. While an acre of beef production yields about 20 pounds of useable protein and an acre of soybeans yields about 400 pounds, typical seaweeds like nori can yield 800 pounds per acre of tidal zone, and spirulina can yield a whopping 21,000 lbs of useable protein per acre of cultivation.

Spirulina and *chlorella* are leading microalgae foods. The market for these algae has been flourishing for over three decades, with a number of successful companies harvesting and freeze-drying algae in food-safety controlled environments. The largest spirulina producers cultivate spirulina in large fertilized ponds. A few years ago spirulina received U.S. GRAS status for use as a good source of carotenoids, vitamins and minerals; supporting healthy eyes and immune function. That approval has sparked renewed interest in spirulina as an important food. Spirulina nutritional content is impressive, containing all the essential and most non-essential amino acids, with 55-65% protein by weight. It also has a variety of minerals, vitamins and phytonutrients such as *zeaxanthin, myxoxantho-phyll* and *lutein*. A number of clinical studies have indicated lower levels of inflammation, brain damage from stroke, and anti-cancer effects from spirulina consumption.

Chlorella pyrensoidosa, or simply chlorella, is also cultured in outdoor ponds. With over 800 published studies verifying its safety and efficacy for a number of health issues, chlorella enjoys a strong consumer-base among healthy consumers. Chlorella's ability to detoxify heavy metals and other toxins make it a favorite of health practitioners. Chlorella is a great source of dietary fiber, and it binds to heavy metals and other toxins. This allows chlorella to expedite the process of heavy metal and other toxin detoxification. Chlorella's phytonutrients include C.G.F (Chlorella Growth Factor), beta carotene, and various vitamins. It is also a complete protein (40%-60%+ by weight) with every essential and non-essential amino acid. Clinical studies have shown that chlorella contributes to increased cell growth; stimulates T-cell and B-cell (immune function) activities; increases macrophage function; and contributes to improvement in fibromyalgia, hypertension, and ulcerative colitis (Merchant and Andre 2001). Its cell wall is tough, but most producers have developed ways of pulverizing or crushing the cell wall, allowing assimilation of its nutrients. Apparently, the polysaccharides and fiber from its broken cell walls give chlorella its unique ability to bind to toxins in the body (McCauley 2005).

Aphanizomenon flos-aquae or simply 'AFA,' is an algae that grows on the pristine volcanic waters of the Klamath Lake of Oregon. The species does grow on other bodies of water, but the Klamath Lake has been known as the cleanest and most sustainable source. Commercial AFA harvesting began in the early 1980s. Although contamination was once a concern, today companies are micro-filtering AFA for potential contaminants. A number of commercial supplements and nutraceutical foods include AFA in their formulations. Unlike chlorella, AFA's nutrients are readily available because of its soft cell wall. The rich volcanic lakebed of Klamath Lake renders it an available source of all the essential and non-essential amino acids, making it, along with spirulina and chlorella, a complete food. AFA is 60% protein by weight and is packed with many vitamins and other phytonutrients. AFA also has 58 minerals at ppm levels and high chlorophyll content.

Another exciting green microalga is *Haematococcus pluvialis*, the highest known natural source of *astaxanthin*. Astaxanthin is an oxygenated carotenoid with significant antioxidant properties. It is thought to have hundreds of times the antioxidant value of vitamin E. Recent studies have shown astaxanthin to be effective for reducing inflammation and stimulating the immune system. Studies have also shown anti-tumor effects, as well as effectiveness in preventing and treating retinal oxidative damage and macular degeneration (Guerin 2003). The antioxidant effects and cancer inhibitory action of astaxanthin has been shown to be greater than beta-carotene. Reports from marathoners and tri-athletes indicate it increases recovery rates after exercise. Astaxanthin is now an ingredient in multivitamins, specific-issue formulations, and various cosmeceuticals.

There are about 1,500 species of sea vegetables, many of which flourish in the cold waters of the North Pacific and Atlantic oceans. Well known sea vegetables include *nori, wakame, dulse, kombu, Irish moss, sea palm,* and several species of *laminaria.* While not particularly correct, most sea vegetables are commonly referred to as kelps. Wild sea veggies are harvested periodically and regrowth is managed carefully—easy to do since the kelp beds are stationary. Out of necessity, kelp farmers have a sustainable supply, and most areas have more than enough to supply market growth. Careful harvesting of kelp around the world will preserve continued supply and functionality within the ocean.

Sea vegetables like kelp have an impressive array of vitamins—more than most land-based vegetables, with A, B1, B2, B5, B12, C, B6, B3, folic acid, E, K, and a steroid vitamin D precursor. Nori and dulse have beta-carotene levels as high as 50,000 IU per 100 grams, for example. Certified organic kelps are showing 60 minerals at ppm levels. They are also good sources of calcium and magnesium. The brown algae also contain all the essential aminos and are high in protein by weight. Nori has 30% protein by weight and the others average about 9%. Laminaria algae also produce the sugar substitute mannitol (McCauley 2005).

Sea veggies also contain a number of beneficial polysaccharides and polyphenols. One such sulfated polysaccharide, fucoidan, has been shown to have anti-tumor, anticoagulant and anti-angiogenic properties. Studies show it also down-regulates Th2—inhibiting allergic response, inhibits beta-amyloid formation (potential cause of Alzheimer's), inhibits proteinuria in Heymann nephritis, and decreases artery platelet deposits (Kuznetsova *et al.* 2004; Nagaoka *et al.* 2000; Berteau and Mulloy 2003).

Red marine algae have become exciting nutraceuticals with confirmed health benefits. Attracting prominent interest is *Dumontiae*, a larger-leaf *Rhodaphyte* typically harvested in colder oceans by either wildcrafting or rope farming. *"Rope farming is highly-sustainable,"* Bob "Desert" Nichols has said. Mr. Nichols practically single-handedly developed the first commercial market for red marine algae in the mid 1990s. *Dumontiae* and other *Rhodophytes* have been confirmed to inhibit growth of several viruses, notably herpes simplex I and II, and HIV. Most studies have pointed to their heparin-like sulfated polysaccharide content for antiviral effects, blocking both DNA and retroviral replication (Neushul. 1990). *"We worked a lot with AIDS groups,"* said Mr. Nichols. *"Word of mouth got out and we were able to help many AIDS and HIV sufferers."*

Now other *Rhodophytes* are being studied for antiviral effects. Michael Neushul, PhD from University of California Santa Barbara's Biology Department has reported antiviral properties among all of 39 California red marine algae varieties tested. Sulfated polysaccharides such as carrageenan were pointed to as the central antiviral constituents, as well as dextran sulfate and other heparinoids. Retrovirus inhibition as mentioned above and murine leukemia inhibition properties have also been shown (Neushul 1990; Gonzales *et al.* 1987; Straus *et al.* 1984).

Red algae have a number of food uses as well. Gelatinous polysaccarides agar, carrageenan and funoran all come from red algae and are used extensively in the food business as stabilizers. Agar contains calcium, iodine, bromine and other trace minerals. Some red algae also produce sorbitol—used as a sugar substitute.

Other nutrient-rich algae from nature include *Dunaliella sp.*, a potent producer of lutein, beta-carotene and zeaxanthin; *Porthyridium sp.*, one of the few Rhodophye microalgae; and *Gigartina*—noted for varied nutraceutical and nutritional properties.

Harmonization: Therapeutic Waters

Water is not just a hydration element. Water has been used in medicine for thousands of years. In fact, it is one of the oldest healing agents documented. Early hydrotherapy practice has been documented in ancient Ayurvedic medicine and Chinese medicine; as well as the medicines of the Greeks, Polynesians, Aborigines, North American Indians, Incas, Mayans, Japanese, and Egyptians. Hippocrates was a proponent of hydrotherapy, and hydrotherapy treatment centers were very popular in Europe and the United States in the eighteenth, nineteenth, and early twentieth centuries. Vincent Priessnitz from Austria popularized many types of water treatments, including water compresses, cold-water therapy, and warm baths in the early nineteenth century. Dr. Wilhelm Winternitz, an Austrian neurologist, observed one of Priessnitz's treatment centers and became one of the most celebrated proponents of water treatment. Dr. Winternitz designed a number of different water treatments and influenced American physicians such as Dr. John Harvey Kellogg, Dr. Jethro Koss and Dr. Simon Baruch. Dr. Kellogg operated the famous Michigan Battle Creek Sanitarium for many years until it burnt down in 1902. The center utilized hydrotherapy as a key healing agent. Dr. Kloss ran his own clinic and worked closely with the Battle Creek Sanitarium. He documented many hydrotherapy treatment methods in his 1922 classic _Back To Eden_.

Despite its history of success, opposition to hydrotherapy came from the pharmaceutical medicine circles in the decades following. Water cures became a target for the new medical establishment, and many hydrotherapy treatment centers were shut down between 1920 and 1950. Hydrotherapy experienced a resurgence in the U.S. following the post-World War II popularity of swimming pools and whirlpools. Today hydrotherapy is widely used in various modalities, treatments, and centers. Many new wellness centers are unmistakably similar to the therapeutic "sanitarium" hot springs, and today draw millions of people seeking both therapy and relaxation. Physical therapists use hydrotherapy to retrain and nurture injured or post-surgical limbs. Water aerobics has become the quintessential form of exercise for the elderly because of its reduction of stress on joints. Colon hydrotherapy is now used throughout the world to cleanse the colon of putrefied waste. Hot baths are regularly recommended by conventional doctors for relieving stress and muscle fatigue, and many Americans sport hot tubs in or around their homes. While the sophisticated use of hydrotherapy is still considered alternative, most conventional physicians agree that water therapy can stimulate circulation and healing. A recent Italian study (Municino _et al._ 2006) of eighteen advanced heart failure patients using hydrotherapy illustrates this effect. After three weeks of daily water training, cardiopulmonary tests and quality of life all increased substantially.

Most of us have experienced the soothing nature of warm water, or the energetic nature of cool or cold water. Water carries electromagnetic, thermal and of course fluid rhythms that balance the nervous system and the cardiovascular system. While water is by no means a cure-all, when used conjunctively with other natural healing processes, water has a powerful effect upon the rhythmic body.

❖ Hot and warm water increase circulation, relaxation (thus decreasing stress), joint movement range, and detoxification efforts. Hot water calms the body and slows the heart rate, as the body's blood vessels relax and dilate in response to warmer thermal rhythms.

❖ Cold water constricts the blood vessels and leads to involuntary muscle contraction. This type of muscle contraction increases the body's immune function by pumping the lymphatic system. Lymph flow is circulated by movement and muscle contraction. Lymph circulation distributes macrophages and B cells throughout the body, enabling them to break down invading bacteria, viruses, and chemical toxins. In a German study (Goedsche _et al._ 2007) on twenty patients with chronic obstructive pulmonary disease, cold-water hydrotherapy was tested for immunostimulation, maximal expiratory flow, quality of life, and level of respiratory infection. After ten weeks of three cold effusions and two cold washings on the upper body per day, IFN-gamma lymphocytes increased, quality of life increased, and the frequency of infection decreased throughout the study group.

❖ For tight joints or sprains, we can alternate hot water and iced water. This is also called a *contrast bath*. The sprained or injured area may be placed into a bucket or pan of iced water, holding until the area begins to feel numb. The injured area can then be put into a tub or pan of hot water (test to prevent scalding). This can be held for several minutes, until the skin feels hot. The process can be repeated a number of times. If the injury is not a muscle or tendon tear—slightly moving the joint a little while numb may help speed healing. This will increase the microcirculation at the capillary level. As reported by Preisinger and Quittan (2006) from the University of Physical Medicine and Rehabilitation in Wein Germany, muscle spasms were reduced both by the application of heat and cold waters. While cold was more effective for reducing spasticity in motor neuron lesions, joint stiffness was reduced with hot water application. They also reported that pain threshold—the ability to endure pain—was increased by combined hot and cold-water therapy.

❖ Daily cold water showers or a quick cold water rinse off after a warm water shower is invigorating and stimulating to the immune system and nervous system. It also helps balance the body's thermoregulation systems, cooling the body in hot weather and heating the body (through muscle contraction) in cold weather. A cold shower also causes the skin pores to close. This leaves the body prepared to step out of the shower or bath. This reduces the potential of a basal cell chilling, which can stress the body. A cold shower will better prepare the body for the temperature change. Blood vessel constriction from cold water also stimulates the contraction of blood vessel walls. This serves to increase blood vessel wall elasticity, especially when the cold shower follows a warm or hot shower.

❖ Another way to accomplish this is the Finnish *sauna and plunge* system as described further below. The Fins are famous for their wooden saunas, built outside, near a cold-water plunge. A vigorous sweat in the sauna is immediately followed by the cold plunge. This ritual has also been a part of other cultures, including many American Indian tribes, who used *sweat lodges* with great health benefits. These cultures have a tradition of living long lives with strengthened immunity.

❖ The hot bath is an incredibly healing proposition. Hippocrates, known to western medicine as the father of medicine, stated that the hot bath *"...promotes expectoration, improves the respiration, and allays lassitude; for it soothes the joints and the outer skin, and is diuretic, removes heaviness of the heat and moistens the nose."* A hot bath will open skin pores, allowing a detoxification and exfoliation of skin cells and their contents. For sore or damaged muscle tissues, the dilation of capillaries and micro-capillaries speeds up the process of cleansing the muscle cells of lactic acid—the byproduct of the energy production in the cell. Our ability to recover from strenuous activity should increase through the use of hot water therapy. It should be noted that too much hot water for too long of a period can lead to cardiovascular stress should one not be accustomed to it. Hot water can also lead to heat exhaustion. For best results, hot water applications or saunas should be limited to about 10-15 minutes at the hotter levels. This can be increased for lower temperatures. In a 2006 review by French *et al.* from the *Monash Institute of Health Services*, hot- and cold-water treatment studies were compared and analyzed for the treatment of low back pain. They concluded nine different trials on 1117 patients. They concluded that heat therapy more significantly reduced acute and sub-acute low-back pain. Hot water also slows internal organ activity. This provides a soothing effect upon the innervations of these organs. The result is a calming of nervous tension. Hot water is also *hydrostatic*—it gently massages the dermal layers. This effect increases micro-circulation while relaxing and soothing muscles and nerves. This effect is increased when hot water is in motion, for example in a hot tub or whirlpool. Hot water can also relax key intestinal issues. In a study of 18 healthy volunteers and 28 patients with anorectal disease, 10-minute baths at 40, 45 and 50 degrees Celsius were tested for rectal neck pressure and internal sphincter stress. The hotter bath temperature produced significantly better test results and better pain scores (Shafik 1993). (Always test hot water to prevent scalding.)

❖ Certainly a hot mineral springs bath would be recommended, but not all of us have the advantage of bathing in hot mineral springs often. For those of us with typical hot water tanks and baths, there are a number of additives that can re-polarize and mineralize the water. We add a natural mineral compound such as *Epson salts* to the water. Epson salts—originally named after the magnesium-rich waters of Epson, England—are primarily magnesium sulfate, which will ionize in the water. Ionic magnesium is beneficial to body tissues, relaxing skin and muscle tissues, and lowering stress. A drop or two of an essential oil such as lavender or rose oil will further support relaxation—with the aromatherapeutic effect of the lavender oil.

❖ The skin is said to be the largest organ of the body and absorbs water quite readily. The skin is similar to a mucous membrane. The ancient seamen understood this fact well. When water ran out, they would soak their garments in seawater. Should the water be filled with toxins such as chemical bubble baths, chemical-laden perfumed soaps, or heavily chlorinated water, the body can readily absorb these toxins putting an extra burden upon the liver and detoxification systems.

❖ The *sitz bath* is sometimes thought of as a foot bath. This is actually a feet and hip bath. The sitz bath stimulates detoxification and relaxation for injuries and cramping of the abdomen, hips, legs, and feet. The sitz is accomplished by sitting in a large washtub while putting the feet into another. The foot tub is heated about five degrees F hotter than the larger tub. The water level can be about belly button high. A towel or blanket is laid over the exposed body to increase sweating. Best to cool the water for a few minutes before exiting the bath or get out when it naturally cools to room temperature. Epson salts and a drop or two of lavender can be used to provide additional therapy.

❖ The *foot bath* is a great way to ease sore feet, increase circulation, ease chills, and ease abdominal cramping. Tub can be ankle to knee deep. Water is heated gradually from warm to very hot while feet are in. A blanket or loose clothing can cover the rest of the body up to the neck. A cold compress on the nape of the neck provides an extra therapeutic effect. 15 to 30 minutes is the recommended duration. Again, Epson salts may be added for additional therapy. Eucalyptus can be added along with a blanket wrap to ease congestion.

❖ The *hose bath* is performed with one person spraying a hose from about ten to fifteen feet away from the subject, spraying up and down the spine, throughout the body. This can stimulate the immune system and decrease pain. It also can cleanse infected areas hard to get to with showers and baths. The water temperature can be contrasted or be at tap temperature.

❖ The *nose bath* or *lavage* is very good for sinus congestion and allergies. While specialized *neti pots* can be used, we can also simply pour room temperature water in the palm of our clean hand, place the nose into the palm and raise the palm as we *snuff* the water up the nose. It can either be pulled through to the mouth or held for a moment in the pharynx and pushed out through the nose. Neti pots are helpful because we can lean back and snuff through the sinus and pharynx more easily when congested. The lavage can be done one nostril at a time or both at the same time. Alternating nostrils has the bonus of being soothing upon the psyche as well as decongesting.

- The *steam bath* is especially helpful for respiratory congestion, sinus congestion and sinus congestion. The steam bath can be accomplished by the formal steam room, by the use of a humidifier or simply with a hot bath in an enclosed room. A cool or coldwater rinse is important for immune stimulation. Eucalyptus can be added to increase the expectoration effect. Again, water intake should increase dramatically depending upon duration.

- The *medicated bath* is a bath with specific botanicals or essential oils. A hot water bath may be medicated with any number of herbal or aromatherapeutic agents. An herbal or aromatherapeutic manual and health practitioner can be consulted for specific botanicals to use for particular issues, as there are many options. For a relaxing medicated bath pour one to three drops of lavender, rose petals, or mint oil into the bath. Alternatively, an herbal tea can be made on the stove and added to the water. Botanicals can also be added directly to the water in fresh or dried form. Oatmeal, starch, Epson salts, and even special waxes can also be added for additional benefit. Prior to taking a bath with an additive, that additive should be skin tested by applying to a small part of the skin and letting it sit on the skin for 10-15 minutes.

- *Mud baths* are often natural warm springs with clay mud zones, but they can also be made with warm water and natural clays. Mud baths are extremely detoxifying because clay draws toxins from the skin cells, while producing an alkalizing effect. Essential oils or botanicals can be added to the mud water to increase the medicinal benefit.

- *Sponge baths, spray baths* and *hot water bottles* are extremely beneficial for specific injuries and cleansing. A hot or cold sponge, rubber water bottle, or spray bottle can be used therapeutically to target specific areas of pain or injury. Relaxation and dilation occurs with heat, and stimulation and constriction occurs with cold. Contrasting can be done for a particular region or injury as well. The bottle, sponge, or spray can be applied directly onto the areas of pain, congestion, or inflammation. Open wounds or burns should be kept dry and clean, so they are not good candidates for this type of therapy.

- *Water wraps* are best done with linen cloth or a cotton washcloth. The cloth is wetted in either cold for immune stimulation; or warm water for respiratory or sinus congestion. A linen can be wrapped around the body. The water also may be medicated, as described for the medicated bath. The wrap can be kept on the skin for thirty to forty-five minutes.

❖ *Cold compresses* or packs are clothes soaked in hot or cold water (depending on the circumstance) and applied directly onto the skin to reduce inflammation (cold) and stimulate the immune system (hot and cold). Cold can be established by crushed ice in a bag or cloth or a frozen ice chest pack as well.

❖ Water is the universal hydrator. What's the rush to towel off so quickly? Rinsing our skin with clean water without toweling off will naturally increase skin's moisture. Alternating cold-hot-cold water with a washcloth cleanses the pores and hydrates the epithelial cells, while stimulating circulation to keep the facial skin vibrant.

❖ We can gargle and swish with water often, alternating cold with warm salt water to stimulate the gums. Water piks are also good for this, as they shoot water into the gums. Swishing vigorously can accomplish this to some degree while exercising key jaw muscles.

❖ Internal fluid balance can be accomplished by drinking approximately ½ ounce of filtered tap or ground water per pound of body weight each day, give or take activity level, diet and environment (fresh food diets can supply 25 ounces). Ceramic or charcoal filters use earth's magnetic nature to screen out toxins while retaining key minerals.

❖ We can drink water with natural cups like glass and copper. Copper adds necessary copper, assisting the body's detoxification. Clean cupped hands are nature's ultimate design for drinking water. This warms the water as it begins to resonate with the body to increase absorption. This can also help us stop to check its clarity and smell it in an attempt to avoid drinking polluted, toxic water.

❖ We can become stewards of the earth's waters. We can give water the respect it deserves. We can stop our dumping of toxic wastes, plastics, and various chemicals. The dumping of toxic waste into our water systems is the same as drinking toxic waste. When we watch a river flow downstream, we are witnessing an exhibition of a conscious source. The living organism we call the earth circulates vital waters and petroleum just as blood and lymph are circulated around our bodies.

Chapter Five

Rhythmic Air

Every day we breathe about thirty-five pounds of air, translating to about five thousand gallons. This is about six times our consumption of food and liquids.

The composition of the earth's atmosphere is precisely suited for breathing. Atmospheric air is composed primarily of nitrogen as N_2 (about 78%); oxygen as O_2 (roughly 21%); carbon dioxide (roughly .03-.04%, depending upon location); argon (almost 1%), water vapor (ranging from 1% to 4%); and smaller/trace amounts of neon, helium, hydrogen, methane, xenon, ozone, nitrogen dioxide, iodine, ammonia, carbon monoxide, nitrous oxide and let us not forget krypton. The earth's atmosphere weighs approximately 5,000 trillion tons.

There are a number of other atmospheric layers covering the living earth's 'lungs.' These layers each have slightly different composition; and vary to temperature, pressure, and activity. Moving upward from the earth's surface, the first layer of air—the *troposphere*—ranges from about seven to seventeen kilometers thick, and thinner at the poles. This is the warmest layer. It reflects and retains the heat of the earth along with the sun's radiation. The Greek word *tropos,* meaning "mixing," aptly describes one of the functions of this layer. This layer is constantly in harmonic circulation as it responds to the relative rhythms of the earth's environment and the sun's solar activity. Because its temperature ranges from warmer at the earth's surface to cooler above, it is constantly rotating air between these warmer and cooler regions. The conveyers for these temperature gradients include various wind channels, jet streams, and water vapor. These conveyers all rotate and gyrate around the earth, guided by mysterious rhythmic currents, which also create our weather systems.

As researchers have been observing the patterns of the troposphere, three distinct cyclic patterns have been identified: These are the three major *cells:* the *Polar cell, Hadley cells,* and the *Ferrel cell.* The Polar cell is a thermal loop of rising convective air that takes warm air at the lower latitudes and moves it to the poles to cool and descend. This descending cooler air produces an easterly wind flow. It also produces a harmonic wave front called *Rossby waves,* with ultra-long wavelengths that funnel moving air into channels. The jet stream is an example of one of these channels. The Hadley cell was identified by the eighteenth century British meteorologist George Hadley to explain the movement of the trade winds. As warm air travels from the sun's zenith point, it moves from the hottest regions on the planet—usually nearest to the equator—upward. The warm convection air is then carried either north or south towards the poles. By the 30th latitude both north and south, this airflow descends with increased pressure. As this happens, some of the flow moves along the earth's surface as *tradewinds*.

Hadley's proposals had several weaknesses, however, because his description of airflow was linear while the earth's surface is curved and irregular and subject to magnetic flows and rotation. This brings into context the conservation of angular momentum. The spherical relationship between airflow and temperature was first demonstrated by William Ferrel in the nineteenth century. Mr. Ferrel related the atmospheric conveyer system to the *Coriolis Effect* and the need to maintain equilibrium between pressure and temperature. The *Ferrel cell* was first calculated by Mr. Ferrel as a blending area for low and high-pressure areas, working conjunctively with the Polar and Hadley cells. The Ferrel cell channels these currents below into the *westerlies*. The flow of the westerlies is blocked by the existence of other pressure differentials, unlike the closed loops of the Polar and Hadley cells. All of these cells and their movements cycle with temperature, pressure, wind, seasons, storms and other rhythms of mother earth.

The *stratosphere* lies outside the troposphere and extends a good fifteen to thirty-five kilometers from the earth's surface. There are a number of wave mechanics aspects of the stratospheric layer. Its various gases apparently react with radiation to create a subtle circulation loop called the *quasi-biennial oscillation*—a layered conveyer of stratospheric wind. The stratospheric wind oscillates between the easterlies and the westerlies in the tropical region. The quasi-biennial oscillation has a period occurring about every 850 days, with a vertical downward path of about one kilometer, ending at what is called the *tropical tropopause*. It has been suggested that these mechanical occurrences are related to gravity waves, although many also connect to atmospheric waves.

This layer contains higher levels of ozone. As a result, some refer to this layer as the *ozone layer*. Actually, ozone is not a layer. The percentage of ozone at this level is simply higher than other layers. Ozone will readily absorb the electromagnetic energy in the ultraviolet-B (270-315 nanometers) wavelength range. This absorption by ozone effectively shields the earth from much of the sun's ultraviolet output—at least in areas where the ozone density is considerable.

The ozone's absorption apparently makes the upper area of the stratosphere warmer than the lower area—the opposite of the troposphere. Researchers theorize the warm layers at the top and bottom of the stratosphere are responsible for the relative lack of turbulence in this layer of atmosphere. Thus jetliners tend to fly within this layer for safety.

Here oxygen O_2 gas is converted to ozone O_3; then back to oxygen gas and oxygen atoms only to recycle once again back to ozone. This recombination loop between oxygen and ozone allows this layer to flex or adjust to ultraviolet conditions and other atmospheric situations like smog. Ozone is thus a key element facilitating atmospheric homeostasis.

Most of us have heard news reports about the ozone layer thinning. Stratospheric ozone levels fell close to 4% per decade since 1980 as the hole widened to a peak of 29.4 million square kilometers in 2000. This is from a size of just over a million square kilometers in 1979—when measurements began. This amounts to a total loss of about 10% of the total ozone in the stratosphere in two decades. Since 2000, measurements had indicated the hole appeared to be shrinking. At least until 2006, when NASA and European Space Agency satellites confirmed the hole hit a record 40 million metric tons and 29.5 million square kilometers. Then in 2007 the hole dropped off again to 27.7 million tons covering an area of 24.7 million square miles.

So is the ozone hole going down—getting better—or not?

The conventional theory was that ozone loss was related to the use of *chlorofluorocarbon compounds* (also called *freons*): Methyl chloroform, carbon tetrachloride and trichloroethane—commonly used in dozens of applications including refrigerants, aerosol propellants and solvents. Their industrial use is based upon their stability, which is why they can float undamaged into the atmosphere. This stability ensures that they can endure the path up to the stratosphere. Once there, these stable molecules are theoretically broken apart by unfiltered ultraviolet radiation into chlorine and bromine ions. These two ions in turn break apart the three-oxygen ozone combination. As a result, oxygen O_2 molecules tend to convert less to the ozone configuration O_3 in the absence of the sun's rays. The remaining oxygenated no-ozone areas theoretically vortex into holes, leaving the classic ozone hole. While chlorine and bromine have been observed as catalytic ozone destroyers for many years, conclusive proof that the ozone layer was disintegrated only came in 1984, and the theory emerged in the years following.

However, this chemical response to humankind's planetary mismanagement is only part of the picture. Peering back to each year's satellite measurements of the ozone, even in the late eighties and nineties when the hole doubled in size, shows a variance of ups and downs. In 1987, the hole was measured at 22.45 square kilometers. In 1988, it dropped down to 13.76. In 1989 and 1990, it popped back up to 21.73 and 21.05 respectively. In 1992, it grew to 24.90. In 1993, it sagged a bit to 24.017; and in 1994, it sagged again at 23.429. Was the issue resolving during this era?

Not likely. In 1996, it jumped again to 26.96. In 1997, it sagged again to 25.13 but in 1998 it jumped again—this time to 28.21. It fell another couple of square kilometers in 1999 and jumped again in 2000. In the years following 2000, it jumped from 26 to 21, to 29, to 23, and to 27, respectfully. Then it jumped to a record 30 again in 2006, and then fell again in 2007 back to 24.

It appears there is more involved than simply chlorofluorocarbon compounds. There are cyclic variations in temperature, pressure, and magnetic fields to consider. The latter relates to the fact that the hole positions toward the annual ozone hole that forms in the troposphere around the South Pole on a seasonal basis. Researchers now believe that the catastrophic ozone depletion in the lower stratosphere may also relate to our increase in carbon emissions, which tends to cool the stratosphere and provide a blanketing effect. This in turn shields heat from rising—resulting in the well-known *greenhouse effect*.

The mild decrease after 2000 was thought to be due to the reduction of CFCs—as mandated by law in many countries. The World Meteorological Organization states that should our CFC use continue to fall, it may take another 50 years or so to bring the ozone levels in the stratosphere back to normal. Understanding the greenhouse effect, this is unlikely unless we get our carbon emissions down immediately.

Just as a skin puncture is a symptom of an injury, producing an inflammatory healing process, the ozone wound on the planet's atmosphere is a symptom of a chronic injury induced by humanity. Now we are witnessing the planet's inflammatory healing process as we experience increased global temperatures and wild weather conditions.

Above the stratosphere exists yet another layer of atmosphere called the *mesosphere*, which ranges from about 50 kilometers to about 80 kilometers above the earth's surface. Here temperatures get colder, alternating from the warmer layer of the stratosphere. In the mesosphere, once again we have observed a circulation of atmospheric tides intermixing with gravity and planetary waves.

Above the mesosphere layer exists the *thermosphere*, ranging from 80 to about 650 kilometers from the earth's surface. This layer again reverses from the temperature gradient of the previous layers, moving from colder to warmer with the distance away from the earth's surface. This takes place through a process of ionization: An interaction between the sun's ultraviolet radiation and gas particles within this layer.

Above the thermosphere layer lays the *ionosphere*, which appears to interact with the sun's solar flare x-ray radiation. This interaction appears to precisely influence radio wave frequency absorption, creating high-energy proton releases. These protons appear to penetrate the atmosphere near the poles. It is this ionosphere proton-wave interaction with the magnetic field pulses of the earth that are thought to cause the magnificent auroras we see in various places on the earth—particularly towards the northern latitudes as the *borealis,* and towards the southernmost latitudes as the *australis.* Recent research reveals these greenish swirling side-spirals of beauty intensely pulse with the rhythms of the solar wind. The solar winds pulse with sunspot cycles, as the heartbeat pulse of the sun synchronizes with the magnetic field of the earth.

Beyond the ionosphere exists the *exosphere.* Here the envelope between the openness of space and the atmospheric layers exists. In this layer, lighter gases such as hydrogen, carbon dioxide, and helium exist. In this layer, the rhythms of the solar system and galaxy pace the pulses of the atmospheric lung of the earth.

What we apparently have above the earth's surface is a series of layers of subtle gases each forming barriers or linings providing for pressure gradients and temperature differentials. These layers are not layers in the sense of clothing layers—with a consistent thickness throughout each layer. These are moving, adjusting, and interactive layers. They are intermixing in areas, while having distinct barriers in other areas. The resulting structures function very much like a cell membrane or epithelium cells of the skin's surface in mucous membranes. They offer channels or pores to allow particular waveforms in while blocking waveforms possibly dangerous to the earth's health.

This organic gaseous system could also be compared with the lungs of a mammal. Instead of air being transferred to the bloodstream through the lung's aveoli sacs and membranes, the sun's electromagnetic radiation is being filtered and partially channeled through the atmospheric membranes. Once these rays are channeled and filtered through to the earth's surface, they can be utilized for photosynthesis, vitamin D synthesis and other waveform processes such as wind and weather creation.

The electromagnetic properties of our atmosphere have been the focus of research for a number of years. This is because our atmosphere is filled not only with various stable molecules such as oxygen and nitrogen, but also with a huge volume of electrically charged particles in the form of positive and negative ions. Positively charged ions have absent electron-waves and negatively charged ions have an abundance of electron-waves. These various energetically charged wave moments are created through their interactions within other waves pulsing through our environment. Cosmic rays, ultraviolet radiation from the sun, electrical storms, wind, and the earth and water itself all interact to create an array of electronically charged wave moments. Friction—closely interacting waves—is the principle relationship creating interference structures. As higher energy waveforms interact with molecular combinations, ions with increased electron-wave moments are produced. Unbelievable as it sounds, this array of interaction is organized into patterns and structures. This is because the earth is a living organism.

Positive and Negative Ions

A decayed radioactive atom is thought to produce 50,000 to 500,000 ion pairs—positive and negative. About 40% of the atmosphere is thought to be ionic. Airborne radon theoretically comprises another 40%, producing in the neighborhood of 250,000 ion pairs per radon atom. The remaining twenty percent of ions is thought to be produced by cosmic proton radiation.

Tests have indicated that indoor ion levels are slightly lower than outdoor levels. This is thought to be because outdoor ions tend to last longer than do their indoor cousins. This also correlates with the existence of the various electromagnetic fields existing within the home due to the use of various electronic appliances.

Outdoor ion counts in rural areas among good weather conditions can range from 200 to 800 negative ions and 250 to 1500 positive ions per cubic centimeter. Positive ion count can increase to over five thousand ions per cubic centimeter during an incoming storm front. During a rainstorm, the level of positive ions falls quickly, and negative ion levels rise.

Ion levels also tend to vary depending upon the proximity to air pollutants. Pollution dramatically reduces total ion count. This is thought to be because both positive and negative ions will attach to pollutant elements. Natural settings typically have more ions than urban areas. For example, a waterfall might have as much as 100,000 negative ions per cubic centimeter. Ion levels around crowded freeways tend to be quite low, often below 100 negative ions per cubic centimeter.

Negative ions appear to form easily. One pass of the comb through the hair can create from 1000 to 10,000 ions per cubic centimeter. The living organism is a tremendous ion producer. Assuming adequate grounding onto the earth or a grounding metal, a typical human exhalation will contain from 20,000 to 50,000 ions per cubic centimeter.

Positive Ions are typically generated with a decrease of atmospheric pressure; an increase in wind and temperature; and a decrease in elevation. This is particularly noticeable in *Foehn winds*—warm winds that descend from mountainous areas down to areas of lower elevation. Wind patterns known to be Foehn include several winds that blow around the world. The dry southerly wind (meaning it comes from the south) blowing through the Alps, Switzerland and across southern Germany is a Foehn wind. The *Sharav* or *Hamsin* winds blowing though the desert of the Middle East are Foehns. The *Sirocco* that blows through Italy and the *Mistral* that blows through southern France are both considered Foehns. The *Chinook* winds of western Canada and NW United States, and the *Santa Ana's* that blow through southern California from time to time are also considered Foehns. Foehn winds have also been occasionally spotted around various mountain ranges such as the Colorado Rocky Mountains and Tennessee's Smokey Mountains. Foehn winds tend to funnel between mountain ridges, which accelerate their gusting speeds to an excess of 50 miles per hour. The other notable characteristic of a Foehn wind is it usually accompanies dramatic increase in temperatures.

Foehn winds are also known for their ultra-low humidity and propensity to cause erratic fires. They also apparently cause a number of negative physical and emotional effects in both humans and animals. For these various reasons, the Foehn is often referred culturally as an 'ill wind.' While some have disputed the effects of Foehn winds, both research and observation has indicated otherwise. Research performed by Sulman (*et al.* 1973-1980) indicated these winds were linked with an increase in headaches, heat stress, and irritability. Others have documented an increase in allergies and sinus ailments. These effects may also be related to the immediate change in temperature, pressure, and humidity. Because positive ions are also linked to these kinds of environmental and physiological changes, it would be safe to assume Foehn winds and positive ions may be harmonically connected with the physical effects of low humidity, higher pressure, and higher temperatures.

Ions thus play a dramatic role within our environment in a number of ways. They appear connected with production of storms, lightning, wind motion and other specific atmospheric conditions. There is some debate among scientists regarding the importance of these ions. The basis for the controversy seems to be that theoretically the interaction between solar radiation and the earth's outer atmosphere results primarily in an equal amount of positive and negative ions. So where are these variances in ions coming from?

Negative ions' effects upon health and behavior have been the subject of intense study for the past fifty years. In 1926 Russian scientist, Alexander Chizhevsky developed research exposing animals to ionized air with negative ions and/or positive ions. In these studies, he found—as other studies have concluded such as Krueger and Reed (1976), that while living in positive ion conditions, rats and mice experienced greater illness and shorter life duration than those living in negative ion conditions. Negative ions have also been linked to health benefits among humans in a number of studies. Increased negative ions were associated positively with better levels of cognition and memory by Delyukov and Didyk (1999), and by Baron (1987). Negative ions were linked with lower levels of aggression by Baron as well. A lower incidence of asthma was observed by Ben-Dou (1983). Lower levels of irritability and higher levels of serotonin were linked with negative ions in two studies by Sulman (1984; 1980).

In Sulman's 1980 study, daily urine samples were taken from 1,000 volunteers one to two days before a storm's arrival, both during Feohn winds and under normal weather conditions. The samples were analyzed for neurotransmitter and hormone levels, including serotonin, adrenaline, noradrenaline, histamine, and thyroxine metabolites. The results concluded that during positive ion conditions, an overproduction of serotonin levels resulted in irritability. In positive ion conditions, Sulman discovered adrenal deficiency and early exhaustion. Positive ions were also associated with hyperthyroidism and subclinical thyroid symptoms considered *"apathetic."*

As Jonassen (2002; 2003) points out, the effects linked with negative ions may not be as simple as the presence or abundance of these ions. Certain atmospheric conditions appear to create pathways through which flow specific currents. As these layered currents flow, various effects are observed. This is compared to the pathway of a river, or the path of current created within the body by an electrode as it is attached to the body. It might also be compared with the pathway of *qi* created by an acupuncturist's needle.

The evolution of the study of ions in the atmosphere leads us to an understanding of wave mechanical currents moving though the atmosphere. We might have a clear notion of how currents move through the solid mediums of wires and metals, or even the ability of fluids to transmit currents. These are visible. Currents moving through the atmosphere are beyond our vision, however. This does not mean we cannot understand their presence and motion through empirical research and instrumental observation. While we might imagine the atmosphere is made up of an unorganized chaotic collection of gas particles bumping around, the organized effects of weather fronts and pressure currents through the atmosphere indicate organized motion.

Organized atmospheric current flows consist of not only air, temperature and pressure circulation, but of waveform pulses of the invisible electromagnetic spectrum. As we can observe through the effects of human-propagated radiowaves, specific types of 'receptor' antennas are designed to read specific types of signals. The circulation of these waveform currents throughout the planet is effectively organized by consciousness. This perspective provides the clearest explanation for Sulman's and others' observations that changing weather patterns (with apparent ion changes) affect some people while not affecting others. There must be additional elements at play, connecting these movements to some physiological metabolisms and not others. As we understand that the basic variance between one living organism is another is the conscious personality living within that body, we can connect the individualist responses to function weather disturbances like Foehn winds to variances in individual consciousness.

To consider otherwise would require that every organism is affected precisely the same by a specific negative and positive ion level. While some people may complain of irritability and/or sore joints prior to an oncoming storm, others of the same age and physical condition have no complaints. We certainly accept the biological individuality at play. That individuality is ultimately tied to the individual personality of the conscious self, residing within the organism.

The Lung Atmosphere

Our lungs are beneficiaries of this elegant interaction of waveform currents within our atmosphere. The lungs undergo a rhythm of their own—a cycle called *respiration*. This cyclic process is divided into two processes: *inspiration* and *expiration*. This of course is not associated with dying, although the body does *expire* as the inner self leaves—hence the association.

Inspiration starts with a process called *pulmonary ventilation*. Pulmonary ventilation occurs with a vacuuming of air through a maze of pipes and filters, while undergoing processes of heating, screening, and moisturizing along the way.

The journey begins as air passes over the bones of the nasal cavity— the *turbinates* located above the palate. Here the air is warmed and humidified. The mucous membrane-lined passageways of the larynx and pharynx also help warm and moisten the air to some degree, but the nasal passages are the primary humidifier system. When the heat-humidifier zone is working right, incoming air is heated to about 95 degrees Fahrenheit and brought to about 97 percent humidity. This moist warm air is important because it now more closely matches the environment of the body. In this condition, the air can keep the rest of the linings relaxed. Should we breathe in colder, dryer air quickly through our mouths, bypassing the body's warming and humidifying system, the lungs will tighten, making it harder to breath. For this reason, it is always better to breath through the nose as much as possible, even when exercising.

As air moves through the nostrils or mouth and through to the *pharynx*, the *larynx* and the *trachea*, it passes over a lining of mucous membrane cells lined with tiny hairs called *cilia*. These cilia capture foreign particles with a web of sticky mucous. After being stuck, the particles are gathered up within the mucous. If possible, the tiny cilia hairs will undulate the foreign particle towards exit points like the throat, nose, and mouth. At these points, the particles can be sneezed or coughed out as phlegm, nose-blown out, or swallowed into the acidic abyss of stomach gastrin. More offensive particles like bacteria and viruses—or those particles not able to be extracted by the cilia—are attacked by the macrophages that line the mucous membrane cell surfaces. The macrophages break down the foreigners and escort their waveform components out of the body through the liver, kidneys, skin, or colon.

Once through this maze of filtering and conditioning, incoming air reaches one of the two *bronchi* on its passage to one of the lobes of the lungs. Because the bronchi fan out through into some eight million tiny branches called bronchioles, this system is called the *bronchial tree*. Eventually the airflow terminates at tiny air sacs at the end of the bronchioles called *alveoli*. These little round sacs are lined with a thin membrane wall covered with a thin layer of water along with a specially designed *surfactant*, which prevents the surface tension of the water from collapsing the alveoli when they expel their air.

There are literally millions of alveoli in the lung. Some estimate three hundred million. If you laid out the surface of these sacs flat, it would cover an area of about 60 square meters. Only about a third of these are necessary to supply enough oxygen to keep the body alive. For this reason, people can lose one lung and still do fine. For those of us with two lungs, it is useful to understand how much breathing capacity we truly have available.

As the incoming air reach the alveoli, the approximately 21% portion of oxygen diffuses through the alveoli membrane into the bloodstream. This process is at least partially driven by the relative pressures existing on each side of the membrane. When the oxygen percentage inside the blood is a lot less than in the alveoli, oxygen diffuses into the blood. This seems a fairly simple process on first glance. However, the process is really quite complicated: It is a waveform interaction of ionic chemistry and atmospheric pressure.

Because air is a mixed gas, the various types of molecules within the air exert pressure against each other. The net effect is a unified atmospheric pressure existing throughout—just as a mix of ocean waves interacts to form a homogenous water surface. When we consider that these air molecules such as oxygen, nitrogen, and carbon dioxide are actually electromagnetic wave interactions, we can better relate to the concept that their wave patterns interfere and interact with each other, creating further relative wave patterns. Each combined resonating wave structure—each molecule—has a distinct internal resonating moment giving it stability. This specific resonance also allows it to repel or attract other resonations. The repelling force provides an interference pattern (or *interference moment*), which creates a specific pressure gradient, temperature and motion. In a mixed gas, interference moments interact to form an overall pressure, temperature, and current. The blended characteristics of the atmosphere are due to the interaction of the waveforms of the various molecular species.

The combination and balance of the combined waveforms is critical. For this reason, our atmosphere can easily be polluted with excess carbon, ozone, and even oxygen. Too much of one waveform can throw off the entire balance, resulting in a potentially dangerous mixture of foreign gas.

Researchers assume that the partial pressure of oxygen in air stabilizes at about 160 mm Hg because oxygen maintains 21% of the total volume of air and 760 mm Hg is the average atmospheric pressure existing at sea level (21% of 760). By the time this air finds its way to the alveoli, however, it is moistened, warmed, and mixed with carbon dioxide coming in from the blood into the alveoli. This air mixture ends up at about 100 mm Hg pressure. At this pressure, carbon dioxide will exist in the alveoli at about 40 mm Hg.

On the other side of the alveoli wall is the blood supply, with immediate access to the rest of the body through the pathways of the arteriole and capillary system. On one side of the alveoli wall is air at one pressure, and on the other side blood containing carbon dioxide and oxygen at another pressure. As carbon dioxide-poor oxygenated blood leaves the lung's capillary system into circulation, it will have oxygen at about 95 mm Hg and carbon dioxide at about 40 mm Hg. As blood enters the alveoli capillaries from the heart after having traveled around the body, the oxygen pressure is about 40 mm Hg and the carbon dioxide is about 45 mm Hg. This is because the oxygen has been depleted by the cells around the body, and carbon dioxide is now being returned as the waste product of *cellular respiration.*

Because the oxygen pressure is greater in the aveoli than in the blood, oxygen tends to diffuse out through the alveolus membrane into the blood. Because the carbon dioxide pressure in the blood is higher than the carbon dioxide pressure in the alveolus cavity, carbon dioxide diffuses into the alveolus. This exchange takes place until the pressure of the two partial gases (oxygen and carbon dioxide) are closer to equal on each side of the alveolus membrane.

Now if we consider what is going on harmonically, we can understand that particular molecular structures stabilize when their wave patterns resonate at particular densities. (Density would be how many resonating wave moments exist within a specific volume.) This would be like comparing how many tennis players a particular series of courts could contain. If five courts had doubles going on, then the total density of players would be twenty (four per court). But if suddenly school let out and a bunch of kids with tennis rackets and balls invaded the courts and started trying to play while the doubles games were going on, balls would be flying all over and people would be running into each other. Some people might even get hurt. For this reason, there are particular rules for tennis courts. One governs how many players can be on each court.

If we consider each court to be a volume of gas, and we observe the rules of the tennis courts, we know that there is a maximum density for any particular court—four for doubles and two for singles. When the courts are cleared of the serious tennis players, the inexperienced kids come onto the courts and break the rules for a while. In this situation, the density of the courts might be eight or ten for a while. Eventually the kids will be warned by an adult, and they will spread out to the empty courts and fill up the two or four-person requirement in each court. This will re-establish the density of each court back to the rule standard.

In the same way, different molecules, with their different waveform characteristics and resonance patterns, fit particular densities within air, as designed through some rules. They may occasionally fluctuate depending upon the circumstances. When circumstances change, density may increase within a particular region of the atmosphere. This change results in a higher level of interactivity, and a greater pressure within that region. This higher pressure in turn forces its occupant waveforms to spread out—most easily to a less dense region. The waveform molecules will be channeled into a less-dense region until their designed (stable) density is found. In the case of the external atmosphere of the earth, these channels create airflow and temperature gradients. In the case of the blood and the alveoli, oxygen is channeled into the less dense bloodstream from the more dense air of the lungs. Carbon dioxide on the other hand moves from the denser blood stream to the less dense alveoli air sac.

This process of diffusion is repeated at the cell level in cellular respiration. When oxygen attaches to hemoglobin in the blood—as oxygen's waveforms resonate and attach to hemoglobin—it is delivered to cells around the body. These cells require oxygen for the process of energy creation and metabolism. Inside the cell membrane of a relatively inactive cell, oxygen pressure is about 40 mm Hg while the carbon dioxide—again, a byproduct of cellular metabolism—is about 45 mm Hg. Oxygen's waveforms are thus drawn into the cell and carbon dioxide's are drawn out into the bloodstream.

We might find nice textbook drawings of tiny holes in the alveoli and cell walls to accommodate "particles" of oxygen and carbon dioxide moving through. These membranes are much more complicated than that, however. They are both made up of complex phospholipids with complex receptors and channels with ionic gates. Like little border checkpoints, these channels allow molecules through under certain rules and inspections. Membrane phospholipids are interlocked in such a way to only allow through molecules that oscillate and resonate at particular frequencies, wavelengths, amplitudes, and angular waveforms. In other words, they are monitoring the *waveform fingerprint* of each molecule, and allowing through only those that fit particular waveform requirements.

Meanwhile oxygen and carbon dioxide molecules are waveforms seeking an environment of stability and balance. They resonate at particular densities and with particular surroundings. These parameters, when accomplished, establish a harmonic at a particular state and environment, and seek to remain in that state.

It is precisely for this reason that exhalation takes place so easily. It is so easy that biologists describe the exhalation process as the *passive phase*. The lungs relax, the diaphragm and rib cage returns to their respective positions, and the carbon dioxide-rich gas easily passes out of the lungs. This is because the gases seek equilibrium and stable densities—quickly found in the outside atmosphere.

About 97% of our oxygen is delivered to the cells by resonating with hemoglobin molecules. After being escorted through the micro-capillaries to the cells, the oxygen disassociates from the hemoglobin. Only about 20% of oxygen disassociates from hemoglobin while we rest. More disassociates as needed. The rest stays in the bloodstream, on standby. This standby oxygen effectively alkalizes the blood, inhibiting oxidative radicals with the presence of O_2.

The cells of the body ultimately utilize a combination of the sun's radiation and the atmosphere's oxygen to produce energy. Plants provide the basic energy reaction, producing glucose from sunlight, carbon dioxide, and water. Our cells process glucose using oxygen as part of the conversion into energy. The process of converting sunlight to glucose is photosynthesis. Photosynthesis is probably more appropriately considered a form of respiration.

The process of photosynthesis is uniquely similar to the process of cellular respiration. In a process called *oxidative phosphorylation*, the two-phosphate ADP synchronizes with oxygen and another phosphate atom to form a triple phosphate (ATP) structure. As with all electromagnetic bonds, waveform energy is held within the phosphate-oxygen bond. Once in ATP form, glucose synchronizes with one of the phosphates to release ADP, glucose 6-p along with heat and energy for metabolism. The process releases energy as the phosphoryl is transferred, and ADP (adenosine diphosphate) cycles with ATP (adenosine triphosphate) utilizing oxygen.

This is not the only exchange cycle the body uses that releases energy. Cellular processes will convert NADH to NAD+, biotin-CO2 to biotin, and acetyl CoA and oxaloacetate to citrate among others in order to release energy. Once the process has been started, glucose provides the final conversion step—often called *glycolysis*.

The Krebs ATP/ADP cycle process releases both metabolic kinetic energy and heat, along with carbon dioxide. As that happens, the carbon dioxide combines with water to form carbonic acid. Carbonic acid in turn releases hydrogen acid (H+) and blood bicarbonate (HCO_3) into the bloodstream. This creates an acidic environment, which means there are greater numbers of H+, which will resonate with dissociated hemoglobin. As this H+ rich hemoglobin and bicarbonate blood reach the alveoli capillaries, hemoglobin (one smart molecule!) releases the H+, which reforms carbon dioxide (CO_2) as it resonates with the bicarbonate. Hemoglobin is then ready for another oxygen molecule and the CO_2 is ready to diffuse back through to the alveoli and out through the lungs to the atmosphere.

It is remarkable that this layered process of gas exchange is so exact. It takes place billions of times each day, through millions of alveoli and millions of capillaries and billions of cells. It is also remarkable that the system of the atmospheric layers circulates the various gasses and to the extent possible, provides a filtering system to clear pollutants with a natural process.

It is important to note that during heavy exercise or minimal breathing the blood will become laced with heavier doses of carbolic acid, carbonates, and acidified with H+ hemoglobin. This acidic situation can become toxic to the tissue systems if it remains too long without clearance. Should this acidic environment remain in the blood for too long it can do some serious damage to blood vessel walls and tissues. These acidic and incomplete waveforms can become more stable should they borrow waveforms from molecules making up our tissues. This leaves these tissues damaged and in need of repair—often referred to as *atherosclerosis* or the hardening of arteries.

Now should there not be enough oxygen available for the cell's energy needs, the process of waveform bonding transfer will still occur. However, it will not have the same efficiency. This is called *anaerobic glycolysis*. The problem with anaerobic glycolysis is that instead of the byproducts being easily exchanged like carbon dioxide, pyruvate produces lactate while NAD+ and NADH become the vehicles for waveform transfer. This process tends to build-up acidic lactates in the bloodstream, which are difficult to clear. While muscle fatigue has been traditionally pegged to lactate buildup in the bloodstream and muscle tissues, there appears to be other factors to consider. Though we can easily link fatigue to oxygen debt in the cellular respiration process, there appears to be other issues involved in muscle fatigue, because muscle fatigue takes place in both aerobic and anaerobic work.

The bottom line is that oxygen is necessary for energy production for all mammals and most other organisms. Our bodies would cease without it. Even fish draw oxygen out of the water they drawl into their gills. Blood rich with oxygenated hemoglobin will provide more efficient energy production. At the same time, there is a sensitive balance in the level of oxygen in the air and in the blood. Oxygen provides the foundation for the body's total air pressure.

When oxygen levels within the blood become too great or too small, trouble is here or on the horizon. Dizziness, lack of balance and even death will occur in places where the atmosphere has varied oxygen densities—such as on mountaintops or deep undersea. As we rise through the troposphere, the atmospheric pressure decreases, allowing oxygen (and other gases) to disperse. There are less air molecules within the same volume—including oxygen. Oxygen density is also reduced with heat. As atmospheric pressures can fall in extremely hot climates, oxygen density will often dramatically fall.

This tendency of air pressure to vary to heat is one way meteorologists predict weather patterns. High atmospheric pressure will usually provide stable weather, while the low pressures of hot, humid weather will typically provide instability and stormy weather. Our body also appreciates higher pressure. When our body has sufficient oxygen within the blood stream, a state of internal higher atmospheric pressure prevails. This is not to be confused with blood pressure, however. Blood pressure is the fluid pressure upon the blood vessel walls. The atmospheric pressure of our body is directly related to its oxygen content, as oxygen is the central gas existing within the body. Of course, there is a limit to this high pressure. Too much atmospheric pressure will cause damage to our various air-filled cavities such as the lungs, the ears, the sinuses, and the arteries.

This is what can occur when we dive into the water's depths. The atmospheric pressure of water increases with depth just as it does as we rise above water level. The atmospheric pressure on the surface of water is typically 14.7 lbs per square inch or 101 kilopascals. At thirty meters below the surface, the atmospheric pressure is four times as great, at 400 kilopascals. These levels begin to put pressure upon the sinus cavities and the eardrums. The ear's tympanic membranes can become damaged easily at this depth. Should we rise too fast from the depth, or fall too quickly from a higher atmosphere to a lower atmosphere—as occurs when deep-sea divers ascend too quickly, or when pilots ascend too quickly in unprotected aircraft—we may experience *decompression sickness*.

Decompression sickness is actually quite prevalent among divers and pilots. At higher atmospheric pressures, gases like nitrogen are absorbed to a greater extent by the cells and tissues. This is because these gases seek greater stability inside the cells. Along with oxygen, the tissues will absorb greater amounts of nitrogen gas. While oxygen is utilized in the energy cycle, nitrogen is not. It will be released back into the bloodstream from the cells as the water or pressure decreases, and the diver rises to the surface or the pilot falls to the ground. As this happens, large nitrogen gas bubbles can form in the smaller capillaries of the body, preventing oxygenated blood to reach the cells and tissues of the body. This can result in fainting, vomiting, dizziness, and even paralysis. Some refer to this state as the *bends*. The bends is usually prevented by rising slowly and carefully, or by transferring through a stepped pressure chamber—often called a decompression chamber.

Now this is extreme, and our normal breathing at sea level is not apt to create the kind of pressure differentials involved in these cases. It provides a useful lesson, however. We can conclude that should we not breathe adequately our bodies will experience a reduction in available oxygen to the cells, along with a drop in our internal atmospheric pressure. A lower atmospheric pressure reduces cellular respiration, lowering energy conversion. With a lower atmospheric pressure, the pressure differentials—driving diffusion between our cell membranes, our hemoglobin and our lungs—will all be out of balance. This effect creates fatigue, memory difficulties, and greater acidic levels within our bloodstream—leading to potential tissue damage.

The primary muscle used for inspiration is the *diaphragm*. The *pectoralis major* and *minor, seratus anterior, sternocleidomastoid, serratus posterior superior, scalene group* and *levatores costarum* are accessory muscles assisting in the process. As the diaphragm contracts it moves downward, causing an increase in the rib cage volume. This lowers the atmospheric pressure in the lungs, and creates a vacuum or negative pressure within the lungs. This effect draws in the air.

The average lung capacity of an adult is about 6000 cubic centimeters. This of course can be increased with cardiovascular exercise. When we breathe unconsciously while relaxed, we might breathe 500 to 700 cc in and out. Typically, about 1200 to 1500 cc will be left in the lungs during this breathing. If we should completely exhale, we have the capacity to move out 4800 to 5200 cc of air. Obviously, this would easily be enough to clear out much of that stale air and replace it with fresh air.

Looking at these numbers it is not difficult to utilize the lungs better than we typically do. With better breathing, we will not only bring more oxygen into the bloodstream but we will push out more stale, acidic carbon dioxide as we exhale. This lowers the carbolic acid and carbon dioxide levels in the bloodstream, forcing a greater level of oxygen replacement to associate with hemoglobin. By bringing in more oxygen, the acidified H+-hemoglobin is replaced with oxygenated and alkalized hemoglobin.

There are various negative consequences of having poorly oxygenated blood and/or highly acidic, carbon dioxide-rich blood. Poorly oxygenated blood can cause or contribute to brain fog, poor memory, fatigue, restlessness, nervousness, anxiety, indigestion, cardiovascular diseases such as *cor pulmonale,* hypertension, atherosclerosis, and angina. Poor oxygenation can also contribute to intestinal challenges such as colitis and heartburn. Poorly oxygenated blood can lead to low healing response, poor sleep, and increased inflammatory response, all of which are linked to a number of autoimmune diseases.

Oxygen is by far the greatest nutrient of the body. Oxygen is vital to the operation of every organ and tissue system. Oxygen also helps provide an environment among the blood and tissues less hospitable to bacterial or viral invasion. With a poor supply of oxygen—caused by either a lack of breathing or contaminated air—the body operates at less than maximum efficiency, leaving our systems open to all sorts of disorders.

There are several types of air pollution. The first is *air particle* pollution—when particulate size ranges from .1 micron to 10 microns. This type of pollution is from soot, typical of automobile exhaust, industrial smoke stack exhaust, fireplace smoke, and smoke from forest fires, barbeques and other combustion. Soot is also called *black carbon* pollution, because the excess carbon burn off from burning fossil fuels is the primary component.

Noxious gases are another type of pollution category. These include gasses such as carbon monoxide, chlorine gas, nitrogen oxide, sulphur dioxide and various other chemical gases and vapors. CFCs are a type of noxious gas.

While these two pollution types are distinct, they are often tough to differentiate. Many pollutants are actually vaporized liquids. Molecules become suspended in vapor, appearing like gases or liquid vapors. They are airborne elements from the solid plane.

We should also specify a difference between *indoor* and *outdoor* air pollution. There are a number of pollutants specifically pervading building ventilation systems and indoor facilities. To this we can add outdoor pollution entering the house. Examples of indoor pollutants include formaldehyde emitted by foam, treated wood plastics, and chlorine gas emitted in indoor swimming pools.

Indoor Pollution

Indoor pollutants arise from a number of sources. Appliances driven by oil or gas can emit exhaust of either noxious gas or soot. The amount of pollution largely depends upon the appliance and the fuel utilized. Propane burns very clean and efficient as opposed to oil, for example.

Various scents and fragrances used in deodorizers, decorations, soaps, and furniture can be surprisingly toxic. While a fragrance might innocently smell like flowers or delicious foods, the typical commercial fragrance contains at least ninety-five percent synthetics. One perfume may contain more than 500 different chemicals. Benzene derivatives, aldehydes, toluene, and petroleum-derived chemicals are just a few synthetics used in fragrances. Toluene is notable in that it has been linked to asthma among previously healthy people. For this reason, we should carefully consider any product with an ingredient called "fragrances." This includes laundry detergents, dishwashing and other soaps, shampoos and other types of hair products, disinfectants, shaving creams, fabric softeners, fragrant candles, air fresheners and of course perfumes. Care also should also be given to the word "unscented" as this still may have some of the same synthetics, but used as fragrance-masking elements. In one study (Anderson and Anderson 1997), mice were submitted to breathing with a commercial air freshener for one hour at different concentrations. A number of concentrations, including levels typically used by humans in everyday use, caused sensory and pulmonary irritation, decreased breathing velocity, and functional behavior abnormalities. Another study performed by the same researchers (Anderson and Anderson 1998) and published a year later revealed that mice who were subjected to five commercial colognes or toilet water for an hour suffered various combinations of negative effects, including sensory irritation, pulmonary irritation, decreased airflow expiration, and neurotoxicity.

Fabric softener emission is also a dangerous source of air toxicity. Several known irritants and toxins are typically found in fabric softeners, including styrene, isopropylbenzene, thymol, trimethylbenzene and phenols. In yet another study, Anderson and Anderson (2000) subjected mice to five commercial fabric softener emissions for 90 minutes using laundry dryers. The results clearly illustrated that fabric softeners significantly irritate their air passages. These negative health effects were also seen from the emissions of clothing dried with fabric softener pads. Anderson and Anderson also found pulmonary toxicity from commercial diapers (1999), adding that a number of chemicals were found in diapers that were known pulmonary and sensory irritants.

The same authors (2000) also found the same types of sensory and pulmonary irritation effects and decreased lung capacity from the use of synthetic mattress pads, and identified respiratory irritants such as styrene, isopropylbenzene and limonene among polyurethane mattresses. When subjecting organic cotton mattresses to the same test, the results were quite the opposite. Increased respiratory rate and tidal breathing volumes were observed with organic fiber mattresses. The authors noted on each of the above studies that any of these air poison sources could be at least a contributing factor in the rampant rise of asthma among our modern society.

Second-hand tobacco smoke is also an important source of indoor pollution. The American Cancer Society in 2004 estimated that 160,000 Americans die each year from lung cancer caused by smoking. Lung cancer alone has between an eleven and fifteen percent chance of survival beyond five years. It should be noted that the highest rates of global lung cancer occur for both men and women in North America and Europe (Parkin *et al.* 2005). Of course, these are countries where indoor smoking rates are the highest.

Recent research has indicated that not only is second-hand smoke dangerous to non-smokers, but it has more than twice the amount of tar, nicotine and other toxins than the smoker inhales. This is because the smoker will inhale the smoke through the filtering mechanism provided by the packed tobacco inside of the burning, and many cigarettes will also have additional filters to screen out toxins. Second-hand smoke contains five times the amount of carbon monoxide—the lethal gas that deoxygenates the blood—than the smoker inhales. Second-hand smoke also contains higher levels of ammonia and cadmium. Its nitrogen dioxide levels are fifty times higher than levels considered harmful, and the concentration of hydrogen cyanide approaches toxic levels. Constant exposure to second-hand smoke increases the risk of lung disease by 25%, and increases the risk of heart disease by 10%. Second-hand smoke exposure has also been positively linked to emphysema and chronic bronchitis, asthma and other ailments (Makmomaski and Kaiseman 1998).

There are also a number of indoor pollutants caused by various indoor organisms. These are typically called *biological pollutants*. They include bacteria, viruses, fungi, mold, dander, and dust mites. Bacteria can come from rotting food, plants, people, or pets. Dander typically comes from the hair or skin of pets or other creatures, as do many bacteria, viruses, and fungi.

Mold, fungi, and bacteria can grow on anything wet. For this reason, it is important to make sure that spills or wetness below or in the house be dried up immediately. This is because almost any type of sitting water will grow mold. Mold spores reproduce and float through both the outdoor and indoor air. Once they land, they can begin multiplying into larger cultures. While there are several species of mold, almost all require moisture to grow and populate. A concentration of spores and their riding the indoor air currents can cause various sensitivities, allergies, and sickness. The spores also produce a number of substances called mycotoxins, which can create health concerns if inhaled. Homebuyers should be aware of this, and make sure that houses they are buying have no water entry into the basement. Moist appliances like air conditioners, bathtubs, bathroom carpets, or air ducts can also grow mold. For these reasons, we might want to frequently check the various corners and dark places in our houses and workplaces for moisture, because the fungal or bacterial populations growing on these surfaces will only get bigger with time— disbursing toxins into the air as they grow. Once found, the area should be dried, the mold or bacteria should be cleaned up (water with a few drops of vinegar or chlorine will kill most mold or bacteria) thoroughly. If a mold growth is over several feet, a mold specialist may need to be called in to eliminate the mold (EPA 2003).

Dust mites are tiny creatures that live in beds, pillows, carpets, and other fabrics. There are two prevalent species: The American house dust mite (*Dermatophagoides farinae*) and the European house dust mite (*D. pteronyssinus*). They both feed on organic matter—and they love dead skin. The primary allergens that dust mites create are the skin moltings they shed as they grow, and their feces. Because of their propensity for eating skin, mites are usually not found in vents or ducts. Their favorite places are beds, pillows and chairs—and they like humid locations more than dry locations. Chemical pesticides do not help, because they will just grow back. The best defense for dust mites is to keep the house dusted and all fabrics cleaned as frequently as practical (Boyd *et al.* 2006).

Asbestos indoor air pollution is becoming less likely with time since the *Environmental Protection Agency* passed the *Asbestos Ban and Phase Out Rule* as part of the *Toxic Substances Control Act* in 1989—which was for the most part overturned in 1991 by the U.S. *Fifth Circuit Court of Appeals.* What have remained are various specific bans such as those from the *Clean Air Act* and remnants of *Toxic Substances Control Act*, including some continued restrictions supported by Congressional rulings. The CAA has stimulated various bans since 1973. The bottom line is that although paper- and cardboard-based asbestos has been banned along with certain spray-on substances, many products still contain asbestos. These include cement sheets, clothing, pipe wrap, roofing felt, floor tiles, shingles, millboard, cement pipe, and various automotive parts. Beyond the banned items, there is no ban preventing manufacturers from using asbestos. The important thing to remember is that the EPA does not monitor manufacturers for their ingredients. In general, asbestos inclusion into today's building materials should be considered a given.

Formaldehyde falls in the same category. Today so many building materials and furniture are built using formaldehyde. These include pressed wood, draperies, glues, resins, shelving, flooring, and so many other materials. The greatest source of formaldehyde appears to be those materials made using *urea-formaldehyde* resins. These include particleboard, plywood paneling, and medium density fiberboard. Among these, the medium density fiberboard—used to make drawers, cabinets and furniture tops—appears to contain the highest resin-to-wood ratio. Another sort of resin called *phenol-formaldehyde* or PF resin apparently emits substantially less formaldehyde than the UF resins. The PF resin is easily differentiated from UF resin by its darker, red or black color. Incidental *off gassing* of formaldehyde into the indoor environment from these fabrics or pressed woods can result from the application of sun, heat, sanding or demolition. It should be noted that cigarette smoke is also a source of airborne formaldehyde.

There are many other chemical toxins—including many yet to be discovered—in our building materials. Knowing some of this information, it is probably safe to say that any kind of building or decomposition of a modern building will likely impart various hazardous chemicals, most likely including asbestos and formaldehyde. This means that the air during any kind of sanding, crushing, fire or demolition should be treated with extreme caution. Using a particle or gas mask is more than a good idea under these circumstances, though it should be noted that most particle masks do not form a tight enough bond with the face to filter much at all. Best is to use a gas filter or a mask with a rubber barrier fitting onto the face.

With regard to off gassing, prior to bringing in any type of new furniture or wood into the house, it is best to off-gas the product by setting it in the sunshine for a couple of days or at least for a full day. As the sun's resonating waveforms connect with the material, various toxins are disassociated and released. Not such a good thing for the environment, but at least it will disburse outside of our immediate breathing environment. Through off-gassing we can possibly avoid a number of sensitivities and allergenic responses associated with immediately incorporating foreign toxins into our indoor environment.

Carbon monoxide is a substantial indoor air toxin. Carbon monoxide is released by burning gas, kerosene, or wood. It can thus arise from the use of wood stoves, fireplaces, gas stoves, generators, automobiles, kitchen stoves, and furnaces. Low concentrations of carbon monoxide in the indoor environment might cause fatigue and even chest pain. Higher concentrations may result in headaches, confusion, dizziness, nausea, vision impairment, and fever. This is due the association of *carboxyhemoglobin* in the bloodstream, which takes place when carbon monoxide attaches to hemoglobin instead of oxygen. This will in effect starve the body of oxygen, and higher concentrations can easily lead to death.

Acceptable carbon monoxide levels in households are about .5 to 5 parts per million. Levels near a gas stove might be 5 to 15 ppm. An improperly vented or leaking stove might cause 30 ppm or more near the stove, which becomes hazardous. The U.S. National Ambient Air Quality Standard for maximum carbon monoxide levels outside is 35 ppm for one hour and 9 ppm for eight hours. Standards for indoor carbon monoxide have not been determined.

Making sure that every appliance is vented properly is task number one in avoiding carbon monoxide poisoning. The appliance should also be checked for leaks, and those leaks should be sealed prior to use. Central heating systems should be inspected for leaks and be cleaned periodically. Idling the car in the garage is a no-no. Open fireplaces should be avoided indoors, and wood stove doors should be kept closed. We might not want to rely upon the draft up the fireplace for the escape of carbon monoxide.

Radon is another indoor pollutant worth consideration. It is estimated that almost 12% of American cancer deaths from lung cancer (about 19,000 deaths a year) result from radon exposure. Radon is caused by various soils and rocks from the earth. The problem with radon poisoning occurs when a house is not properly ventilated, especially in cold weather.

Cold outside weather combined with an unventilated warm indoor environment creates an energy vacuum, drawing radon into the house and keeping it there. Houses without adequate coverage over ground soils may have greater radon exposure.

Many other indoor pollutants exist, depending upon the structure and condition of the environment. For example, automobiles, trains, planes, or buses can provide a whole range of dangers, from carbon monoxide to lead, formaldehyde, and plasticizers, which can off-gas, especially when the weather gets warmer. This is especially the case for new cars. That "new car smell" may actually be toxic off-gassing of a mixture of plasticizers and formaldehyde. In the case of older cars, air vents may be clogged with a number of molds, dust and bacteria, which may spray out when ever the "air" is turned on. In these cases it might help to periodically clean out the filters of an older car, and in the case of a newer car, we might consider leaving the windows cracked while parking in sunny location between driving for a few weeks to let the various materials outgas.

Over the last thirty years, researchers have become increasingly aware that certain buildings, especially older ones with older ventilation systems, can make people sick. This effect is often termed *sick building syndrome*, or SBS. The major symptoms reported in SBS include chronic fatigue, brain fog, headaches, allergies, nausea, chronic sore throat, asthmatic attacks, and bronchial congestion, among others.

SBS is indicated when multiple workers or inhabitants of a building complain of one or more of these symptoms soon after beginning to work or live in the building. Sometimes, however, SBS can develop over time, or directly after or during an extreme change in the weather. A humid summer or heavy rainfall period, for example, may stimulate a growth in mold in the ventilation system. After a smoggy summer in a city or during a fire season, the ventilation system may become clogged with soot. Workers or occupants of the building need to speak up and request the filters and vents are cleaned and flushed.

For these reasons, we would suggest that in all cases, indoor environments be properly ventilated. Windows should be open in the house, apartment, and workplace as much as possible. When driving, the windows should at least be cracked. Air conditioning the car or home with the windows shut, though seemingly uneconomical, will cost our bodies in other ways—often years later—with various lung disorders and cancer. Better to be a little warm with fresh air in the summer than cool and toxic.

The best remedy is to allow natural airflow into all of our indoor environments. While most might assume air is just a gas floating around aimlessly, we are learning that air has various flow and detoxification properties. As we discussed above, air moves in channeled flows of temperature, pressure, and radiation currents. Air also carries with it a number of electromagnetic and waveform properties. We can see this effect when we consider the power of oxygen to deliver transferable waveforms enabling phosphorylation inside the body's cells. We can reduce all this into particles if we want to, but we will be missing an entire dimension of airflow should we limit the potential of natural, rhythmic air. As air flows naturally, it cools and cleanses the immediate vicinity. Air also provides a filtering effect. The mixture of elements in air enables the breakdown of toxins through vaporization and diffusion. While many molecules such as fluorocarbons damage our atmosphere and trap warm air, the combined elements in air work together to breakdown these elements, gradually eliminating and dispersing their toxic effects.

Outdoor Pollution

According to the 2007 *American Lung Association State of the Air* report, 46% or about 136 million Americans live within a county having "unhealthful" levels of either ozone or particle-based outdoor pollution. Over 38 million Americans live in a county with "unhealthful" levels of both ozone and particle pollution. A third of Americans live in "unhealthful" ozone level counties. Interestingly, this is substantially better than a 2006 report indicating almost half of Americans lived in ozone-rich areas. It is unlikely that the source of ozone pollution—carbon emissions—went down this much in a year. Most likely—just as the ozone hole is fluctuating with some unknown atmospheric rhythm—there is some unknown relationship with the weather, temperature, pressure of all of the above. Meanwhile, more than ninety-three million Americans—about one in three—live in areas seasonally high in short-term particle pollution and about one in five Americans live in areas of high *year-round* particle pollution. Unlike the fluctuating ozone levels, the number of high particle pollution areas has definitely increased since last year.

Not that ozone is such a bad gas. Ozone is used in a number of positive ways. Ozone therapy is gaining recognition among the alternative medical community. Ozone is used in hot tubs as a purification measure. It is also used as an industrial cleaning and sanitizing substance. Ozone is part of our natural mix of gases in our air. Typical levels are between 25 and 75 parts per billion. At higher levels in our immediate atmosphere, however, ozone is seen as a pollutant. This is because ozone levels typically increase with higher levels of carbon monoxide levels, sulphur dioxide, and other pollutants. Like the canary in the coalmine, ozone levels provide a good indicator of unhealthy air. When combined with smog, ozone is thought to have a detrimental effect upon the body. Alone, ozone is an oxidizer. This means it will release unpaired oxygen radicals, which can interfere with biomolecular bonds within the body. While oxidation is a natural part of our metabolism, too much oxidization—especially when an abundance of radicals are released—can damage tissues, blood vessels and organ systems.

The EPA's Air quality index rates 85 parts per billion of ozone as unhealthy for sensitive people and over 105 ppb as unhealthy for anyone. As ozone levels rise, it can irritate the lungs and throat, increasing the potential for inflammatory throat and lung infections. As a result, indoor ozone generating machines—which have become popular over the last decade—are discouraged by both the EPA and the FDA if they emit levels higher than 50 parts per billion.

Higher ozone levels in the air simply reflect its cleansing action upon other pollutants. Ozone is part of the atmosphere's normalizing systems. The very same cleaning and antibacterial effect we have begun to utilize are part of the earth's detoxification mechanisms to clean and break down particulate pollution.

PM(10-2.5) refers to a pollution particle size from 2.5 micron to 10 micron. PM(2.5) refers to particle sizes below 2.5 micron. 2.5 micron or less is considered *fine particulate*, while less than .1 micron is considered *ultra-fine*. Ultra-fine particles may also be considered as noxious gas pollution, because their molecular size is small enough to pass through the alveoli into the bloodstream.

Particles above 10 micron in size are usually trapped by the cilia and mucus membranes in the nose, throat, or mouth. As mentioned earlier, these are usually disposed through the movement of mucus or broken down by immune cells.

Fine particles are small enough to escape this labyrinth and get into our lungs, but they are usually too large to get into the bloodstream. These particles can become lodged in the tissues of the lungs and bronchi and slowly breakdown. As they breakdown, various toxic substances can become absorbed into the tissues of the lungs and dumped into the bloodstream. This accumulation of toxins can quickly overburden the liver and bloodstream. In other words, the pollution of the air quickly becomes the pollution of our lungs, bloodstream and liver.

The chemical makeup of these toxins depends upon the source of the pollution. Particle pollution caused by automobile exhaust will have various fluorocarbons and nitrate particles, while coal fired power plants will emit large levels of sulphur dioxide within their soot. Course particulate pollutants are typically caused by mining, construction, or demolition.

Building demolition can rain fumes of various dangerous substances a mile or more from the demolition site.

The most blatant illustration of the demolition effect is the World Trade Center bombing of 2001. The collapse of the towers caused such a toxic fume that there are still thousands of people suffering from a slow poisoning of the lungs. Dangerous chemicals like asbestos, formaldehyde and many others were breathed by thousands of people. This effect was not limited to rescuers and those who escaped from the building. Many others who happened to be in the vicinity are now suffering. One of the more prevalent diseases contracted was *sarcoidosis,* a life-threatening inflammation of the lungs. As toxin deposits build up and damage the lungs, scar tissue forms. This scar tissue affects the elasticity and efficiency of the lungs, causing a life-threatening lung collapse. A recent study released by nine doctors (Izbicki *et al.* 2007) researching the delayed effects at Ground Zero reported that firefighters and rescue workers were diagnosed with sarcoidosis at a rate of five times higher than the incidence rate prior to 9/11.

Cancer has also been an unfortunate result of the Trade Center bombing. While we know asbestos is toxic to the lungs, it is still a popular building material. Many structural components of new buildings use asbestos as an ingredient. Because it is a cheap fire retardant material, buildings still go up using it. Asbestos has been linked to thyroid cancer and lung cancer in many studies, which caused a number of high profile lawsuits and residential building restrictions.

Another carcinogen released when buildings are demolished or collapse is benzene. As we have mentioned before, benzene is identified as a carcinogen, and certain types of leukemia are associated with benzene exposure. Other toxins thought to be released by the building collapse are mercury, lead, cadmium, dioxin, polycyclic aromatic hydrocarbons (PAHs) and polychlorinated byphenyls (PCBs). Many of these components are still used to build buildings. All of them were used in buildings a few decades ago, when many of our current skyscrapers were built.

Besides cases of cancer and sarcoidosis, ailments associated with the WTC bombing include reactive-airways syndrome, asthma, chronic throat irritation, gastroesophageal reflux disease (GERD—associated with heartburn) and persistent sinusitis. Other cases thought to be associated with the WTC have included miner's lung and thyroid cancer. While many thought these disorders were temporary, many new diagnoses are being made. This is because toxins can build up and reside in lung tissue cells, affecting future lung capacity for many years to come. Studies on firefighters involved in the rescue report an average loss of 300 milliliters of lung capacity (Senior 2003).

It is safe to say that exposure to these sorts of chemicals through air pollution may not be symptomatic for years after the exposure. This was certainly the case with those exposed to Agent Orange in Vietnam. Cases of prostate cancer, skin cancer, and chronic lymphocytic leukemia did not appear in some veterans for decades later (Beaulieu and Fessele 2003).

There are several types of outdoor pollution to consider. *Smog* is primarily made up of ozone. This form of O_3, gas is caused not by a nature's interaction between radiation and the atmosphere, but through the reaction between fuel vapors from automobiles and sunlight. For this reason smog levels tend to peak during warmer periods. Smog levels are also higher in warmer regions like Southern California and urban areas in the southern U.S, such as Atlanta. Other cities also experience greater smog levels during the summer months. Wherever higher concentrations of vehicles combine with warm sunshine, smog levels go up. The 'perfect storm' of almost constant sunshine, warm weather, and vehicle concentration makes Southern California one of the worst smog and ozone regions in the United States.

Ozone will oxidize on a cellular and internal tissue level when breathed in. These tissues will be damaged in almost the same way that other oxidized radicals damage tissues. Ozone will readily be absorbed through the alveoli as a gas. From there it can enter the blood stream and readily damage artery walls and tissues. Ozone is also a major lung tissue irritant, causing inflammation and epithelial cell damage, resulting in decreased lung capacity and/or decreased lung growth development among children. Ozone is directly linked to the incidence and worsening of asthma, bronchitis, and COPD (chronic obstructive pulmonary disorder). Recent research indicates that this mechanism is the oxidation of lung surface lipids lining the cells of the lung. These oxidized lipids stimulate inflammation as the body seeks to mitigate ozone's damaging effects. This launches scavenger macrophages, which bind to the oxidized lipids in an effort to reverse the damage (Postlethwait 2007).

While great emphasis is put on the link between heart disease and diet, exercise and smoking, little attention is put on the effects of air pollution on the heart. In one study of 11,167 coronary events, long-term air pollution was linked with a significant increase in heart disease (Rosenlund *et al.* 2008). The level of toxic NO_2 exposure seemed to be the primary marker.

Ozone pollution appears to have a very similar affect upon the earth's forests. Apparently, high ozone concentration creates a nitrification effect, causing high levels of nitrogen oxides, nitric oxide, and nitrate particulates. While the higher levels of nitrogen may initially benefit plant growth, they also damage the leaf surfaces, reducing these plants' ability to photosynthesize. Higher nitrogen levels also stimulate increased nitric oxide emissions from the soil. This affect has been observed in the San Bernardino Mountains in California—an area saturated with the highest levels of smog in the country. High nitric oxide emissions increase soil leaching and erosion, robbing the soils of their nutrient content. This is exasperated by rain run-off, which further pulls nutrients from the soils, robbing the plants of needed nutrition. The result is a further deterioration of the ecosystem, creating more dry brush and increased fire risk (Bytnerowicz *et al.* 2001; Wood *et al.* 2007, Sakugawa *et al.* 2007).

Particulate pollution or soot is the most dangerous form of outdoor pollution. Auto exhaust, aerosols, and chemicals from power plants and wood burning are the major sources. While the particles themselves are too small to be seen by the naked eye, they can be seen as a whole in the form of a haze in the sunlight. While the body's cilia hair and mucus membranes in the nose and throat filter and catch some of these particles, many still make it into the lungs.

Like any living organism, the earth and its atmosphere have means to cleanse toxins out of the system. The atmosphere conducts a number of self-cleaning currents, which move and break down pollutants. One such mechanism is the hydroxyl radical (OH) molecule. The immediate effect of the hydroxyl radical is to form photo-oxidants like nitric acid, which undergoes a photolysis reaction with nitrous acids. The process cycles with the formation and evaporation of dew droplets (Zhou *et al.* 2002; Kleffmann 2007).

The earth itself can create hazardous effects in response to our altering of her cycles. This was observed in Owens Lake, California. For thousands of years, this lake offered respite from the heat and inhospitality of the California desert region known as Death Valley. The size of Owens lake was nearly 28,500 hectares—a huge body of water fed by the Sierra mountains. In 1900, the waters from the rivers feeding the lake were diverted into the Los Angeles basin to feed a growing population. Within ten years Owns Lake, a huge body of water some compared to the Aral Sea, turned into a dustbowl. Now the silt covering the water basin is coated with arsenic. Arsenic is a natural element, and in this case, it is providing resistance and restructuring for the soil at the lake bottom. This arsenic is also blowing around the region with the winds, substantially increasing the cancer risk for local residents to about 1 in 40,000 (Raloff 2001).

The earth's responses in particular are seen in increasingly unpredictable weather and climactic disturbances, increased tectonic movement—resulting in earthquakes and volcanoes—and destructive fires caused by a loss of moisture in certain regions. All of this related to our society's interference with of the earth's breathing apparatus. Like a spreading bacterial infection, human society is pushing the earth towards a major detoxification event.

Breathing Issues

In holographical irony, the human body is also showing signs of breakdown caused by the gradual destruction of the earth's atmosphere. The massive increase in allergies, asthma, COPD and lung cancer are only a few of the symptoms increasing among humans. We are also seeing a gradual loss of breathing capacity, in particular among those living in counties with higher ozone and air particle pollution. Breathing capacity losses as high as 15-18% have been observed among teenagers raised in Los Angeles County, for example. As the third world countries fully embrace petrochemical use, this problem will only get worse. This will become an even greater issue around the world.

Like the atmosphere, the human biology also works systematically to clean itself of air particle pollution and ozone pollution. The ciliary hairs on the mucous membrane lining the entrances to and the surface of the bronchial tree beat rhythmically with the expansion and release of the lungs. This expansion and contraction increases the surfactant film as well. In the alveoli, the surfactant works to help transport oxygen across the tissue-blood barrier. In the rest of the airways, this surfactant provides a transport mechanism for foreign particles. Once particles are captured within the surfactant, they are carried back by the rhythmic movement of the surfactant gel towards the trachea, where they are met with immune system cells such as macrophages. These macrophages break down the particles through *phagotosis*.

Should the particles remain airborne, they will likely be moved back to the trachea by the rhythmic ciliary hair undulations. The ciliary hair move in waves, very similar to what we might see among kelp beds as they move with the undulating sea currents. This wave-like action of the ciliary hairs acts as an effective transport system.

These transport systems explain how we often gather an accumulation of mucus within the esophagus. Most of us clear our throats and/or hack out mucus daily without a second thought. Little do we realize that much of that mucus is the lungs' self-cleaning mechanism as it moves out particle pollutants. This interruption mechanism prevents polluted air from being absorbed into our blood. Those particles not tossed out with the mucus get phagotized and discarded through the lymph or bloodstream, and out the colon, urine, or sweat glands.

Extreme cases of lung invasion by toxins or particles from the air may cause an inflammation of the airways. This inflammatory response is also an attempt to clear the body of toxicity. Should the transport systems of the bronchi fail to remove the particles—perhaps because too much has come in all at once—then the body will launch an immune response, sending natural killer cells, histamine and extra blood and lymph to the site, in order to conduct a superfund-like clean-up process.

Unfortunately, this heightened response almost invariably creates an expansion of the tissues, obstructing airflow during the process. A cross-section of an inflamed bronchial airway looks remarkably similar to an artery with atherosclerosis. The cross section of the airway channel reveals a thickened lumen, replete with a sticky, mucoid plaque appearance. This thickened lumen restricts airflow through the bronchial airways. Wheezing, coughing, and even asthmatic symptoms are often the result of this decreased airflow.

This is all part of the body's natural detoxification mechanisms. While coughing and wheezing seem like the enemy, the airflow reduction during a lung cleanup process is an important signal of trouble in the air. Part of the body's natural defense system is to partially shut down while it performs repairs. When we consider this tremendous effort, we must remember that our physical bodies were simply not designed for the kind of constant onslaught our lungs are now being exposed to: Persistent smog, soot, chemical irritants in our homes, chemically manufactured clothing, sick buildings, and toxic cars. These all deliver substances introduced to our bodies only over the past fifty to one hundred years. Our bodies therefore treat them as foreign invaders. The body must reject them by launching an immune response. Furthermore, as these chemicals invade our bodies and become resident in one respect or another in our cells, those cells become genetically altered as they attempt to adapt. This is one of the central mechanisms involved in cancers: genetically defective cells that keep dividing.

This mechanism is also directly related to the modern-day epidemic of the *autoimmune disorder*. This disorder is nature's mechanism to minimize cellular interference. As airborne foreign intrusions overload the body, they interfere with cellular metabolism. As the cells adapt they become altered. The immune system senses this alteration of the cells, and launches an immune response in an attempt to remove altered cells. This immune response in turn results in inflammation, producing the various allergic symptoms many in our society increasingly have. This can also be seen in the dramatic rise in cancer over the past fifty years. It can be seen in the rise of so many other disorders as well, at least partly explaining why many industrialized countries like the United States have some of the highest disease rates despite the greatest spending on medicine.

It can also be seen in the form of asthma, a disease increasing around the world, especially among industrialized nations. There are two types of asthmatics, and we can see the first type simply with in the statistics: Asthma is twice as likely to occur in infants living in a smoker's house. A premature baby, who also likely put in a respirator after birth, is four times as likely to have asthma compared with a term-birth baby. Eighty percent of asthmatic children also have additional or multiple allergies. Other asthma cases follow a viral respiratory infection. In many of these cases, the child will cease having asthmatic symptoms during elementary school (Plaut 1999).

Meanwhile, adult cases of asthma are increasing. Between 1980 and 1996, adult asthma increased by almost 74%. Rates of other adult respiratory diseases such as sinusitis and ear infections have paralleled these rates. Asthma also appears to be more prevalent among obese adults (Simon 2007). Meanwhile, Australian research indicates that asthma amongst children is highest among indigenous peoples in urban settings (Poulos 2005).

Various studies confirm that adult onset asthma caused by occupational hazards is substantially greater in industrialized countries as compared to less developed countries. While developing countries have lower than two cases per 100,000 of population, industrialized populations have up to 18 cases per 100,000 (Jeebhay 1997). This illustrates the connection between asthma and our industrialized technologies.

It is quite easy to link asthma with the body's immune response to foreign air particles, soot, fragrances, or other airborne chemicals for any duration of time. The response from contact with the airways could be immediate or it could be delayed, however. In the case of the later, a foreign pollutant may irritate the lungs over a period of time with no real outward symptoms. The immune system may work to remove the toxin without a major inflammatory event. Then at some point, there may be an overload signal, or the body may run low of other immune response mechanisms. Now the body will begin launching 'all the missiles.' With a heightened inflammatory response, the body may have the greatest opportunity to rid the body of the invasion. This should of course also help inform the inner self that the body's environment needs to be corrected.

It is this last step that our society as a whole is failing to respond to appropriately. At this point we should be investigating what parts of our air may be irritating the lungs—and removing them. Instead of spending billions of dollars looking for a magic pill or breathalyzer chemical to remove the symptoms of asthma, we should be removing the cause. We can simply isolate all of the chemicals or toxins that exist in both our indoor and external environments, and work immediately to remove them.

The Environmental Protection Agency reports that the air inside the average home is from two- to five-times more polluted than the outside air. Therefore, we have a lot more control over our environment than we might think. We might want to look below the sink at the various cleaning agents we are using to clean the house, or the corrosive elements that may be seeping into the air when we use them. We might examine carefully what kinds of fragrances we adorn our bathrooms and bodies. We may want to carefully consider the perfumes in our laundry soaps and that wrinkle-free pad we might be dropping into our dryer, which will not only reside on the clothes we are wearing but leak into the air, affecting everyone else. While many attribute allergies to natural issues such as pollen, in reality we may be reacting to pollen simply because our immune systems are already overloaded with so many chemical allergens. The pollen may well be that *one more thing* the body has to deal with, triggering the inflammation.

Pollen allergies in themselves are actually quite easy to deal with. In most cases, simply ingesting the honey and bee pollen from local hives—especially those of the same flower allergic to—will train and adapt the immune system to that particular pollen. Other strategies would include clearing the colon of putrefied waste, exercising and eating more vegetables and fruits, and of course, decreasing our exposure to chemicals. All of these measures will decrease the toxic burden on our bodies.

Asthma and allergy victims typically look to pharmaceuticals as the solution. The typical measure for bronchial obstruction is to administer bronchodilators such as theophylline. Other pharmaceutical agents used include *corticosteroids, beta2 agonists, leukotriene antagonists, anti-IgE therapies* or a combination thereof. While some are used as stopgap measures to obtain normal airways—especially theophylline—many are used on a long-term basis. The more prevalent prescriptive solution is cortisone, a synthetic version of cortisol, which our bodies naturally secrete from the adrenal glands to quell inflammation. It should be noted that such corticosteroid inhalants such as fluticasone propionate, when used over a period of several months, have been shown to suppress the body's own adrenal cortisol inflammation-reduction mechanisms. At higher doses, these pharmaceutical anti-inflammatory measures effectively suppress adrenal activity, leaving the body immune-suppressed. In some cases this can lead to a health crisis, and among children, stunted growth (Mahachoklertwattana *et al.* 2004; Macdessi 2003).

A newer drug being prescribed for asthma is Adair®, which contains salmeterol, a long-acting beta-agonist, or LABA. In a 2006 report published in the *Annals of Internal Medicine*, Salpeter, *et al.* determined that 19 studies indicated that LABAs actually increase the risk of asthma exacerbations, some with fatal consequences. The study estimated that some 5,000 deaths per year may be associated with the use of LABAs.

It should be noted that higher doses of corticosteroids should not be curtailed abruptly. Changes should be made gradually, and under a health professional's supervision (Drake *et al.* 2002, Todd *et al.* 2002, Dunlop *et al.* 2002).

Ironically, rather than removing the unintended elements we have added to our air, medical doctors and pharmaceutical companies are working to add additional chemicals into our lungs and bodies, further overloading our chemical loads. Instead of eliminating the cause and lightening our toxic burdens, we prefer to ignore the cause and even add to it.

The other type of asthma is a seemingly autoimmune disorder possibly combined with deep-seated psychological factors. Research has indicated that up to 25% of asthmatics fall into this category. As research has continued in asthma cases not set off by specific allergens or activity, the body apparently seems to be responding to its own lung cell tissues as though their mechanisms are foreign. This attack by the immune system is orchestrated through the activity of mast cells, neutrophils and Th2 T cells (T helper type 2 cytokine cells). They launch an attack upon the lung cells in almost the same way as a foreign agent is attacked. As this inflammatory cascade progresses, the airways become inflamed, and can seize up. T-cell *chemokines* communicate a threat level attack by stimulating receptors to invoke the immune response (Palmqvist 2007). Histamine levels are also commonly heightened among asthmatic attacks, consistent with allergic reaction (Duke 1997).

As we look at our society's response to the rise in allergic responses to our pollutants as a whole, we see a policy of avoidance and further industrialized production. The production of further pharmaceuticals to "combat" asthma incurs an additional burden on both our atmosphere and body. The pharmaceutical solutions themselves may appear noble, except for the fact that nature already produces various botanicals that can quell the inflammatory response and increase healing of the lungs in cases of viral respiratory disease. So why are our healing professionals not focusing on such low-impact, low-risk solutions? Why hasn't our society focused on such natural and safe healing agents such as *grindelia, sundew, coltsfoot, hops, licorice, stinging nettle, gingko, skullcap* and *chamomile?* Why has our society decided to ban such potent anti-asthmatic botanicals used for thousands of years by Chinese and Ayurvedic healers like *lobelia* and *ephedra* (Mabey 1988, Duke 1997)? Both of these later botanicals were demonized because of a few deaths and illnesses that were primarily unrelated to their medicinal or prescriptive use, while hundreds of thousands of people die every year of pharmaceutical medicines.

There are also natural substances containing generous amounts of natural theophylline, also balanced with other natural constituents. These include *cocoa beans, coffee beans, herba mate, ephedra* and others (Duke 1997). Why have we not extracted these natural constituents instead of using risky synthesized forms?

Simply put, the answer to these questions comes down to economics. What is more profitable? Which method guarantees a patent? Because natural products cannot be patented, there is no way to restrict competition. Simply put, nature is diametrically opposed to restrictive profitability. Nature proposes that herbal therapies are available to everyone, regardless of their economic status or insurance provider.

While doctors who prescribe pharmaceuticals are given licenses to deal lightly tested drugs with uncertain and sometimes deadly side effects, these low-impact alternative healing agents are being attacked as heretical and dangerous. Herbal remedies have thousands of years of safe use by billions of people. While pharmaceutical manufacturers poison our air and water with chemical manufacturing waste streams, herbal growers work to sustain threatened species and preserve poisoned land. The situation presents an upside-down equation.

In addition to the botanical solutions, the body's digestive and probiotic system is a key component to consider for allergic or asthmatic situations. Research over the past two decades indicates that 80% of the body's immune system lies within the digestive system. The composition and condition of our lower colon has been linked with issues of sinusitis, allergies and other immune responses. Should the colon be slow or occluded due to low-fiber, high meat protein diets, various putrefaction byproducts formed by the action of anaerobic microorganisms can be absorbed into the bloodstream from the colon.

These byproducts include ammonia, indole, skatole, hydrogen sulfide, neurine, tyramine, phenol, and histamine. Most of these are derivatives of undigested amino acids such as tryptophan and tyrosine. The complex protein polypeptides in animal meat are more difficult to break down, ending up in the colon. A couple of the more implicating byproducts of this mechanism are tyramine and histamine. These two chemicals have been linked with sinus and lung irritation, allergic response and inflammation (Dotolo 2003).

This all means our airways are closely tied closely with our diet, the activities of our intestinal tract, and the probiotics that colonize our intestines and colon. Probiotics help the body in many ways. They repel and kill pathogenic bacteria. They help digestion. They help break down toxins.

Probiotics secrete antibiotic substances that kill pathogenic bacteria. This reduces immune response associated with these bacteria or their waste stream. In one study on newborn mice, those who received either *Lactobacillus rhamnosus GG* or *Bifidobacterium lactis* (Bb-12) for 8 weeks following birth had substantially-reduced Th2 cytokine production and increased TGF-beta-secreting CD4+/CD3+ T-cells, which damper and balance the body's allergic and asthmatic immune response (Feleszko *et al.* 2007). This illustrates a probable link between asthmatics and appropriate probiotic colonies. Connecting the dots, we can easily see how the body's populations of symbiotic bacteria provide at least part of the solution to the mystery of asthma.

We must ask ourselves how these tiny living bacteria somehow produce the right chemicals to assist the immune system? As many of us learned in high school biology, this is a prime case of symbiotic behavior. The organisms living within our bodies are just as reliant on oxygen (at least the aerobic ones—which are the substantial majority) as our own cells are. As a result, the same oxygen that we use to energize our cells is providing fuel for up to three trillion symbiotic bacteria. Again, we might compare this situation with the planet earth, supporting the so many populations of (hopefully) symbiotic creatures that live upon or within it. The stirring question we must now contend with is: Are humans a symbiotic or pathogenic species with respect to the living earth?

Research has established that bacteria make decisions to avoid life-threatening situations, and conversely, are attracted to life-sustaining situations just as we are. These tiny living organisms are thus working to promote sustainability, which means they will contribute to an environment offering better oxygen supplies.

This also coincides with research linking *lactobacillus* probiotics with a reduction of diabetes among rats eating a high-fructose diet (Yadav 2007). Blood sugar disorders seem to be related to asthma through a mechanism relating to insulin sensitivity (Mineev *et al.* 2002). This indicates a deeper relationship between the insulin- and glucose-reception and asthma. Could the relationship be connected with the utilization of oxygen inside the cell? Looking at the evidence, this appears likely.

Contrary to popular belief that dairy consumption is associated with asthma, a recent study of 14,893 children between five and thirteen years old across Europe confirmed that significantly lower rates of asthma were associated with farm milk (Waser *et al.* 2007).

Research has also confirmed that living near or exercising in water significantly lowers the risk and incidence of asthma. A number of studies have confirmed swimming reduces asthma occurrence. In two studies investigating whether swimming itself reduced asthma or whether weather or greater humidity near the water reduced asthma incidence in children, it was determined that just the act of swimming (over other forms of exercise) reduced asthma incidence in children (Inbar *et al.* 1980). However, it should be noted that swimming in heavily chlorinated indoor pools might have quite the opposite effect. A 2006 study compared the frequency of middle-aged adult hay fever and indoor swimming pool frequency as children. In one study, indoor swimming pool use as children increased the risk of hay fever as adults (Kohlhammer *et al.* 2006). This risk was further illustrated in a 2007 (Nickmilder *et al.*) study showing the risk of asthma and other bronchial issues increases substantially for children with greater indoor pool frequency. In 2004 Williams *et al.* illustrated that indoor pools with one parts per million of chlorine concentration caused 60% of the 41 swimmers (half with a history of exercise-induced asthma) to suffer from airway constriction. Meanwhile only 20% suffered from constriction in a pool with .5 parts per million.

Research has also confirmed that breastfeeding through at least the first year decreases not only asthma rates but all types of allergic activity among children. Apparently, breastfeeding increases IgE levels— antibodies noted for their low-allergic response (Halken 2004). Again, we are reminded that nature has created a balancing of chemistry and organisms to bring about the best potential for health. Why mess with it? It is evident that nature has an organized rhythm of synchronized activities that promote health. Interfering with these rhythms adds stress and creates disease. We must synchronize with these rhythms if we expect long-term survival for our species.

Another effect is epitomized in the mechanism of fear-type asthma. The seemingly unprovoked identification of the lung cells in this case indicates an inconsistency within the genetic information map of the cell: the DNA. DNA is a holographic program reflecting the status of the inner self in relationship with the physical environment. In an asthmatic attack responding to a fearful situation, we are dealing with *expression*. The attack response, complete with raised histamine levels during a fearful situation relate to a prior event or activity recorded into the cellular DNA.

Often an experience where ones lungs became restricted during a frightful or life-threatening event will create this recording. This might be compared to the famous Pavlov's dog experiments, where the dogs related eating to the sound of a bell, stimulating their appetites.

Over time, this behavior relationship will not only be reflected by conscious activity, but the DNA will begin to express the relationship as the body's cells begin to line up behind the behavior. This is illustrated by obesity: The genes begin to express overeating with leptin regulation tangential to PPAR-gamma repression in SIRT1 pairs. Sirtuins have been implicated in the diabetic mechanism because of their ability to affect genetic expression related to issues of degenerative glucose and insulin metabolism. SIRT1 appears to balance glucose levels by modulating the PGC-1alpha molecule. PGC-1alpha is a transcriptional co-activator. When SIRT1 is inhibited, increases in hypoglycemia, glucose and insulin sensitivity, free fatty acids and cholesterol have been observed. On the other hand increased SIRT1 expression reversed these effects, but only in the presence of PGC-1alpha during fasting (Rodgers and Puigserver 2007; Das 2005).

Subtle Winds

The wind is mother earth's form of breathing. We have discussed many aspects of the wind with regard to its dynamic motion among various temperatures and pressures. The bottom line of this information is that wind is a critical waveform element. Without the winds, we see no circulation of temperature and weather systems. In the same way, without the breathing winds of our body, we cannot circulate critical nourishing elements around our own inner atmosphere. Although we consider oxygen to be the more important of these, by circulating the air within our bodies we also circulate other cyclical elements including olfactory waveforms, water vapor waveforms, thermal waves, and a number of subtle pranic rhythms existing within the gas layer.

It is for this reason that the ancient Ayurvedic and Chinese sciences introduced the functions of wind as an elemental nature of our physical body. In Ayurveda, wind is seen as *vatic*. It is drying, and stimulates movement. It thus stimulates reflexes, and obviously cyclic breathing. Being a subset of the element of air, the vatic body constitution is dry, slender, expressive, creative, high energy, active and sometimes nervous. Yet at the same time, because by nature wind is drying, *vatic* types may tire quickly, especially when they have a lot of nervous energy.

Wind and breathing are also detoxifying for these same reasons. The body detoxifies many types of toxins through the lungs. Many wastes vaporize into gas form. As this vaporization takes place, the alveoli can exchange these waste products via the lungs. This effect is immediately noticeable when someone consumes an abundance of alcohol, or even garlic. The metabolites of these substances are almost immediately recognizable on the breath. We can almost immediately identify someone who is drunk—even hours after they have had their last drink—simply by smelling their breath.

As many of our normal metabolism byproducts—as well as toxic waste byproducts—get broken down by liver enzymes and lymphocytes— many of these byproducts can also easily be vaporized into gas form. As this takes place, they can be escorted out of the body through the lungs.

Chinese and Ayurvedic medicine describes pathways for the course of oxygen and other gases as channels. Modern medicine has also documented biochemical pathways working within the lungs, bloodstream, liver, digestive tract, and cells. These include ion channels, ligands and receptors, and neurotransmitter connectivity. These pathways also illustrate a more subtle channeling of gases. The cyclic pathways connecting the air in our atmosphere between energy production within the cells and the various detoxification systems working throughout the bloodstream illustrate parts of this biochemical channeling process. Certainly, the bronchial tree and alveoli present a channeling system for incoming air. Once this air is translated through alveoli membrane the bloodstream piggybacks oxygen to hemoglobin along the pathway of the arteries. Various subtle channels for gas exist as well. An example is the olfactory-turbinate-nerve channel we will discuss shortly. Various other channel systems also cycle gases through the body, including the digestive system, the nervous system, and the subtle chakra/*nadi* channels.

Information Airs

Once thought of as a theoretical fancy, aromatherapy is now a functional health science, as researchers have been able to verify that particular aromas stimulate various physical responses, mood changes, and cognitive responses. Aromatherapy in clinical research has confirmed improved moods and cognition, lower agitation and anxiety using lavender (Lin *et al.* 2007; Muzzarelli *et al.* 2006), decreased cortisol levels using lavender and rosemary (Atsumi and Tonosaki 2007); and increased free radical scavenger activity in the body using rosemary (Atsumi and Tonosaki 2007). Other studies have linked aromatherapy to improved therapeutic responses among various cardiovascular diseases, cancers, bacterial and viral infections, diabetes, and dermal issues (Edris 2007).

The mechanism of aromatherapy involves a complex interaction of brainwaves, the mind's screen, and the neural system. When an aroma— an electromagnetic wave moment with a particular rhythmic fingerprint—

interacts with the olfactory bulbs lining the nasal cavity, olfactory nerve pathways are stimulated. That stimulation sends information pulses to the olfactory cortex in the temporal lobe, which then reflects and transduces those rhythms into the limbic system and to other cortices, which may include the motor cortex, the prefrontal cortex and others, depending upon the odor. These combined rhythms create a brain mapping as documented by researchers like Zao *et al.* (2005). This mapping is reflected onto the mindscreen, while at the same time the body negotiates electromagnetic resonance among the meridian, nadi, and chakra channeling systems.

Beyond the nasal olfactory system lies another more subtle receptor system called the *Jacobson's organ*. This *vomeronasal* organ has a number of receptors that detect more subtle molecular moments, notably molecules such as pheromones and other non-odorous molecules. Pheromones have been the focus of study in both animals and humans for many years. The subtle molecular wave moments contain information. For example, a plant may produce a pheromone, which drifts to other plants. As the pheromone connects with another plant, it stimulates the production of pollen. This jump-starts the process of reproduction. Animals will also produce various pheromones that communicate to potential companions that they are 'ready' for reproductive activities. Other types of pheromones indicate where food is found. For example, ants leave hydrocarbon-composed pheromones to mark the trail where food was found, and repellent versions for trails no longer offering food. Various species of plants and animals also produce territorial pheromones, which help dissuade others from encroaching. Dogs' urine contains this type of pheromone. Many trees will produce territorial pheromone-containing oils, which drop from leaves or drip from barks.

For a number of years, researchers thought pheromones only existed in plants and animals but not in humans. While there is still some controversy, it is generally accepted today that menstruation accompanies some pheromone activity. Research has confirmed that menstruation stimulates synchronization between females living within close proximity. Other pheromone communications appear to be exchanged between men and woman on this subtle level. Many of these subtle non-odorous communications are translated through an interaction between the vomeronasal system and the receptors of the olfactory system when interpreting odors (Johnston 1998).

As we consider this more subtle system of pheromone exchange, it is obvious these olfactory wave moments contain specific information, enabling precise communication. The fact that an organism can communicate territory or reproductive timing to another organism indicates these chemicals contain precise information. The fact that the ant's trail can sometimes indicate a positive signal for food and sometimes a negative signal indicating the food is gone tells us these biochemicals carry specific messages. These messages are decoded by the nervous system of the organism and translated to particular activities. In other words, these pheromone messages carry information from the consciousness of one organism to the consciousness of another.

In the ancient Ayurvedic medical science, the evaluation of informational airs moving within the body was used for diagnostic purposes. These *prana* were perhaps a bit more complex than simply the oxygen air that we breathe. Each type of informational air or *prana* was associated with a particular elemental level, such as earth, air, water, fire, or ether. The *prana vayu* is considered the gaseous air, which was described as moving linearly, up and down the length of the body. Its energy is considered nourishing. *Vyana vayu* is the air of the liquid layer, moving with the waters of the body. The *apana vayu* is the earth or solid elemental air, and tends to move downward, detoxifying. *Samana vayu* is the thermal air, circulating primarily outward from the abdominal region. *Udana vayu* is the electromagnetic or ethereal air, moving upward and outward, and considered the expressive information air of the body. These various airs or *vayus* are described as harmonically moving through the healthy body, and imbalanced when disease is present. The ancient Ayurvedic scientists also considered methods of breath control critical to the process of bringing the various informational airs, or *prana*, into balance and harmony.

Harmonization: Air Therapy

(Note: Increased oxygen levels within the body can lead to light-headedness. It is recommended that anyone with a breathing disorder or in any way physically challenged perform these slowly and gradually, preferably with the supervision of a health professional in the beginning and with the presence of a partner.)

❖ *Shallow breathing* is what most of us do during our waking hours. This is a chest-oriented, top-lung capacity breathing. We raise the top portion of our chest as we suck in a bit of air, filling approximately a quarter of our capacity. While a good portion of this breathing is autonomic, the follow-through is probably more contraction and tension-oriented. In other words, our autonomic nervous system stimulates the initial breath, while the status of the sympathetic central nervous system determines how deep we breathe. For most of us, this is not much. Most of us simply have not been taught to breathe.

❖ The second form of breathing is *belly breathing* also called *core breathing*. This is a form of deep breathing that most of us use unconsciously when we sleep. This is also the form of breathing often taught and exercised by singers and public speakers. Belly breathing uses the greatest capacity of the lungs of the other types discussed here. Belly breathing also tends to bring about a deep state of relaxation, encouraging beta and theta wave activity in the brain. Because core breathing provokes the lower energy centers, it is probably best suitable for sleeping or recovery from a traumatic situation.

❖ The first step in belly breathing is to incorporate the core muscle group into our breathing process. This is the *transversus abdominis, internal* and *external obliques*. The abdominis muscles are located in the lower belly above the hip, angling towards the pubis in sort of a V-shape. These muscles support the abdominal region, the diaphragm and the lower back with a sort of girdle. These core muscles (with supporting muscles like *linea alba, rectus abdominis, serratus anterior* and a few others) can be strengthened with abdominal exercises. Strengthening these will also support our back and other muscle groups.

❖ The easiest way to perform *core breathing* is through visualization. First, we can focus our attention onto the girdle of muscles below and above the belly button described above. This girdle can be seen as wrapped around a balloon about the size of a soccer ball sitting just inside our belly button. As we breathe in, we can focus our attention on expanding the balloon.

❖ As this first balloon expands and fills, we can inflate a second balloon sitting right on top of the belly button balloon. This second balloon sits at our solar plexus. It is expanded just after the first balloon is filled, in a one-two step.

❖ After the second solar plexus balloon rises and fills up, we are finally ready to lift and comfortably expand our rib cage, as the air continues to fill into our lungs. This is the third step.

❖ The entire inspiration can take about five seconds or more—depending upon the situation (If we are meditating we might opt for 6-8 seconds.) We do not need to force the rib cage open. Once we have expanded the two balloons of the abdomen, the rib cage can open to a comfortable level with only a small amount of effort.

❖ Once we reach the full inspiration volume, we can hold the air in for just a couple of seconds. Then, as we relax our chest, the air will begin to release automatically. As our rib cage slowly contracts naturally, we can then slightly contract and empty the two balloons, starting from the top balloon—pushing the solar plexus and then the belly button inward. This will empty the bottom of the lung and prepare us for a more complete refill.

❖ The third type of breathing is *diaphragm breathing*. This breathing uses a much greater capacity of the lungs than shallow breathing, but not as much as deep core breathing does. However, diaphragm breathing is a practical way of breathing deeply, and is the preferred method of meditative breathing. The central difference between deep core and diaphragm breathing is that while core breathing begins from below the belly button, diaphragm breathing focuses around the diaphragm's ability to collapse and relax during inhalation, while tensing and contracting during exhalation. While it would appear that there would be a far less use of the lung without the wind-up from the core, the fact is that diaphragm breathing is the most efficient and flexible way to breath while awake, because it can accompany both shorter, quicker breaths during exercise along with the longer, slower breaths needed during meditation or relaxation.

❖ The key is in the ability to relax the diaphragm when inhaling. This takes an inner focus towards the solar plexus and a drawing of our energy into the center of the body. The location to bring this focus is just below the bottom of the rib cage, at the center, below the sternum. As we relax the diaphragm, our upper abdominal region will expand outward both forward and sideways—more precisely at the top of the obliques. The filling the lungs through this method can be either be slowly and calmly, or quickly and energetically, depending upon the physical activity.

❖ At the end of the filling cycle, after we have relaxed the diaphragm outward and upward completely, we will feel fullness, not only within our lungs, but also within our upper abdominal region. At first, this may not feel natural. This is actually the feeling of the diaphragm expanding and the bottom of the lungs expanding with it into the upper abdominal region.

❖ At the end of this filling cycle, where the expansion is greatest, the breath should be held for a moment. The length of this "moment" will depend upon what our physical and mental condition is. In slow, meditative breathing, the moment should last about three seconds or so. In faster, energetic breathing, this moment should probably be more like a second. If we are exercising, the moment may be a little more fleeting. The key is to be conscious of that expanded moment, however long it is. When we are conscious of that moment, we will be bringing intention into our breath. Bringing intention into our breath will in turn deliver a soothing energy throughout our body's energy channels. (By the way, "being in the moment" is no big achievement. It is the knowing *why we are here in this moment* that brings enlightenment.)

❖ When we breathe out during diaphragm breathing, we do not have to focus at all on the activity of the diaphragm or the lungs. After the diaphragm has been fully relaxed during the filling cycle, it will automatically contract back when we breathe out. If we are trying to maximize oxygen potential—for example if we are not feeling well and we want to energize our bodies—we can push out that last section of air just before we relax the diaphragm to bring more air in.

❖ For best results, we may want to breathe in through our noses and out through our mouths. This may not always be comfortable. There is absolutely no problem with breathing both in and out through the nose. In addition, this may not always be practical when we need an increased amount of oxygen when we are exercising. Although there is nothing wrong with breathing in and out through the mouth during strenuous exercise, we should know that he nose will not only cool or heat the air before it enters the lungs as needed, but will filter and screen out particles like soot and dust before they enter our lungs.

❖ As the air is being brought into the nose, we can comfortably rest the tip of the tongue against the inside gums that meet our front teeth. This positioning of the tongue keeps the air passages open, also allows an easy release through the mouth as we breathe out. We might consider keeping our lips parsed as we channel the wind out.

❖ The key element in breathing therapeutically is to establish a consistent rhythm. Whether we are core or diaphragm breathing, we might want to find a comfortable pace we can maintain over the duration of our activity, rather than attempt an unrealistic pace. Establishing a rhythm is easy—we can find something around us with which to pace our breathing, such as a ticking clock, some music, the pace of our walking or running, the sounds of songbirds outside, or any other rhythmic noise. Pacing with natural rhythms is preferable. By pacing our breathing with natural rhythms, we will be harmonizing our breathing rhythms with nature. This will have a calming effect upon us, helping us to better adapt and thrive within our environment.

❖ We can do these breathing exercises anytime: Whether we are sitting, standing, walking, or lying down. While core breathing is probably better for lying down and relaxing deeply or sleeping, diaphragm breathing is probably better for waking activities. The more supported our body is, the slower our pace can be.

❖ As we practice these methods, we will begin to strengthen and coordinate the various supporting muscles allowing our bodies to breathe deeply. As these muscles become more coordinated, we will begin to breathe deeper without consciously thinking about it. At first we can practice a few minutes each day, and progressively increase the time.

❖ Along with the healthy feelings our bodies will experience from deep rhythmic breathing, we should experience a soothing of nerves and better moods. Deep breathing allows us to approach stressful situations with more focus, clarity, and feelings of increased relaxation.

❖ There is an anatomical connection between the nose and the health of the rest of the body. This was confirmed with a diagnostic machine that was first developed in the mid-twentieth century by Dr. Maurice Cottle called the rhinomanometer. Dr. Cottle is best known for his work in the development of the electrocardiogram. Dr. Cottle was a physician and specialist in rhinology, a branch of otolaryngologists focused upon the nose. An English-trained physician, Dr. Cottle was the founder of the *American Rhinologic Society* in 1954 and was instrumental in advancing many nasal surgical procedures. Dr. Cottle wrote two books on the subject of rhinomanometry and as professor and head of the Otolaryngology at the Chicago Medical School, he introduced the process of diagnosing various ailments simply by measuring the swelling and shrinking of the nasal passages as one breathed. The tissues throughout the mucus membrane, especially around the turbinates, have invisible energy channels connecting various organ systems and processes. The delicate membranes of the turbinates are not just mucus membranes: They are erectile, with thousands of tiny receptors. They respond to stimulation just as other erectile parts of the body do—such as the penis, clitoris, and nipples. However, instead of responding to external stimuli, these erectile tissues respond to the subtle harmonics of the quality, quantity, and timing of our airflows. For example, Dr. Cottle could diagnose coronary heart disease and depression through rhinomanometry. A swollen, over-engorged lining indicates possible sexual overstimulation. A blocked airway deep in the lungs is indicated through unique pressure changes in the nasal membrane. Allergens can be challenge-tested with rhinomanometry. Colds and other viral infections can also be diagnosed using rhinomanometry (Cottle 1968).

❖ Dr. Freud and Dr. Wilhelm Fleiss, who was an otolaryngologist, also worked with the nasal passages. They also found associations between various physiological processes in the body and the turbinates. They found that numbing one particular spot with cocaine would decrease the intensity of menstrual cramps. They found a swelling in another spot on the turbinates indicated an ulcer. It is said that prior to World

War I, an extensive mapping of the turbinates with respect to particular organs and issues was in development. This work apparently came to a halt when the war started.

❖ Should we focus on our breath as we breathe deeply as discussed above, we can become aware of many of our body's own internal issues. Through our focus upon the temperature, smell and pressure differentials of the air we are breathing in and out, we can easily pick up our body's energy levels—whether we are tired or sleepy for example. We can usually tell if our body is feeling stressed, or whether we might be on the verge of an infection, or heartburn, for example. As we practice our deep breathing over time, we will develop a greater ability to read our emotional moods through our breathing. Through deep breathing, we can also become increasingly aware of our body and mind as an observer. We can watch our body and mind from a distance, assessing their situation and behavior. With time and practice, we can begin to see how the consciousness of our inner self can watch the waveforms of the body. Through the manipulation of waveforms, our inner self can utilize the physical mind and body as a vehicle rather than be lost within them.

❖ Deep breathing is also an ancient art. The Greeks, Egyptians, Mayans, Polynesians, Chinese and about every other culture have practiced breathing techniques for thousands of years.

❖ The most ancient, and now probably the most famous, is that of the ancient Indian art of *pranayama*. There is a widely held misunderstanding that pranayama simply means "deep breathing." While pranayama is related to breathing, this is not true. Pranayama means to control the *prana,* which is the vital current of consciousness expressed through the body by the atma, or inner self. For this reason, some refer to prana as the life air or vital air. Put another way, this prana is the reflective essence of the inner self circulating through the body. This is also expressed in Chinese medicine as the *qi* or *ch'i.*

❖ The deeper understanding of pranayama as taught from the ancient Sanskrit language is part of the practice of *yoga.* Yoga—also widely misunderstood—literally means to link or connect with the Almighty. Thus yoga is quite practically, a practice of devotion rather than a series of stretches as most consider it. *Hatha yoga* is one branch or subset of yoga according to the ancient Vedic texts. It is a focus on bringing about a cleansing within the body and mind for the purpose of devotional growth. Other forms of yoga include *astanga yoga, karma yoga, tantra yoga* and others. The highest perfection of yoga according to the ancient texts is the practice of *bhakti yoga,* which means quite literally, "devotional action on behalf of the Supreme Being."

❖ Pranayama plays a part of many yogic practices because during physical exertion, study or devotional acts, controlling the movement of the life air helps control the senses. The art of moving and controlling the life air is connected to the breath because through a rhythmic and paced breathing—especially during that moment between breaths—one can begin to focus internally and towards the Creator.

❖ The momentary pause between breaths is key to focus the prana. This pause is maintained only momentarily, and done twice during the full cycle of breath for two to three seconds each: After the air is pushed out of the lungs and after the lungs are full.

❖ Pranayama breathing is assisted by a half-closing of unfocused eyes. While some suggest the nose can be focused upon as the eyes are half closed, this comes with the risk of crossing the eyes. Closing the eyes is the best alternative if the eyes cannot be unfocused. The idea is to remove focus from the material and meditate onto the higher realm.

❖ The asana—meaning devotional posture—suggested to accompany meditative pranayama is the traditional lotus posture. However, pranayama can be practiced with any of the *hatha yoga asanas* or while sitting in a chair or even walking. Experts in pranayama have gained the ability to control internal functions such as heart rate and blood pressure. As evidenced in biofeedback, this is not a difficult thing to master as long as we have a reference such as a heart beat to watch.

❖ In the Ayurvedic teachings of pranayama, the breath out was to be accompanied with a *mantra.* The yogic mantra such as *Om,* (pronounced "Aumm") is a precise sound intended specifically to invoke the Creator according to the ancient Sanskrit texts. This element of the mantra as devotional seems to have been lost during the translation of "meditation" to the western world.

❖ The ancient Chinese also practiced special breathing exercises in relation to spiritual growth. The tradition of meditative focus was to some degree promulgated through the teachings of Lord Buddha, as a method to assist one to reach the inner *Buddhahood* or enlightenment. This art was later passed through Taoism monks who taught enlightenment through the focus upon the *Tao*—translating quite literally to "the Supreme." The monks utilized a combining of breath control and disciplined movement—perfected over the centuries— now referred to as *qi kung* or *quigong.* This was popularized during the seventeenth century by Master Li Ch'ing Yuen, who apparently lived to the age of 256. Qi kung (pronounced "chi gung"), focuses the ch'i or qi. The ch'i or qi is seen as the expressed radiance of the inner self circulating through the body. Thus chi and prana might be considered synonyms.

❖ Hatha yoga/pranayama and qi kung are closely related. They both combine movement with breath control, though qi kung appears to involve greater motion. A qi kung master becomes skilled at coordinating the breath and movement to enable a focus inward, and the ability to channel the qi into different parts of the body at will. Like experts in pranayama, many qi kung masters have gained the ability to consciously control the heart rate and blood pressure, along with other physical functions. During a qi kung session, a number of very slow and deliberate movements are incorporated with diaphragm breathing. This also is a process of balancing the polarities between the yin with the yang. Diaphragmic qi kung breathing allows the focus to be put onto the channeling of qi throughout the body, just as pranayama allows us to focus onto the channeling of the prana of the inner self through the body.

❖ *Tai chi* applies the methods of qi kung to a martial art—or fighting movement. Contrasting to the western thought of fighting requiring rough and fast movement, tai chi is a very slow and controlled arrangement of movements applied to self-defense. While most of us have seen tai chi movements practiced in a park or other natural setting, to see tai chi applied in an actual fight is an astounding display of calmness and physical control. Tai chi is probably the form of qi kung most recognized in America. Tai chi was apparently introduced in America by Master Da Liu—a student of Li Ch'ing Yuen.

❖ We can easily incorporate some of the tools of *pranayama* and even *qi kung* by diaphragm breathing evenly and slowly during as much of our day as possible. We can take a few minutes once or twice every day to sit or move intentionally while performing these our breathing exercises. We can perform these while walking, sitting, or exercising. We can also take up tai chi, or hatha yoga to discover increased vitality, energy and immunity, while raising our inner awareness.

❖ We can also try to locate our living situation around air that is as free as possible from particulate and smog pollution. This may be difficult for many of us at the moment. If we live in an urban area, it may be difficult to avoid the industrial and automotive pollution. Certainly all of us can restrict our indoor pollution as far as possible by removing the fragrances and chemicals in the air we breathe.

❖ We can also restrict the level of formaldehydes and pesticides by choosing to furnish our households with natural products free as possible from chemical preservatives. This means using cotton clothing; natural flooring and furniture, along with natural fragrance-free soaps and cosmetics. This means buying fewer plastics and more natural fibers. Should we change our purchasing behavior, we will also alter the behavior of those companies manufacturing these items.

❖ If we live in an urban area with poor air quality we might consider exercising or doing our activities in the morning, when the air quality is typically better. As temperatures rise, ozone levels increase. We might also consider exercising near a lake, river, or ocean. Polluted air around water tends to disperse quicker due to wind, humidity, pressure gradients, and temperature differences around water.

❖ *Steam therapy* can significantly help decongest the lungs. Humidifiers and steam rooms are good sources for steam. *Eucalyptus* can be added to increase the steam's cleansing effects.

❖ When it comes to freshening up stale indoor air, pressurized aerosol air fresheners are not the way to go. They may contain various synthetic fragrances, benzyl ethanols, naphthalene, and formaldehydes among other undesirables. Various micro-particles are also created by these aerosols. Candles may smell nice, but they are often made with synthetic fragrances, hydrocarbon-based paraffin waxes, and lead or other heavy metal wicks. The combination often releases unhealthy black soot into our air. *Beeswax candles* with *essential oils* can provide a good alternative for the candle-loving household, combined of course with fresh air.

❖ There are a number of air purification systems available commercially now. Some work better than others and many may be over-hyped and even misleading. *Electric ion generating machines* and *ozone generating machines* typically generate both ions and ozone. Ionizers have been reputed to remove dust and bacteria, yet this remains controversial. While dust and soot may be attracted to negative ions, they are likely to remain airborne, or possibly end up on floors or furniture. The amount of ions generated by these machines may be of concern as well. While outdoor air may range from 500 to 5,000 negative ions per cubic centimeter and indoor air may only have a couple hundred per cc, negative ion generators can easily pump out from ten thousand to ten million negative ions per cubic centimeter. At these higher levels—especially over a million—negative ions may become irritating to mucous membranes. They can irritate the throat, the eyes, and the lungs. Quite simply, our bodies were not designed for this level of negative ions. Ozone generators may be effective at removing bacteria, because bacteria require oxygen to live, and ozone depletes their oxygen levels. For this reason, a good ozone generator will often create a fresher smelling indoor environment. While ozone is not the same as smog, which has carbon or sulphur molecules connected to the molecules, ozone is an atmospheric response to smog, as ozone helps stabilize oxygen levels and clear out other molecules. At the same time, higher ozone levels may also result in lung irritation and allergic response. As mentioned earlier, the FDA limits ozone generating machines in medical devices (used in hospitals and clinics) to emit no more than .05 parts per million. The bottom line with both ionizers and ozone generators is that the atmosphere contains a fragile balance of components. This is not a random mixture. The level of ozone present in outdoor pollution reflects the atmosphere's cleansing process. In other words, the decision to use these machines should be made carefully, and in full knowledge of the current air condition and the effect these machines may make.

❖ It should be noted that neither the ozone nor ion generator systems have been shown to effectively remove dust, dander or allergens (unless the allergens are bacteria, which can be removed with ozone). A HEPA (stands for *High-Efficiency Particular Air*) filter is a better strategy for removing dust, dander, or other particulates. HEPA filters are designed to pick up over 99.9% of particulate sized .3 micron in size. Putting a HEPA filter on a simple air ducting system with a fan or heater will do wonders for removing dust, dander, and allergens.

❖ An *electrostatic air filter* for forced air systems can be a good choice. These filters typically filter between ninety and a hundred percent of dander, mold, mites, dust, soot, and bacteria. Many of these filters come with the ability to clean and reuse, allowing us to clean them as often as needed—which should be monthly if use is constant. *Environmental filters* can also be installed for urban environments, which will bring in outdoor air while warming to the indoor temperature.

❖ Healthy alternatives to freshen the air include cleaning surfaces with *lemon* and *vinegar,* and/or *baking soda.* Rotting food and unclean surfaces create mildew and bacteria quite quickly. Fresh air is quite easy to achieve with clean surfaces. Dusting and wiping down flooring with vinegar can be a good strategy. A small cup of vinegar, placed in a corner can also absorb odors. Fresh cut flowers or herbs like *basil* or *lavender* from the garden can provide a wonderful indoor fragrance. Essential oils from *peppermint, cloves, cinnamon,* and *eucalyptus* are also good strategies to freshen the air. These are also antibacterial and antifungal to some degree, so they can reduce infectious air. These botanicals can be applied to cotton, or put into humidifiers or boiling water on the stovetop. Cooking with fresh herbs can also add wonderful fragrances into our indoor air.

❖ As for indoor air quality, a house full of indoor plants will do wonders for raising oxygen levels and lowering indoor carbon levels. Placing between two and five plants in a hundred square foot room—depending upon the outside air—can significantly remove carbon and raise oxygen levels. Research from the *Mississippi Stennis Space Center* concluded that indoor plants absorbed and broke down formaldehydes, tricholoethylenes, benzenes and zylenes. Better performing plants included the lady palm, the rubber plant, English ivy, and the areca palm plants. The removal rates can range from 1000 to 1800 micrograms per hour. One study done in Norway found 23 percent less complaints of fatigue and sinus congestion among workers working around plants. Studies performed in Texas and at Washington State University found that cognition response and problem solving were also higher in people working around plants (Wolverton 1997).

❖ We can also create forests around our homes. This means planting and maintaining indigenous trees to enable them to share our immediate space. Nearby outdoor plants can do wonders to increase the oxygen and decrease the carbon in our immediate environment. This also means instead of paving our courtyards, we can leave the trees or replant them. By doing this we will not only be giving our lungs more oxygen, but we will be assisting the planet in its current asthmatic condition—a condition we have caused as we have *"paved paradise and put up a parking lot."*

Chapter Six

Therapeutical Sun

The energy of our entire ecosystem is due to the sun's electromagnetic radiation, thought to be hurtling at us 186,282 miles per second from a heat source estimated to be 93 million miles away.

The sun is one of hundreds of billions of stars that encircle our Milky Way galaxy at a speed thought to be about 500 miles per second. One rhythmic revolution of the sun around the center of this galaxy is thought to take about 250 million years.

Scientific analysis has estimated the surface of the sun is only about 5,760 degrees Kelvin; yet its corona—or radiance—is estimated to be between 1,000,000 and 5,000,000 degrees Kelvin. The sun's radiance will deposit between 1,000 and 1,300 watts of power onto the earth's surface per square meter. This problem is also referred to as the *coronal heating problem:* Why is the sun's corona so much hotter than the surface? While some researchers conjecture the sun's radiation comes from nuclear fusion reactions occurring deep within the sun's core—the mystery is by no means solved.

The sun is also magnetic. Its magnetic field fluctuates to up to 3,000 times its normal range—so it has extreme ups and downs: Rhythmic occurrences. The sun also emits solar flares, which appear to our eyes and instruments to be massive eruptions spewing up to 20 billion tons of matter into space. On a mass basis, researchers estimate the sun represents about 99.8 percent of the total mass of this solar system. That makes the rest of the planets dwarf-like compared to the sun.

The sun is truly an enigma of massive proportions—a subject of continuous conjecture by researchers.

Much of the nutritional energy utilized by life on this planet comes through the thin leaves of plants: with a mechanism called photosynthesis. Through microscopic pores in the leaf called *stomata*, the plant absorbs carbon dioxide. From the roots below, water is absorbed and brought through tiny veins to the leaf. Tiny *chloroblasts* of multiple chlorophyll molecules drink specific rays of the sun, utilizing those waveforms to split water into hydrogen and oxygen atoms. The resulting hydrogen atoms combine with carbon dioxide to form carbohydrates (CH_2O, leading to $C_6H_{12}O_6$, for example) while oxygen atoms are released into the air as oxygen gas. The carbohydrates form sugars, starches and cellulose within the plant, directly providing fuel to plant-eating species, and indirectly providing fuel for those species that eat plant-eaters. None of this could be possible without the energy of the sun. It is the sun's violet-blue and lower red-orange wavelengths the leaf utilizes. The green wavelengths are reflected back—giving the leaf its characteristic green color.

The molecular structure of the chlorophyll molecule is daisy-shaped—called a *porphyrin ring*. Its molecular shape almost precisely mirrors the radiant orb from which it converts energy. This ring-shaped chlorophyll structure (sometimes connected with *phytol* chain) allows the radiation to freely migrate, enabling an electron transport process through *fluorescence resonance energy transfer*. This process converts specific wavelengths of radiation from the sun into an excited state within the molecule. The excited state is a constructive interference between the radiation waves from the sun and the energy waves of the electrons (standing waves) within the chlorophyll molecule. This coherent interference creates an energy transfer chain very similar to the Krebs energy cycle occurring within our cells. Both are considered electron-transport cycles. As the electron-waves are transferred, NADP+ is reduced to NADPH, which enables the conversion of carbon dioxide to sugar—the fuel of choice for other organisms.

The process of photosynthesis is incredibly complex. We merely summarize it here. Within the process exists many catalysts and enabling compounds gathered from the earth's soils, such as magnesium, which lies in the center of the porphyrin structure. For example, carotenoids assist in a number of ways, by both buffering the process when too much radiation is involved, and alleviating barriers to the process at lower levels. Carotenoids such as vitamins A, C and E function to protect the plant from the effects of too much radiation. The carotenoid vitamin A works in a similar way in our eyes—helping prevent degeneration from extreme radiation. Other compounds involved in the photosynthetic process such as *phycobilins, ferredoxin, adenosine triphosphate, fucoxanthins* and so many others quite simply boggle the mind with such an array of inter-functionality.

About three-quarters of this planet's photosynthesis takes place within the oceans, thus involving the *cyanobacteria* and *rhodophyta* families of the plant kingdom. Quite simply, the complexity of photosynthesis and its ability for providing the fuel needed for life is a miraculous array of waveform precision and design.

Solar Waveplex

The sun produces several types of radiation. They are categorized as *visible light rays* (from 400 to 700 nm wavelength), *infrared* radiation (from 750 nm to 1 mm), *ultraviolet rays* (280 to 400 nm), *x-rays* (about 10 x -5 nm), *gamma rays* (10 x -11 to 10 x -14 nm), *cosmic rays* (10 to 12 cm) and *microwaves* (1 mm to 30 cm). The earth's atmosphere blocks a good amount of the x-rays, cosmic rays and gamma rays and almost all of the ultraviolet rays (depending upon the levels of ozone), allowing a good portion of the sun's visible and infrared radiation to hit the surface. In all it is estimated that the atmosphere blocks about 40% of the sun's total radiation. About half of the heating of the earth's surface is thought caused by infrared radiation from the sun. The rest is thought caused by visible wavelengths of light from the sun being absorbed and radiated to the surface in longer wavelengths. We would propose that these guesstimates do not include the thermal heat generated by the earth itself, which is difficult to measure.

While we can easily separate the colors of the visible radiation into reds, yellows, blues and violets, there are many more colors we cannot see in between and around these wavelengths. Each different waveform stimulates specific sensors in our retinal cells and optic nerves, within a limited range. Different organisms see the sun's light differently as well. For example, bees see ultraviolet light, which assists them in seeing flower pollen.

In contrast, a typical incandescent light bulb will emit visible light by heating a filament inside a bulb of ionized gas. The filament is usually made of tungsten in an atmosphere of halogens such as nitrogen, krypton, or argon. The incandescent bulb will typically produce three or four-colored visible light together with near-infrared waveforms, which create the mild heat typical of a light bulb. Though the bulb will produce visible light, here are various wavelengths missing when comparing to the visible light from the sun. An incandescent bulb typically discharges visible light in the red, green, blue, and violet waveforms. The sun on the other hand, will emit these along with various others, including yellow, purple, turquoise, and so on. For this reason, visible light from the sun allows us to see more crispness and more colors. As these additional waveforms are reflected or absorbed by the objects around us, our eyes see the result.

In addition to these other bands of color waveforms within the visible spectrum, sunlight also emits a full (near and far) ultraviolet and infrared spectrum, along with the other waveforms as mentioned above—many of which are filtered as they pass through the atmosphere.

We have had little success trying to duplicate the sun's waveform complex. *"Full spectrum lighting"* is a far cry from sunlight. Currently there is no real standard for what is called "full-spectrum lighting," so its reference is often misused. There is a quantifying standardization called the *General Color Rendering Index* (or CIE). This gauges the intensity of the color temperature. While red and infrared waveforms are "hot," blue and violet waveforms are considered "cool." A CIE rating that approaches 100 is considered cooler, in that it covers not only some of the ultraviolet waveforms, but also emits at least a range of the main colors such as red, blue, green, and violet.

Tanning system lighting typically provides primarily ultraviolet radiation. Many tanning lights emit about 95% ultraviolet-A and 5% ultraviolet-B in order to produce the tanning (and burning) equivalent of the sun. Indoor "full spectrum" lights will usually provide some ultraviolet waveforms as well, usually near-ultraviolet, which covers part of ultraviolet-A as well as a decent part of the visual spectrum. Many of the newer "full-spectrum" lamps provide cooler CIE numbers, reflecting a good dose of blue and green spectrum additionally.

Research performed by Rosenthal and Blehar (1989) concluded that lighting with blue and green wavelengths has therapeutic value for *seasonal affective disorder* (SAD). There have also been a number of inconclusive studies on therapeutic "full-spectrum" light. In one review by McColl and Veitch in 2001, research from 1941 through 1999 concluded that "full-spectrum lighting" research has for the most part, not shown any positive effects on either behavior or health—inclusive of hormonal and neural effects. Most researchers have also concluded that there is only marginal vitamin D production from "full-spectrum" lighting. Vitamin D deficiency is one of the more prevalent issues in light-deficiency disorders.

Recent research has confirmed that our bodies also produce light. Dr. Fritz-Albert Popp's work with biophotons over the past few decades has illustrated this effect. Russian researchers V.P. Kaznachejew and L.P Michailowa conducted tests to quantify the transmission and reception of these ultra-weak waveforms. Their conclusions showed that both cells and organs produced and received these emissions, and illustrated that information was exchanged through a multiplex of waveforms (Schumacher 2005).

A Toxic Sun?

With all the warnings about the sun's harmful effects, we might consider the sun was more of a toxin than necessary for life. Today modern culture fends off the sun as if it were toxic enemy number one. Today the sales and use of sunblock, sunglasses, and wide-brim hats are ubiquitous. Can the sun be this dangerous?

What were people doing before sun block came along? Did not most humans work primarily outside all day—in the fields growing and preparing food, washing clothes, bathing, and so on? Now as we sit indoors in front of our computers, for some reason the sun has suddenly become toxic.

Most humans have experienced the positive effects of sunlight. This is why humans tend to vacation in sunny areas. A vacation to a tropical or sunny destination typically results in increased feelings of well-being and relaxation. Many also experience a decrease in allergies, headaches, joint pain and backaches. While there is certainly the element of relaxation and lowered stress during vacation, it is unmistakable: Sunlight contributes to the health of the body.

The sun benefits us in a number of ways. For humans, plants and most animals, the sun's resonating energies deliver heat, raising core body temperature. Higher core body temperatures facilitate increased cell function and greater energy. This increases our metabolism, providing increased pathways for the body's various detoxification and purification systems. Boosted core temperatures increase cortisol levels during the day, which are balanced by increased melatonin levels, ushering greater relaxation and deeper sleep during the night.

Sun also regulates our natural biorhythm cycles. The sun's waveforms stimulate the body's pineal gland, which stimulates a number of key master hormones. These in turn synchronize the master body clockworks triggered through the suprachiasmatic nuclei cells. These cells mark the passage of time for the body, regulating cellular and endocrine functions. A day without sunlight will leave these cells confused. This is illustrated by the common experience of disorientation after sleeping in late on a Saturday morning. On an extended basis, a day without a good portion of natural sunlight—even if just seen through a window screen—will leave the body's natural rhythms in a state of disarray.

The reception of light through the pineal gland is a biological process. Iguanas for example, have a parietal (third) eye centered a little higher and between the eyes, which can read and distinguish between light and dark wavelengths. When light radiation hits this area, the clock is reset. Lizards have this same mechanism and apparently other animals and humans have derivatives of this parietal eye—perhaps playing a much greater role than we have ever figured.

A lack of natural sunlight will disrupt the body's endocrine systems. For example, German ophthalmology Professor Dr. Fritz Hollwich published in 1979 that subjects working under white fluorescent tube lighting for a period of time will result in a production of significantly higher levels of stress hormones such as adrenocorticortropic hormone and cortisol. He also found full-spectrum or natural light results lower levels of these stress hormones. Dr. Hollwich's research was pivotal in the decision to ban white fluorescent bulbs from German hospitals.

In another study by Dr. Hollwich (Hollwich and Dieckhues 1989), 110 cataract patients underwent metabolic testing (primarily blood) before and after surgery. Prior to surgery, the opacity of their cataracts significantly reduced the amount of light, down to about 10% of normal. Testing prior to surgery showed slowed metabolism, adrenal insufficiency and hormone imbalances. After surgery—the removal of the lens opacities—metabolism and hormone levels returned to normal. These results were confirmed with another study (Hollwich and Hartmann 1990) performed shortly thereafter on fifty cataract patients with the same results. This later study also looked at water balance, blood sugar and blood cell count—all of which improved following surgery. Dr. Hollwich describes a retino-hypothalamic pathway connecting to the endocrine-visceral system.

Dr. Niels Finsen was awarded the 1903 Nobel Prize in medicine for revealing that sunshine was extremely therapeutic for a number of infectious diseases, including lupus vulgaris, small pox, and Pick's disease. Dr. Finsen's famous sunbaths and separated light colors became known as *Finsen Light Therapy*. In the early 1900s, two Swiss physicians Dr. Bernhard and Dr. Rollier found that the sun in thin atmospheres such as the Swiss Alps provided an effective therapeutic protocol for surgical tuberculosis and lung tuberculosis. Sunlight has since been shown effective as a part of treatment for various other conditions.

About ninety percent of humans in modern society now work indoors. One hundred years ago, this statistic was reversed. At least ninety percent if not more, of humans lived and worked outside or in locations where natural light was directly present. While there are many warnings present in the medical literature to stay away from the sun as though a toxin, The *National Institute of Mental Health* in Bethesda, Maryland included the following statement in a 1988 report (Skwerer *et al.*) on seasonal affective disorder: *"Along with food, air and water, sunlight is the most important survival factor in human life."*

While millions of people have been diagnosed with *Seasonal Affective Disorder* (or SAD) over the past few years, some estimate a good 25 million Americans are afflicted with some form of the disorder—at least the milder yet more pervasive "winter blues" version of SAD. According to Norman Rosenthal, M.D., who led the above study and most noted for many others on *seasonal affective disorder*, the necessity of sunlight to mental health, about 6% have SAD in the U.S. and 14% have winter blues. For some, a move from the southern latitudes to the northern latitudes precipitates the disorder. For others, there seems no connection except with other ailments, such as depression.

As the fall and winter descend upon us, sunlight hours decrease and melatonin levels increase. This is part of our natural biological cycles. However, some of us are more sensitive to melatonin for some reason. SAD and 'winter blues' can become a problem in these more sensitive folks. It seems that women are also more likely to come down with the disorder.

A number of diseases are attributable to lower levels of sunlight. Hypertension, atherosclerosis, early dementia, Alzheimer's, multiple sclerosis, psoriasis, fibromyalgia, depression, arthritis, and low back pain are but a few of the issues which have been linked to a lack of, or ameliorated by sun exposure. A lack of natural sunlight depresses the immune system, weakens eyesight, lowers endocrine activity—lowering and/or disrupting hormone secretion—lowers concentration, increases stress and contributes to depression. In addition, sunlight has been shown to decrease symptoms and duration of several cancers, especially as they relate to melatonin levels. Studies have also shown that natural sunlight decreases hyperactivity in children.

In the converse, an increasing amount of research is indicating that a lack of sunshine during the day combined with the lack of complete darkness during the night—as our environment has become increasingly lit up at night—reduces melatonin availability. This reduction in melatonin has been linked with various types of cancers, including breast cancer, prostate cancer, colorectal cancer and endometrial cancer (Reiter *et al.* 2007).

In a study done by the Heschong Mahone Group (1999), students learning within environments with the most natural sunlight tested better and exhibited faster rates of learning. Another study supporting this was conducted earlier by Anderson *et al.* (1991).

Energy conversion from the sun is not limited to plants. From the sun's ultraviolet B rays, our bodies synthesize vitamin D_3. By far the most important of the D_3 is produced when ultraviolet-B in wavelengths of 270-290 nanometers enters our epidermis. Here a derivative of cholesterol called *7-dehydrocholesterol* undergoes a *conrotatory electrocyclic reaction* to produce a pre-vitamin D. This later molecule undergoes hydroxylation in the liver and kidneys to convert to the final D_3 structure, 1,25 dihydroxyvitamin D (some refer to this as 25-OHD). The conrotary reaction illustrates another synchronous circular reaction as atoms and their bonds rotate around the ring.

In the environment of the 7-dehydrocholesterol molecule within the epidermis, *melanin*—the element which makes the skin turn brown—provides a filtering mechanism for the rays. The greater the melanin level, the fewer ultraviolet-B rays reach the 7-dehydrocholesterol molecules, and thus the less vitamin D_3 is produced.

Vitamin D is most known to regulate calcium levels and absorption. Without proper vitamin D_3 production, calcium will be absorbed into our bones and teeth. Vitamin D is also critically important for healthy immune function, nervous function, cardiovascular health, mood regulation, pain regulation, insulin/blood sugar balance, as well as numerous endocrine and digestive functions. Vitamin D is a necessary component for good health, and its most natural form (D_3) comes from natural sun exposure (Lehmann 2005).

Because vitamin D_2 will also convert to 1,25 dihydroxyvitamin D; until recently it was assumed D_2 and D_3 were functioned identically within the body. This assumption has since been proven to be incorrect. In a study done at the Creighton University and the Medical University of South Carolina (Laura et al. 2004), twenty healthy males were measured following supplementation of D_2 and D_3. While both converted to 1,25 dihydroxyvitamin D, D_2 converted at far less levels and fell off much earlier. The study's authors concluded that D_2 has only about one-third the potency of D_3.

In a Boston University study published in 2008, forty-five nursing home patients took a multivitamin containing 400 IV of vitamin D_2. During the study period, their 25-OHD levels registered deficient from a range of 49% to 78% in measurements over an eight-month period.

In 1998 (Thomas et al.) a study of 290 medical ward patients at the Massachusetts General Hospital in Boston showed 57% were vitamin D deficient (164) and 65 of those (22%) were severely deficient. 46% were deficient despite taking the recommended dosage of supplemental vitamin D. The elderly and those in pain are most often vitamin D deficient. One study (Al Faraj and Al Mutairi 2003) found 83% of 360 low-back pain patients had vitamin D deficiency and another (Plotnikoff and Quigley 2004) showed that 93% of 150 nonspecific pain patients had vitamin D deficiency. In the latter study nearly all of those patients declared pain relief after three months of vitamin D supplementation, and in the former study, of the 360 low-back pain patients, 95% showed clinical improvement after treatment with supplemental vitamin D_3.

Clinical studies have also indicated a 20-30% increase in breast cancer incidence and a 10-20% increased fatality rate for breast cancer among vitamin D-deficient women (Nielsen 2007). Chronic kidney disease is also more prevalent among those deficient in vitamin D (Khan 2007). Vitamin D deficiency was linked to tuberculosis by Sita-Lumsden et al. in 2007. Vitamin D has also shown to protect against macular degeneration by Parekh in 2007. Cognition and mood related disorders are also linked with vitamin D deficiency (Berk 2007). Vitamin D deficiency was linked to a higher incidence of osteoporosis by Barone (2007). Fetal diabetes, preeclampsia and fetal neurological disorders were connected to vitamin D deficiency by Perez-Lopez in 2007. Serum vitamin D was also connected with higher insulin sensitivity by Kamycheva *et al.* in 2007.

Vitamin D, when taken together with calcium, was linked to greater glucose metabolism by Pittas *et al.* in 2007. In a Creighton University study (Lappe *et al.* 2007) of 1179 postmenopausal women, cancer rates among those supplementing with calcium and vitamin D_3 had an almost 60% lower cancer rate than the control subjects. A number of cancers are apparently prevented or ameliorated by vitamin D.

In a review of cancer studies from 1970 to 1992 (Grant, 2002), the American Cancer Society reported that UV-B solar radiation is associated with a reduction of cancers of the breast, colon, ovaries, prostate, bladder, kidney, lung, pancreas, rectum, stomach, uterus, and esophagus. Dr. William Grant, the author of the research, reported that the reduction of some cancers range from 30-50% with adequate UV-B solar exposure.

The active and most therapeutic version of vitamin D is 1,25-dihydroxyvitamin D_3, most effectively produced through the conversion of sunlight. It is also thought that the availability of 1,25-dihydroxyvitamin D in the tissues and bloodstream is raised—and possibly somewhat regulated—by the levels of isoflavones in the body. Isoflavones are nutrients available in various plant foods, with higher levels in grains and beans (Wietrzyk 2007).

In the Boston University 2008 press release noted above, tanning beds produced therapeutic levels of 1,25-dihydroxyvitamin D_3. While this information might be considered useful for extreme therapeutic cases of deficiency, other studies have reported higher incidence of melanoma among tanning bed users. In one study of 551 persons, those who used tanning beds more than 20 minutes per session had significantly higher rates of malignant melanoma (Ting *et al.* 2007).

Not many foods contain vitamin D. It is found in various dairy products such as cheese, butter, and cream, and in some fish and oysters, but the sun or supplements are the most reliable sources. The average American diet will only provide about 250-300 IU. Dr. Grant reports that 2,000 to 4,000 IU is required for significant cancer rate reduction.

Conservative sunlight exposure provides--primarily through UV-B radiation--sufficient vitamin D. Just twenty minutes of sunlight on the arms, hands and face will produce about 400 IU of vitamin D on the average skin type. A day of summer sun in a bathing suit until the skin is pink will produce as much as 20,000 IU. A day in the tropics could easily result in 100,000 IU/day. Although the RDA is 200-700 IU (700 for elderly adults), many nutritionists believe that 1,000 to 5,000 IU per day is optimal. Note that an SPF-8 sunscreen will block about 95% of vitamin D synthesis.

The human body requires sunlight levels above 2500 lux before it can stimulate any significant levels of endogenous vitamin D. Indoor lighting simply does not supply enough of the correct waveforms to stimulate significant conversion of vitamin D. Indoor light ranges from 60 lux at low lamp level to 200 lux in average indoor lighting. The brightest indoor lighting might produce about 1000 lux. Levels of over 1000 lux typically require some sort of daylight.

Worse, many indoor office and lighting systems have poor quality fluorescent lamps. Older models with conventional ballasts tend to flicker at lower frequencies. Their blinking will range from 60-100 times per second. Research performed by Küller and Laike (1998) in Sweden, has illustrated that fluorescent lighting powered with conventional ballasts increase stress and lower accuracy. This study, done on adults working in laboratory offices, found that the conventional ballast increased alpha wave activity among subjects. The newer technology of high frequency electronic-ballast compact fluorescent light bulbs have increased frequencies in the 20-40 kilohertz range—over 2000 times faster than the older models. Higher-speed flickering seems to have a far less negative affect.

The typical American gets about 100 lux per day on average, experiencing only quick bursts of any higher lux, and then barely over 1000 lux. 1000 lux is about as much light as is available in the twilight period—just after sunset or just before sunrise. Research on shift workers has confirmed that light above this level for three hours will re-establish sleep cycles within 48 hours, while other techniques such as coffee and alarm clocks can take up to eight days to re-establish ones sleep cycle. Levels of over 4000 lux are needed for most endocrine-stimulating functions.

Most of the benefits of the sun seem to occur through the stratum corneum of the epidermis. This is the outer layer of the skin. Research has attempted to quantify the dose needed for health benefit—referred to as the *minimal erythemal dose* (or ME dose). This minimal effective dose is considered the minimum dose required to produce a slight reddening of the skin. However, the amount of sun required to accomplish this dose is different for each skin type. A darker-skinned person, for example, will require several times the amount of sun than a fairer-skinned person might.

There are six basic skin types. They range from the fairest skin type at number one to the darkest skin with the number six. While the number one skin type might quickly burn and require only about 10 minutes of full summer sun to establish an ME dose, a number six skin type will take about four times longer or 40 minutes, with little chance of burning. Most

Caucasian people fall into the number three skin type, which might require about 20 minutes before establishing the minimum dose level.

This minimal dose might even be too much for someone not accustomed to the sun—at least until some tolerance is developed. Sunlight studies from Russia have indicated that about 30% to 60% of the ME dose levels are advisable until this tolerance is built up among the skin cells. This tolerance comes from a build up of melanin in the skin cells. Melanin effectively reflects and shields the skin from the sun's more harmful waveforms. Once repeated exposure is accomplished for a few days, exposure can safely go up to 1.5 times the ME dose.

Melanin protects the skin from the longer wave UV-B. Longer wave UV-B has been shown to produce higher levels of free radical damage to cells. Longer wave UV-B also appears to damage folate reserves within the cells—an important anti-cancer B vitamin. When humans lived primarily out-of-doors, the build-up of melanin was incidental, and naturally protected us from these longer waves. On the other hand, occasional sunlight--such as vacation exposure--exposes the skin unprotected by melanin.

A fair-skinned person should receive about ten minutes of arm-leg-face overhead sun in the summer, or better yet, about 30-45 minutes in the early evening or morning sun to achieve a therapeutic dose of sun. A medium-skinned person with more melanin should get about 15 minutes of overhead sun or better 45-60 minutes of morning or early-evening sun. This increases to 20 minutes overhead or 75 minutes early/late sun for darker-skinned people and 30 minutes overhead sun or 90 minutes of early/late sun for Negroes and Aborigines each day to achieve ME.

These are summer levels and they also vary to latitude. These also assume moderate north or south latitudes. It will take longer to achieve a therapeutic dose in an extreme north latitude location. It will also take longer during the winter months. For those living in these regions or seasons, it is important to spend a significant amount of time outside during the winter months. Roughly an hour in the middle of the day with the face and hands exposed should be adequate in northern or southern latitudes during the winters. The skin will have less melanin in the winter, so there will be increased absorption.

Regardless of how much vitamin D is synthesized, natural light of any kind can have positive affects upon ones mood and energy levels, deferring many of the symptoms of vitamin D deficiency, especially if exercise is involved.

For a number of decades researchers have been observing the effects a lack of natural sunlight has upon various plants and animals. In the

1950s, while producing the famous Walt Disney time-lapse photography film, Dr. John Ott discovered that for flowering pumpkins grown indoors under fluorescent light, the male flowers would blossom but the female flowers would not. Meanwhile outdoor pumpkins typically produce both male and female flowers.

Dr. Ott filmed and photographed flowers and plants indoors and outdoors. His time-lapse photography sessions with various flowers and plants revealed that plants without a full spectrum of infrared and ultraviolet light become in one respect or another, disturbed. Dr. Ott's research on light and color continued into the 1980s and 1990s. He published numerous articles documenting his research—spanning over forty years—on the human body as well as plants an animals. These studies showed various negative effects resulting from a lack of natural sunlight upon the human body. They included mood disorders, learning disabilities, increased stress, abnormal growth, and poor eyesight among others. Without regular sunlight, humans—like plants—will begin to degenerate in a host of ways. As many of us have observed, should a plant be shut into a room with nothing but artificial light, its flowers and/or fruits will soon wilt and whither.

The effect of sunlight on animal and plant fertility has been known by farmers and ranchers for many years. The poultry and hog industries have known for years that full-spectrum light increases egg production and larger hog litters respectively. It has also been observed that human fertility also improves with sunbathing. We also know the increase in the warmth and duration of the sun stimulates the production of pollen and flowers among plants.

Dr. Ott's experiments with sunlight and rabbits also revealed that rabbits raised under artificial light—especially among males—became more aggressive even to the point of becoming cannibalistic. Meanwhile rabbits raised in the daylight showed none of these tendencies. Male rabbits raised in natural sunlight not only were less aggressive, but were observed graciously tending to the litter in the absence of mother rabbit.

The Skin Cancer Mystery

Interestingly, skin cancer rates have gone up dramatically over the past four decades, at close to the same rates as other cancers have gone up. Does this mean more people have gone out into the sun? While some researchers propose a sun tanning trend began in the 1960s, if we consider this issue in relation to the industrial revolution, over the past century more people are working and living indoors than ever before in history. Hence the assumption people have been receiving more sun is perhaps an over-simplification.

In a study done in Australia of 1014 patients and controls, for example, superficial spreading melanoma rates increased among those with low total outdoor exposure in early adulthood and increased boating and fishing exposure later (Holman *et al.* 1986). The same study showed no relationship between melanoma and sun exposure. Only Hutchinson's melanoma illustrated a link with sun exposure.

Research has confirmed lower melanoma rates among blacks and Hispanics, but by no means do they not get skin cancer—despite the fact that few blacks and Hispanics sunbathe. Also, the increased level of melanin in black and Hispanic epidermal layers screens or filters more ultraviolet radiation, putting these populations at decreased sun exposure in general.

Furthermore, recent research has confirmed that cancer rates have continued to rise over the past two decades despite a dramatic rise in the use of sun-protective agents like sunscreen and hats; and decreased sunbathing in general. Research has shown modern society's awareness levels of the need for sun protection is at an all-time high. In one study (Stoebner-Delbarre 2005) of 33,021 French adults, 92% understood the sun increased the risk of skin aging and 89% understood it increased cancer risk. With few alternatives, the study's researchers concluded the continued increases in skin cancer are due to this awareness not translating into action. The other possibility might be that there are other factors involved.

There are two basic types of skin cancer: Melanoma and non-melanoma skin cancer. Melanoma grows from dysfunctional melanocytes. Melanoma is the more dangerous of the two, rendering far lower survival rates compared to non-melanoma skin cancer. Yet true melanoma is also rarer. By far most of the skin cancers diagnosed—at least in the United States—are non-melanoma-related. According to the American Cancer Society about 62,190 new melanomas were diagnosed in the U.S. and over one million non-melanoma skin cancers were diagnosed in 2006. The Cancer Society also estimates 10,710 skin cancer deaths in 2006—with 7,910 of those being melanoma.

There are three basic types of cells in the epidermis layer: the *melanocytes*, the *squamous keratinocytes* and the *basal keratinocytes*. Melanoma is associated with genetic damage occurring in melanocytes, and non-melanoma is associated primarily with genetic damage in the keratinocytes. Non-melanoma is divided into two types; *basal cell carcinoma* and *squamous cell carcinoma*. Both of these non-melanoma types are considered benign, however. They both can spread if conditions are right, but in most cases, the spread is either slow or visually obvious. Melanoma is known to spread more quickly, leading to greater fatality rates. It can also easily be excised and stopped if caught early.

Basal cell carcinoma occurs at the deeper basal cell level. A good 70-90% of skin cancer is the basal cell version. As for the squamous cell type, this typically occurs around the face, lips, neck, or ears—yet also interestingly in the genital areas. These may form from *actinic keratoses*—which some refer to as sun spots. There are a number of other non-melanoma skin cancer types such as *Kaposi sarcoma* and *Merkel cell carcinoma,* but these are fairly rare.

The precise mechanisms for both melanoma and non-melanoma skin cancers are still being debated. It is thought that ultraviolet-B rays penetrate the basal cells deep in the epidermis layer, creating genetic damage. Research however, has indicated a more complex multi-step scenario. Both the sun's ultraviolet-B (280-315 nanometers) and ultraviolet-A (315-400 nm) rays, especially intense during the mid-day can produce a *reactive oxygen species* (or ROS) among the tissues in the epidermis. These reactive wave-moments appear to induce genetic mutations among the DNA in the cells—*if they are not neutralized.* Should these reactive radicals not be neutralized by the body's protective mechanisms, then DNA damage may occur amongst the kerotinocyte skin cells. Research presented by the Sydney Cancer Center's Melanoma and Skin Cancer Research Institute at the *Photocarcinogenesis Symposium of the 14th International Congress on Photobiology* (Halliday *et al.* 2005) showed ultraviolet-A causes a similar amount of gene mutations as does ultraviolet-B. While ultraviolet-B mutations predominated in the upper tumor areas, ultraviolet-A damage predominated at the basal (lower) layers.

But while research has resolved that these ultraviolet rays can create mutagenesis—damaging DNA—the picture is still more complex, as there appears to be an increased presence of mutation-suppressors and decreased levels of mutation-repair mechanisms available among ultraviolet-damaged tumors (Nishigori *et al.* 2004). Again, we find other factors evident besides simply the impact of ultraviolet radiation onto the skin.

Melanoma has been the fastest growing cancer in America (Fecher 2007). While it was assumed sun exposure was the general causal agent in non-Hodgkin types of melanoma, three recent studies have revealed that risk for melanoma actually decreased from 25% to 40% with increased recreational sun exposure. One study (Armstrong 2007) indicated that possibly this effect is related to higher-levels of vitamin D intake.

Recent studies are now illustrating that vitamin D reduces a wide variety of cancers, including prostate, colon and breast cancer (Schwartz 2007). Indeed the connection between sun exposure and skin cancer seems to be more related to immunosuppression—deficiencies of the body's immune system and antioxidant levels to regulate and neutralize oxidization and the radicals that cause oxidation within the tissues (de Vries *et al.* 2007).

A recent review done by researchers from the University of Southern California's Keck School of Medicine concluded *"...controversy exists, especially in the use of sunscreens."* (Irvy *et al.* 2006).

The Department of Radiation Genetics at the Kyoto University in Japan (Matsumara *et al.* 1996) studied 32 cases of basal cell carcinomas among Japanese patients; 16 of which were developed in sun-exposed areas and 16 developed in less-exposed areas. In both groups, and consistent many other recent studies, the p53 gene was seen as the primary site of mutation. Furthermore, 75% of the non-exposed group showed transversions—the exchange of nucleotides between purines and pyrimidines. This exchange of bases at the genetic level relates to the bonding moments making up the DNA molecule—creating a switching out between two different sequence types. Once again, the study's authors noted these results indicate a more complex mechanism other than simply sun exposure for skin cancer.

Transversions have been observed *in vitro* primarily within two mechanisms: From the impact of toxic molecules—including free oxygen species and various toxins like benzene— and from the impact of ionizing radiation. Ionizing radiation occurs from various electromagnetic sources such as microwaves and various electronic appliances. While the sun might be considered a potential source of ionizing radiation, a molecular environment capable of creating ions at an intensity of radiation causing that ionizing must exist within the mechanism. More importantly, an environment within the body must exist where those ions—or reactive species—remain un-neutralized.

While a popular myth has spread that all free radicals are damaging at any level, recent research is revealing that exposure to reasonable amounts of free radicals, or oxidative species, is necessary for sustained health. In a study performed at the University of Jena in Germany (Schultz *et al.* 2007), free radicals formed in the cell through glucose inhibition significantly extended lifespan. The effect was apparently because the immune system developed a resistance to the oxidative stress caused by the reactive oxidative species. Consistent with other observations of immune strengthening due to increased resistance, it appears that some free radical exposure is required by the cells.

Yet we also know that an overexposure to free radicals is unhealthy and even carcinogenic. The question is where is the line between a healthy amount of free radicals and overexposure? As the above study illustrates, nature has a design for maintaining a healthy balance between antioxidants and free radicals. This assumption also provides the basis for understanding why skin cancers (and so many other cancers and autoimmune disorders) are largely diseases of industrialized modern culture, where sunlight and other natural inputs have been replaced by synthetic and often toxic versions.

One important consideration in this issue is the impact upon the body by sunscreens and other chemicals we spread onto our skin. Many of the chemical ingredients that are in sunscreen have been identified as carcinogenic. These include benzophenone-3, homosalate, 4-methyl-benzylidene camphor, octyl-methoxycinnamate, and octyl-dimethyl-PABA. These five, in some combination thereof, are contained in about 90% of commercial sunscreens. All five have showed increased cancer cell proliferation both *in vitro* and *in vivo* in a study conducted at the *Institute of Pharmacology and Toxicology* at the University of Zurich (Schlumpf *et al.* 2001). This study also showed negative estrogenic and endocrine effects from several of these as well.

A study done at the University of Manitoba in Winnipeg (Sarveiya *et al.* 2004) reported that all sunscreen ingredients tested—including octy-methoxycinnamate and oxybenzone—significantly penetrate the skin. The penetration of common sunscreens was found to increase the penetration of even more dangerous herbicides—a concern for agricultural workers and non-organic gardeners (Pont *et al.* 2004). Furthermore, it has been established that some organic sunscreens can cause photo-contact allergy (Maier and Korting 2005). Cook and Freeman (2001) from Australia's *Skin and Cancer Foundation* reported 21 cases of photo-allergic contact dermatitis caused by either oxybenzone, butyl methoxy dibenzoylmethane, methoxycinnamate or benzophenone. The Cook and Freeman research has led to a conclusion that these sunscreen ingredients are the leading cause of photo-allergic contact dermatitis.

Contact dermatitis is hardly pervasive amongst the general population. A study at the National Institute of Dermatology in Colombia conducted a study of eighty-two patients with clinical photo-allergic contact dermatitis. Their testing showed twenty-six of those patients—31.7%—were shown to be positive for sensitivity to one or several of the sunscreen ingredients (Rodriguez *et al.* 2006).

The widespread proliferation of these harmful sunscreen ingredients—an occurrence increasing in the 1960s sunbathing era—is a significant factor in the skin cancer epidemic. Their addition to the epidermis layer create an environment of excessive oxidative radicals, as the sun in the presence of oxygen further oxidizes these synthetic molecules without nature's balancing molecules. At the bare minimum these chemicals substantially increase the toxic burden within skin cells. This burden would either minimize the body's ability to neutralize oxidizing effects of the sun's radiation, intensifying the sun's oxidizing factor. This intensification would essentially convert the sun's rays from therapeutic to dangerous.

Prior to the prevalent use sunscreen, sun-worshipers ceremoniously applied sun lotion onto their skin. This was meant to intensify the sun's tanning effects. Many other chemicals were used in this process, including various hydrocarbons, which revert to oxidized radicals when exposed to the sun. In addition to sun lotion, sunbathers also apply various other chemical-based lotions onto the skin, both as a matter of course and following sunburn. These various moisturizing lotions also contain a variety of synthetic chemicals.

Living Thermals

Solar flares and subsequent proton storms provide a periodic and waveform relationship between the sun and its production of thermal radiation. Cyclical solar flares erupt on the surface of the sun, sending various forms of radiation such as x-ray flares and proton storms into its immediate atmosphere. Some of these flares eject out in *coronal masses*, which appear like tentacles reaching out into space. These events hurl electromagnetic waves through the solar system. These projections can create damage for orbiting vehicles and can penetrate the earth's atmosphere, causing power outages and radio blackouts.

In 1843, German astronomer Samuel Schwabe first documented his observations of solar activity, noting that sunspot activity appeared to be periodic and rhythmic. Over many years of subsequent analysis by astrophysicists, it sunspot cycles range from 9 years to 14 years. They average 11.1 years. Within this obvious cyclical behavior, astronomers also noticed other cycles within the amplitude of each cycle. Advances in monitoring technology revealed these flaring explosions—called solar flares—move cyclically with respect to their intensity. When sunspot and solar flare activity are graphed over more than a century, a harmonic butterfly-shaped waveform is revealed.

Russian scientist Alexander Chizhevsky charted the periodic incidence of various societies' historical upheavals from 500 BCE to 1922. Amazingly, some 80% of these uproars—including wars, major riots and other skirmishes—coincided with peak sunspot activity. Research from other scientists has confirmed these correlations, including upheavals after 1922.

While Chizhevsky hypothesized that magnetism was at the root of this relationship, we still have no concrete understanding of this effect.

Recent research has linked a variety of human physical effects with solar storm cycles. Musaev *et al.* (2007), Stoupel *et al.* (2005 and 2007), Marasanov and Matveev (2007), Kinoshameg and Persinger (2004), Vaquero and Gallego (2007), Davis and Lowell (2006), Otsu *et al.* (2006), Ivanovic-Zuvic *et al.* (2006), Berk *et al.* (2006) and others have found strong correlations between solar sunspot cycles and human birth rates, mortality rates, infectious disease epidemics, heart attack mortalities, immunosuppression occurrence, suicide rates, mental disease rates, lung cancer rates, Down syndrome incidence, depression rates, incidences of multiple sclerosis and rheumatoid arthritis occurrence.

The range of effects relating to solar storm activity indicates that certainly the sun's thermal prowess reaches us in a number of ways. For example, as photosynthesis takes place within plants, a transfer of thermal energy also takes place in subtle ways. For example, hot climates appear to support the growth of botanicals that transfer thermal heat to spiciness. One of many examples is *capsicum sp.*, also called chili pepper or cayenne, which grows predominantly in hotter climates. Chili peppers impart a heat that stimulates an increase in metabolism and immune function within the body. This in turn increases core body temperature, and assists the body to purge bacteria, which are also more prevalent in hotter climates.

This same effect is seen with *Piper nigrum* or black pepper, another hot climate plant. Black pepper is recommended by Ayurvedic physicians to increase *pitta* in the digestive tract, stimulating the flow of digestive enzymes and gastrin. This in turn tends to increase appetite, further protecting the body from bacterial infection from water and food. Other hot-weather botanicals that transfer thermal heat include *garlic, ginger, cumin,* and *paprika.* Each of these will thrive in sunny, warm climates, while being functional antibiotics.

The ability of the sun to transfer thermal energy is not foreign to us. Most of us have sat down in a nice sun-warmed chair, or put on a sun-warmed hat. Most of us have also witnessed the heat inside a building in the summer, as the hot thermal waveforms of the sun beat down upon the materials of the outer walls of the building. This transfer of thermal radiation takes place as the thermal waveforms are exchanged from one molecule to the other throughout the material. This begins with the surface molecules and—depending upon the insulative or conductive nature of those molecular bonds—is transferred through to the inner layers of the material. If the molecules are good conductors of thermal waveforms, the summer sun will make the material too hot to even touch. If the molecules have good insulating ability, the material will not feel very hot at all.

There are several ways materials may transfer thermal energy. Radiation is the most efficient transfer. This usually takes place through the atmosphere and through space, but it can also take place through other types of mediums, including water and some types of solids. Thermal waveforms may also be transferred through *conduction*. Conduction takes place when one molecule is hotter than its neighboring molecule. Thermal energy tends to flow from hotter to colder through solids during conduction. Thermal energy may also be transferred through *convection*. Convection is primarily the same action as conduction, but it takes place primarily within fluids. In convection, thermal energy tends to transfer from hotter molecules to cooler ones in an effort to equalize the thermal energy equally among the fluid.

The human body also radiates thermal energy. The human body radiates from about 100 to 500 watts continuously, depending upon the person's age, health, and metabolism. A younger, more active person will typically radiate thermal waveforms at a higher level. During intense activity, a person's body may radiate well over 500 watts of thermal heat. Hugging is a means of transferring this radiating heat from one person to another. As one person's heat combines with another's, the body may begin to sweat. This becomes self-evident in a crowded room, bus, or railcar. As bodies fill the room, their radiating thermal energy increases the room temperature substantially. For this reason, we often see people sweating in a crowded bus or train, regardless of the temperature outside.

If the room temperature is cooler, the body will be working to maintain its core temperature. It will thus radiate less heat. If we were to chart the body's thermal radiation, we would find a cyclical pattern evident. We would find that during the nighttime and during sleep, radiation heat decreases. It then increases dramatically in the morning, rising and falling through the day with the body's metabolic cortisol and thyroid hormone cycles. Typically, we find body temperatures gain through the morning; peaking and descending by around ten to eleven AM. After dipping into the afternoon, body temperatures begin to rise substantially in the later afternoon. During these heightened periods—the morning and the later afternoon—humans tend to perform more activity. Many exercise during these higher morning and afternoon temperature cycles for this reason. After night falls, we begin to radiate less, as our core body temperature supports the lower metabolism of sleep.

Physical radiation cycles tend to be about 24-25 hours in period. There are other body thermal cycles to consider as well. Most mammals tend to produce more heat during the spring and fall. During these periods, core metabolism is up. During the peak of the summer solar cycle, the body will slow down to try to reduce heat. To reduce heat, the body's radiative thermal temperature will increase. In the winter, most mammals tend to get cozy in their hideaways, and perform less strenuous outdoor activity. Their metabolism tends to slow down and they tend to sleep more. Black bears, for example, slow down quite a bit. They sleep for long periods, and eat almost nothing during the winter. Though they do not "hibernate" as we used to think they did—sleeping the entire winter—they do indeed slow down quite a bit, and sleep more. During these 'down periods,' body heat is preserved, and radiative heat is thus reduced.

This cycling pattern of the physical body is not unlike the solar sunspot cycles of about every eleven years. During one part of the cycle the sun's radiative thermal heat is reduced, and during another part of the cycle the sun's radiation of thermal heat is intensely increased. Could the sun's body be much different from our physical bodies? They both radiate thermal energy. They both undergo periodic cycling in their radiation periods. They both tend to have a birth and death—a definite lifespan. They body have periods of growth and periods of declination or aging. The nebula of a dying solar star is not unlike the graying and wrinkling of a once-active physical body.

The Clockworks

The human body is driven by a number of internal and external clockworks that in the end must be synchronized with the sun to enable health. There are four types of basic human cycles: *Circadian (circa*=about; *dian*=a day)—more or less in the range of a day; *Ultradian*—less than a day; *Infradian*—more than a day; and *Circannual*—in the range of a year.

Over the past century researchers have been hunting down a possible source for the body's biological clockworks. In 1929 it was two Harvard researchers John Fulton and Percival Bailey who observed disrupted sleep rhythms among patients with hypothalamus lesions. In 1958 Harvard's Dr. Woody Hastings illustrated how *G. polyedra,* the marine species which lights up the oceans at night, were seemingly cued by the sun's path. Exposing the tiny plant to various light pulses at different times, Dr. Hastings concluded a biological mechanism must be responding to the sun and switching on and off its cell illumination process.

Following this research began a focused hunt to corner the possible biological source of the body's rhythmic behavior. While researchers easily accepted plants and many animals possessing biological clockworks related to light, controversy took hold when researchers from Germany's Max Planck Institute reported the human system did not have a light-driven clock. Charles Czeisler, M.D., PhD, an eminent Harvard sleep researcher for several decades with over 180 research papers under his belt, questioned that research. Upon visiting the Planck facility he found although the subjects' outside lighting was controlled, they were still able to switch on and off indoor lights. As various light studies done elsewhere confirmed humans' clockworks responded to light, the variance provided by the light switch affected the study's results.

While the early *Max Planck Institute* study was questioned, in 1985 *Max Planck's* Dr. Rutger Wever monitored temperature cycles with daily cycles of 21 and 28 hours instead of the assumed 25-hour daily cycle, and found that temperature rhythms quickly adjusted to the new light schedule.

In 1972 University of Chicago researcher Dr. Robert Moore dropped radioactive label material in rats' eyes and traced its pathways between the visual neurons to a two small clusters of neurons deep within the hypothalamus. From this it was surmised that these two centrally-located pinhead-sized clusters of about 10,000 nerve cells called the *suprachiasmatic nuclei* (or SCN cells) were the physical biological clock researchers had been looking for. Over the next few years, Dr. Moore, together with Lenn and Beebe, traced synaptic contacts with retinal afferent dendrites among young rats, confirming the formation of SCN entrainment with light (Lenn *et al.* 1977).

For the next two decades, Moore and other researchers such as Dr. Charles Weitz, Dr. David Welsch and Dr. Eric Herzog investigated these SCN cells from various aspects. It was thought without this pair of SCN cells animals' body rhythms would break down into chaotic patterns. Individual SCN cells were tested. It was found that each cell seems to have its own independent clockworks, yet still synchronizes as a group. The genetic structures of the SCN cells were further investigated, indicating several possible genetic switching mechanisms (Buijs *et al.* 2006).

In 1970, Yamaoka reported finding SCN cells in the region of the thyroid gland. Over the next few years, enough dissecting had been done to demonstrate that SCN cells exist throughout the human body. In 1980 Dr. Czeisler and his associates (Lydic *et al.* 1980) reported the suprachiasmatic nuclei in the hypothalamus of primates and other mammals was homologous (similar in constitution) with those found in human fetal, infant and adult brains.

Further research has concluded these cells also exist in humans and responsive to light as well as selected mRNA and prostaglandins. It was found they primarily exist within the hypothalamus. The SCN cells appear to connect the hypothalamus with the activities of the pineal gland—a small conical structure lying above the posterior end of the third ventricle which receives impulses from the optic nerves. The SCN is implicated in the secretion of most if not all the major hormones and neurotransmitters within the body stimulated by the hypothalamic hormones. The synchronization between these SCN cells and light pulses impacting the pineal gland appeared to take on a double-neuron oscillating mechanism guided by a combination of light and genetic expression (Fukada 2002).

As the research on the genetic side of the SCN cell activity unfolded it became evident the activities of the thousands of oscillating nerve cells making up the SCN cell region are somehow defined by a set of clock genes (Kalsbeek *et al.* 2006). Recent genetic research has identified several clockwork genes. The central CLOCK gene has been identified as 3111T/C; rs1801260 (Benedetti *et al.* 2007) and APRR9 in plants. Other clockworks genes have been identified as BMAL, PER and CRY (cryptochrome-12q23q24.1) and DEC genes (Gomez-Abellan 2007, Kato *et al.* 2006). This gene has been identified in that it is expressed because of light and/or circadian rhythm. Studies on light inducement have confirmed that these genes respond to light, and exist not only in the SCN but also within various cells around the body, notably testes, ovaries, kidneys, and most other organs. They have also been found in adipose cells, various nerve cells, even cartilage cells. Each of these genes has their special expression. The PER expression in the SCN for example, are increased by light exposure during the night, but not disturbed by increased light exposure during the day (Shearman *et al.* 1997).

The work of Dr. H. Okamura (2005) from the Kobe University School of Medicine has revealed that these clock cell genes are located throughout most of the body's major tissue and organ systems. Apparently the genetic expressions of SCN cell oscillations are coupled with the independent clockwork genes in these various locations of the body, and this coupling (or resonating) of genetic expression synchronizes the pacing of the activities of these cell locations.

Through the process of elimination and treatment to animals a number of studies have given us reason to believe damage or mutation of these clockwork genes can result in various disease models.

For example, damage to the Per1 and Per2 genes has been linked with a number of human cancer types (Chen-Goodspeed and Cheng 2007).

The clockwork genes have also been linked to the rhythmic expression of hormone release. These genes are now seen as key regulators (or mediators) for all metabolic processes. Mutations of human clock genes have been implicated in metabolic syndrome (Gomez-Abellan *et al.* 2007), and bone marrow CD34 immune cells (Tsinkalovsky *et al.* 2007), depression (Benedetti *et al.* 2007), glutathione release (Igarashi *et al.* 2007) and various others. Clock genes are disrupted in mania-like behavior (Roybal *et al.* 2007). The clock genes have also been observed mediating expression of the plasminogen activator inhibitor, yielding a greater risk of heart attack (Chong *et al.* 2006). In 2005, researchers from the University of Pittsburgh's School of Medicine found that bipolar disorder was linked to a disruption of clock genetic expression (Mansour 2005).

The clock genes apparently correlate information from the SCN cells with various peripheral oscillating cues from around the body. One of the biggest synchronizing mechanisms for these genes along with light is feeding schedules. There is an apparent entrainment to feeding cycles and energy metabolism connected with the SCN/CLOCK gene interaction (Mendoza 2007). Alcohol consumption also appears to substantially alter the expression patterns of the clock genes. The PER genes—especially among brain cells—are significantly affected by alcohol consumption (Spanagel *et al.* 2005).

Recent research has revealed that the clock genes work by way of a feedback loop using transcription and translation in alternating steps, with protein phosphorylation via kinases reactivating expression loops of the WC-1 and WC-2 proteins, turning on and off through the induction of light to heat (Lakin-Thomas 2006). Cry1 apparently mediates CLOCK/Bmal1 complex repression, which sets up a feedback response (Sato *et al.* 2006). Further research has uncovered the potential of pros-taglandin-2 as an activation-switch for resetting these genes, leading researchers to propose a connection between pain and biological clocks. These feedback loops have also been referred to as rhythmic, with a con-served control of gene transcription regulation (Hardin 2004).

It also appears that these SCN neurons are coupled to other cells, and through a signaling process linked to glucocorticoid production, sympa-thetic nerve activity and other metabolic systems are activated due to the genetic alteration by SCN by light and other oscillators. Harmonically, other organs also have their own macro oscillating mechanisms, which turn on and off the various functions of the particular organ.

One of the key oscillating expressions of these clock genes and SCN cells the rhythmic stimulation of key neurotransmitter/hormones mela-tonin and serotonin. The major on and off switches for these biochemicals include of course light along with another switching (inhibi-tory) neurotransmitter messenger called GABA (Gamma-aminobutyric acid) (Perreau-Lenz *et al.* 2005). Meanwhile a dense network of serotonin neurons connects with the SCN cell area. Light-driven oscillations of the SCN cells thus stimulate rhythmic serotonin release through these neu-rons' activity (Moore and Speh 2004).

The photoreceptor signaling process is still somewhat mysterious, but it appears that a protein called malanopsin is involved in a photo pigmentation process stimulating SCN cells. A signaling transduction pathway has been proposed with respect to the gene phosphorylation system: Glu-Ca2+-CaMKII-nNOS-GC-cGMP-cGK—>—>clock genes.

In 2005, Dr. Erik Herzog and Sara Aton discovered a peptide that lies between SCN neurons, seemingly polarizing their rhythmic oscillations. The *vasoactive intestinal polypeptide,* named because it was found in the gut, is apparently produced by the SCN cells. Dr. Herzog proposes the VIP lying between SCN neurons *"is like a rubber band between the pendulums of two grandfather clocks, helping to synchronize their timing"* (Aton *et al.* 2005).

Dr. Paolo Sassone-Corsi's 2006 studies at the University of California led to an understanding that the CLOCK gene functions more like an enzyme in the switching process. A year later, Dr. Sassone-Corsi's research found that a single amino acid within the BAL1 protein provides the initial switching signal when it undergoes a single modification. That amino acid bonding modification stimulates the rest of the body clock's switching systems (Hirayama *et al.* 2007).

As researchers delved deeper into the mystery of the body clock mechanisms, they have found increasingly complex signaling processes taking place on a myriad of levels. A continuing mystery for example is blind people also have extremely strong biological clocks and they too run on the sun's clockworks. It has been proposed that retinal cells contain another type of neural photoreceptor system—one receiving electromagnetic rhythms outside the visual relay system. Another theory proposes a few photoreceptor cells remain in blind people allowing them to stimulate the pineal/SCN system while not having enough activity to stimulate the LGN/visual cortex.

A bigger mystery is how these various clockwork genes and cells communicate and synchronize throughout the body. In 2004, Dr. David Welsh and a team of researchers observed individual fibroblast cells—which will differentiate into connective tissue cells, osteoblasts and other cells—also contain self-regulating circadian clockwork genes, synchronizing their behavior with the rest of the rhythmic activities of the body. If we could imagine a house full of thousands of clocks, each having different timing mechanisms and different alarms and functions, we might come close to the fantastic synchronization necessary to coordinate all of these various clocks.

Certainly the rhythms of all these clockworks are synchronous. A good example is the cycle between cortisol and melatonin. While driven by different endocrine systems, cortisol (known to increase metabolism) tends to increase and peak as melatonin (known to decrease metabolism and induce sleep) levels recede. Then as cortisol levels recede (for most people it's mid-evening) melatonin levels increase, peaking around midnight, just as the cortisol levels have wound themselves down. While seemingly independent cycles driven by seemingly independent endocrine systems, their cycles synchronize with each other. Nonetheless, they must ultimately be synchronized with the clockworks and daily path of the sun.

Harmonization: Solar Healing

❖ Research has indicated that the visible light spectrum (400-700 nm) received from the sun through both the eyes and the skin increases the body's immune response. As light is received through the retina, its energy is delivered to the LGM and visual cortex through transduction while being delivered to the suprachiasmatic nucleus in the hypothalamus. This stimulates the release of hypothalamic-pituitary hormones. Light also stimulates the pineal gland directly, stimulating a cascade of hormones and neurotransmitters through the pituitary gland. Melatonin, norepinephrine, and acetylcholine secretions (the latter two known for stress response) decrease, while cortisol, serotonin, GABA and dopamine secretions increase with increased sunlight. These latter three are noted for relaxation and calmness, while cortisol is related to inflammation reduction. All are related directly or indirectly to immune response.

❖ Visible light also penetrates the epidermal and dermal skin layers, interacting directly with circulating lymphocytes. Visible light increases the body's immune responsiveness, assuming the body is not immune-suppressed (Roberts 2000).

❖ Natural vitamin D from the sun is extremely important to all of the body's tissues—and an essential part of the immune system. In a revealing statement from a review of studies from the Universitatsklinikum des Saarlandes in Homburg Germany (Reichrath 2006), *"there is convincing evidence that the protective effect of less intense solar ultraviolet radiation outweighs its mutagenic effects."*

❖ We might closely consider the mutagenic effects ultraviolet-B has upon plants. (Yes, plant genes are also damaged by ultraviolet-B radiation.) However, healthy plants also have natural repair mechanisms that mitigate this genetic damage. This is quite evident when we notice that while some plants remain green and healthy as they sit in the intense sun all day long, other plants of the same species—poorly watered or poorly fertilized—will turn brown in the mid-day sun. On the molecular and cellular level, this browning effect would be comparable to the sun's exposure effects among poorly nourished human skin cells (Zaets *et al.* 2006).

❖ Ultraviolet-A has been shown to directly assist the immune system by aiding the repair of DNA damage. This effect was illustrated in a series of studies on tiny unicellular paramecia led by Dr. Joan Smith-Sonneborn, a University of Wyoming professor. While bursts of unscreened ultraviolet-C caused DNA damage, ultraviolet-A exposure reversed the damage. Going beyond the reversal of genetic damage, additional exposure to ultraviolet-A radiation extended the paramecia's life span as much as fifty percent (Smith-Sonneborn 1979; Rodermel and Smith-Sonneborn 1977).

❖ At the same time, natural sunlight helps the body produce hormones and neurotransmitters that encourage the repair of damage done by oxidative radicals. Natural sunlight also warms the body with thermal radiation, stimulating the production of key enzymes that stimulate the immune system and its detoxification routines. These combined mechanisms of the sun's oxidizing effects and the stimulation of the immune system create a natural balance between radicals and antioxidants, leaving the body's immune system stimulated.

❖ These mechanisms can be supported with a healthy supply of *carotenoids, flavinoids, sterols* and other protective *phytonutrients* from a natural plant-based diet. Interestingly, some of these are precisely the same protective agents that also help protect plant cells from sun damage. The immunostimulatory effects of sunlight combined with these *phytochemicals* neutralize the effects of radical oxidative species and help repair any genetic damage before significant mutations can take place. Other important phyto-nutritional components include *beta-carotene, zeaxanthin* and *lycopene* from veggies, carrots, and tomatoes; *anthocyanins* from oats, red berries and cherries; *lignans* found in various grains and beans, and other phytonutrients found in fruits, vegetables, beans and nuts. In addition to these antioxidant foods, a good food-based multi-vitamin can provide an effective insurance policy.

❖ The lowly *eggplant* provides a good example of the balance phytochemicals provide to the skin. Eggplants contain a class of phytochemicals called *solasodine glycosides*. One of these, called *solamargine*, binds to sugar receptors (*endogenous endocytic lectins*) inside of cancer cells. This molecular binding results in solamargine being drawn into the cell's lysosomes, causing the lysosome to rupture. This rupturing causes the death of the cancer cell (Cham 1994). This is one phytonutrient out of many providing anticarcinogenic effects.

❖ Imbalanced oil consumption or an over-abundance of trans-fats can create a weak cell membrane. The stronger the bonds in phospholipid cell membranes, the more protection the cells will have. A healthy combination of DHA, ALA, GLA and omega-6 oils can help.

❖ In a nutritionally deficient system, wavelengths of ultraviolet-B and ultraviolet-A rays further damage cells weakened through contact with toxins or other stressors—creating oxidative radicals. Suppressed immune systems are one of the key markers for skin cancer risk.

❖ Even in a healthy environment, the body is subject to many small cancers on a daily basis. These form from the various enzymatic and metabolic processes of the body, combined with environmental input. A healthy immune system will remove these before they grow.

❖ Sunlight stimulates the production of natural skin oils and *melanin*, which both further the protection of our skin from damage from the sun. An increase in the volume and activity of melanocytes assists the cells in the production of melanin. Melanin acts as a filter for the sun, allowing the body to draw upon the sun's useful radiation with less oxidative damage. For this reason, traditional cultures with more sun exposure—living closer to the equators—typically have browner skin; while those cultures determined to survive in the colder climes became deficient in melanin over time. This situation is easily remedied, however. A gradual increase in sun exposure over a period of time will radically increase the skin cells' melanin levels in most skin types—reducing damage from ultraviolet oxidation.

❖ The sun is also an effective antiseptic. Various studies have shown the sun to be antimicrobial in many respects. Many bacteria and fungi are intolerant to the rays of the sun. Some are overheated by the sun's thermal rays. Many others are destroyed by the sun's infrared radiation (Piluso and Moffat-Smith 2006). These include certain molds and bacteria, which can significantly multiply in a dark, wet environment.

❖ Our body clock harmonizes to the sun through the suprachiasmatic nuclei cells (SCN) cells located throughout the body, with a greater density around and within the pineal gland. This sets our various hormonal and neurotransmitter cycles (such as cortisol, melatonin, growth hormone, thyroid hormones, testosterone, estrogen, and many others) throughout our body on a daily basis.

❖ Reasonable sun exposure can mean simply taking a walk or sitting outside, allowing the sun's rays to shine upon our heads, faces, arms, and/or hands.

❖ As we resonate with the sun's radiation and reflection, we balance the destructive waveforms stemming from the various sources of stress.

Sunshine helps relax the nerves and stimulates the production of various relaxing neurotransmitters such as serotonin and dopamine.

❖ Ayurveda tells us that letting the sunshine strike our faces and necks for a few minutes each day boosts the fifth (throat) and sixth (brow) chakra energy centers—allowing us to better focus on our objectives and relationships. Allowing the afternoon or morning sunshine to strike our eyelids with our eyes closed transmits healing thermal radiation into our eyes without damaging our retinas. This can also help sore sclera by increasing blood and nutrient flow to the eyes.

❖ According to Ayurveda, the sun stimulates the *pitta* quality. This means sunshine is an excellent therapy for persistent chills or sensations of coldness throughout the body. Sunshine is also called for weak digestion, low energy, fatigue, reduced circulation, and dull aching pains. Exposing those parts of the body to sunlight each day—preferably in the mid-morning or late afternoon—will stimulate the body's natural healing response.

❖ Exposure is best without any creams or lotions. Many lotions have toxic ingredients. Even the healthiest oils and lotions can become oxidized or otherwise degraded by the sun's rays.

❖ We can gradually build up time exposed daily to the sun. Initial morning or late afternoon exposures of 15-20 minutes can be built up to 45 minutes to an hour a day or more once melanin levels begin to build. This can easily be observed by a slowly developing brown skin color. Skin type, previous sun exposure, time of day and location are determining factors in how much exposure is healthy. In no case should the exposure result in a pink skin color. This should be considered sunburn, and is not advisable.

❖ Walking in the sun without a hat for reasonable periods with the head exposed to the sun even if through a matting of hair delivers valuable thermal and infrared waveforms through the hair and scalp, radiating thermal energy through to brain neurons. This opens our neural system to the electromagnetic rhythms contained within the sun's waveforms—stimulating the nervous system and calming the mind. For those who have trouble walking or getting outside, the sun can be sought through a window at the appropriate time of day. If it is winter we can still make it outside—sunny day or not—to receive the blessings of the sun. Cloudy or sunny, we still receive most of the sun's waveforms. The sun's ultraviolet-A and ultraviolet-B waves will penetrate cloud cover, just as will infrared rays. Heavier cloud cover will slightly reduce the intensity of all three waveforms, especially the infrared waves. This also means even on a cloudy day we should limit our mid-day sun without hats or clothing in the summer time.

❖ Spending time out of doors in natural light is critical to cognition and moods. The research has illustrated that learning increases in school-rooms with more natural light. Behavior and moods are more balanced and relaxed when we spend a significant time outside. Should we go for a walk in the sun to have an important discussion or difficult debate, we may find solutions easier. Outside, we can better see the bigger picture—and better prioritize things. Under the waves of nature, we can more easily understand just how unimportant our issues are.

❖ Obtaining thermal energy from the sun is essential. When the sun comes up, our bodies need the sun's thermal radiation to maintain many of our metabolic functions. As the day pulses on, we increas-ingly find that our core body temperature relates directly to our radiative environment. A constant indoor environment leads to slower metabolism and a colder core body temperature. This especially holds true during colder weather or changing weather. During the spring with increased radiation exposure, metabolism increases dramatically. In the heat of summer, the body will often slow down in an effort to lower the body core temperature to stay cooler.

❖ Finding a balanced core temperature requires us to seasonally make adjustments between sun and shade, and indoor and outdoor activi-ties. These adjustments provide rhythmic activity to our lives. As we cyclically adjust with the seasons in terms of our outdoor time, we will be in effect synchronizing our body to nature's cycles.

❖ Finding this thermal balance is quite simple. This might consist merely of taking a daily walk through a natural environment of trees and meadows, or working outside in the garden daily. During the hot summers we can walk under trees or in the morning to be partially shaded and cooled. By walking in and out of the trees or as the sun begins to rise into the sky, we fall into a natural cycling of sun with in-termittent shade. During the cooler winter months, we can seek more direct sunlight and less shade—the trees help as many have lost their leaves. Using nature's shading synchronizes our exposure with the sun's intensity and availability, as it is less intense and less available in the winter, and more intense in the summer.

❖ For many, summers bring more sun exposure, and winters bring al-most perpetual indoor activity. This is actually opposite of what is necessary. We can try to spend more time outside during the winter and even in the mid-day, as the sun's intensity is less. Our summer outdoor activities can be focused upon the mornings and early evenings when its cooler and the sun is lower.

❖ The *far infrared sauna* can provide an alternate form of thermal therapy, especially during cold weather. It is also good for chronic pain, fatigue, congestion, muscle aches and of course toxicity. Because it improves circulation it can stimulate the body's healing response. While the Finnish sauna uses a heat element, the far infrared system uses a series of infrared lamps to heat the body. Far infrared saunas are effective because far infrared waves immediately penetrate the dermal layer, expanding micro-capillaries, relaxing muscles and soothing nerves. A cool or cold bath after the sauna can stimulate the immune system, improve cardiovascular function, and increase lymphatic circulation. Increased water intake is vital during and directly following the sauna.

❖ Should we spend a significant time indoors, it might help somewhat to install full-spectrum compact fluorescent light bulbs. These environmentally superior, low energy bulbs emit a greater spectrum of light than do incandescent bulbs. Newer versions have become reasonably priced. They will not provide anywhere near the full-spectrum needed from the sun, but they can help improve moods and color perception. Best if they contain high ballast frequency devices to reduce their flicker frequency.

❖ Sun gazing during sunrise and sunset can have a rejuvenating effect upon the eyes if done correctly. Watching the sunset and sunrise has been a recommendation for eyesight problems among many traditional medicines including Ayurvedic, American Indian, and Greek disciplines. The filtering of the earth's atmosphere during these two times of the day filters the rays considered damaging during the midday sun. Both the ultraviolet and the infrared spectrum are almost completely blocked an hour before the sunset and an hour after sunrise, depending of course depends upon location and cloud cover. We can double-check our local weather station website or newspaper for the hourly ultraviolet index. In addition, inexpensive photo monitoring cards can indicate the safety of any specific location.

❖ Evening or morning sunset/rise gazing is best done intermittently, accompanied by blinking and looking away at other natural images. Initially we can point the eyes and face at the setting or rising sun with eyes closed. After doing this for a minute, the eyes can slowly be opened, while blinking frequently—about once a second. After 10-15 seconds, the eyes can then wander over other images until the image of the sun has disappeared from the eyelids. As we progress over time, we can gradually repeat this process two or three times. Gazing at the setting or rising sun should always be accompanied with frequent blinking and periodic looking away. During the first few weeks or months we can start with only a minute or two, gradually increasing to a comfortable 5-10 minutes. We can consider the strength of our eyes, our body's age and the body's general health. We may want to consult with our eye specialist prior to undertaking gazing.

❖ After a few months of consistent sunset/rise gazing every day we may discover many subtle and obvious benefits. We may discover our eyesight actually improving. We may perceive colors as brighter. We may also see a change in our moods—possibly a greater sense of optimism. We may feel more relaxed and calm. The synchronizing effect between our SCN cells, our pituitary, hypothalamus, and visual cortex may also lead us to sleep better and have a general reduction in tension. Many of these effects are provided by the resonance between the sixth energy center or chakra and the rays of the sun.

❖ These two periods of the day—sunrise and sunset—have been considered crucial times for meditative processes. Prayer, hymns, chanting and general meditation are extremely effective at sunset and sunrise. These activities can be intermingled with the process of sunset/rise gazing as mentioned above, as we appreciate the consciousness pervading our universe.

❖ It is not appropriate to stare at the sun well after sunrise or well before sunset. Blindness and retinal damage can result. We might want to be careful during a solar eclipse as well. Starring at a solar eclipse can damage our eyes as much as looking directly at the mid-day sun.

❖ Moon gazing and star gazing are also considered helpful for eyesight, and can be meditative as well. Most of us have been awestruck by the beauty of the skies on a clear night. Finding that darkened place to spy at the distant galaxies, planets, and stars can provide us a view of that humbling expanse of our universe.

❖ Letting the warmth of the sun's rays strike our feet and legs increases circulation, especially useful in areas of the body where the veins tend to weaken with age. The effects of the sun's radiation can increase circulation and organ health. In addition, we can walk barefoot on an earthen surface heated by the sun. This might include warm beach sand, grass, or even better, smooth rock. The thermal rhythms stored within these elements will radiate up through the soles of our feet, delivering a combination of the sun's and earth's waveforms.

❖ With a healthy diet and good sense of solar timing, our skin type, and length of exposure, sun risk can be minimal. We should try to get sun every day for at least twenty to thirty minutes during the summer and more during the rest of the year. This minimal level of sun exposure can maintain our body's minimum vitamin D levels, assuming we expose our skin to the direct sunlight during this time.

❖ Sun burning should be avoided. The healthiest sunscreen is a hat, clothing or a shade tree. Any chemical sunscreen should be avoided, unless cover and clothes are not possible. If we must wear sunscreen, we can choose natural versions, preferably without PABA. Natural sunscreen lotions with added vitamins and botanicals are now available. These added antioxidants may help neutralize free radicals from sun exposure at the source.

❖ If our bodies get sun burnt, the best agent for relief and healing speed is *aloe vera gel*. Natural aloe is easy to grow in any sunny environment—another synchrony of nature. Aloe is used by breaking off or cutting a small portion of leaf off the plant and peeling back its skin. We can rub the gel right over our skin. The juice of the aloe is not as bioactive as the gel. This milky plant's gel holds much of its active constituents. Natural aloe gel contains *aloectin, anthraquinones, polysaccharides, resins,* and *tannins.* These phytochemicals work synergistically to speed healing, neutralize oxidized radicals, and soothe pain.

❖ Also good for sunburn is the gel of the *cucumber.* Cucumber gel contains a unique blend of *elaterin resin, starches, lignin, saline matter,* and *minerals.* These work together to provide a healing and antioxidant effect on the skin. Fresh slices can be laid onto the skin, or crushed into a lotion or simply rubbed onto the skin.

❖ *Lemon* can also be helpful for sunburn. The combination of *citric acid, citral, hesperidin sugars* and *limonenes* give lemon a soothing and antioxidant effect upon the skin. Freshly squeezed lemons can be diluted in water 50/50 and sponged right onto the skin for immediate relief. The effects of these phytochemicals illustrate the harmonic existing within nature.

Chapter Seven

Electromagnetic Radiation

Today's electromagnetic environment is quite different that the environment we lived in even fifty years ago. While our environment is pulsing with nature's electromagnetic radiation, humankind's electronic revolution has injected synthetically-pulsed alternating electromagnetic radiation into our environment. Today's physical environment is now drowning in supercharged alternating currents.

Almost two decades ago the sounding alarm was made, through such books as Dr. Robert Becker's _The Body Electric_ and Ellen Sugarman's _Warning, The Electricity Around You May be Hazardous To Your Health_ that electromagnetic radiation from alternating current sources has a negative physiological affect upon the body. This fact is not been challenged. The question posed—and the heart of the controversy—is whether and to what degree electromagnetic radiation from alternating current sources is dangerous to the health of the human body. Furthermore, while dangerous effects are important, equally important is what extent electromagnetic radiation from alternating currents may be effecting our vitality and general wellness. This latter question is often missed in the flurry of research probing questions of whether this type of EMR is carcinogenic.

Today our environment is embedded with a plethora of new electromagnetic pulses bombarding our bodies. We are surrounded by electronic appliances that emit varying degrees of electromagnetic pulses. Our buildings are wired with EMR-emitting circuits and breaker systems. Most of us spend multiple hours in front of computers and televisions, absorbing electromagnetic pulses. The battery and transmission systems built into our cell phones, iPods® and laptops bring our skin in direct contact with EMR-emitting appliances.

With billions of us partaking in at least some of these activities daily without obvious negative effects, some wonder what all the fuss is about, and why some health proponents are still arguing the case against EMFs.

There are two basic forms of radiation to consider: _Ionizing radiation_ and _non-ionizing radiation_. According to a 2005 report by the _National Academy of Science_ on low levels of ionizing radiation, about 82% of America's ionizing radiation comes from natural sources: the earth, sun, space, food and the air. The rest—18%—comes from human origin. The bulk of this fabricated radiation comes from x-rays and nuclear medicine. This accounts for close to 80% of the 18%. Other elements like consumer goods, toxic water, occupational exposure, and nuclear power account for the rest of the ionizing radiation exposure according to this report.

Radiation Ionization

Ionizing radiation is typically defined as electromagnetic radiation capable of disrupting atomic, molecular or biochemical bonds. This disruption takes place through an interference of waveforms between the ionizing radiation and the waveforms of atomic or molecular orbital bonds. As this interference is likely to cause the atom or molecule to lose electrons, ions are likely to develop as a result. These ions can often turn to oxidative species or otherwise imbalanced molecular species. Should ionizing radiation with enough intensity impact the physical body, it can result in cell injury or mutagenic damage. Various natural and synthetic radiation forms are considered ionizing. Natural ionizing radiation includes portions of solar radiation, x-rays, cosmic-rays and gamma-rays. Fire can also cause ionizing radiation at high temperatures if the radiation comes close enough. Synthetic versions of ionizing radiation include electrically-produced x-rays, CAT-scans, mass accelerator emissions and a host of other electromagnetic radiation produced through alternating current.

Non-ionizing radiation also can be split into natural and synthetic versions. Natural versions include sound, light and radiowaves. Most natural non-ionizing radiation can also be synthetically produced. For example, sound may be digitally produced through the manipulation of alternating current by stereo receivers and speakers. This effect utilizes electrical semi-conduction. Most scientists also categorize radiation from electrical power lines, electricity generating or transfer stations, appliances, cell phones, cell towers and other shielded electricity currents as non-ionizing radiation. Microwaves are also considered to be non-ionizing. Most assume non-ionizing radiation is not harmful. This assumption, however, has undergone debate over the past few decades.

In June of 2005, the *National Academy of Science*, after studying most of the available research regarding non-ionizing radiation, concluded that even low doses below 100 milliseiverts were potentially harmful to humans and could cause a number of disorders from solid cancer or leukemia. This jolted the scientific community, because for many years researchers thought that small doses of non-ionizing radiation were not that harmful.

A rem is one unit of radiation dose in roentgens. An mrem is one thousandth of a rem. One hundred rem equals one sievert. One sievert equals one thousand milliseiverts.

Ten sieverts (10,000 mSv) will cause immediate illness and death within a few weeks. One to ten sieverts will cause severe radiation sickness, and the possibility of death. Above 100 mSv there is a probability of cancer, and 50 mSv is the lowest dose that has been established as cancer-causing. 20 mSv per year has been established as the limit for radiological workers. About one to three mSv per year is the typical background radiation received from natural sources, depending upon our location and surroundings. About .2 to .7 mSv per year comes from air. Soil sources are responsible for about .8 mSv. Cosmic rays give off about .22 mSv per year. Japanese holocaust victims received .1 Sv to 5 Sv from the bomb.

Our total radiation dose is a thus a combination of natural sources and those emitted by our artificial electromagnetic empire. A report from the *Hiroshima International Council for Health Care of the Radiation-Exposed* noted that the world's average radiation dose from natural radiation sources is 2.4 mSv. However they also noted that Japan's natural radiation average is comparably low at 1.2 mSv, while Japan's average radiation dose from medical radiation is higher than average, at 2.4 mSv. This gives Japan a significantly higher radiation average of 3.6 mSv.

UK's *National Radiological Protection Board* estimates that the national radiation exposure in Britain for the average person is 2.6 mSv, with an estimated 50% coming from radon gas, 11.5% coming from foods and drinks, 14% coming from gamma rays, 10% coming from cosmic rays and 14% originating from appliances—primarily medical equipment.

Recent research indicates that radiation from medical equipment is increasing. This is primarily driven by the growing use of CT scans, which generate a larger dose of radiation than the more traditional x-rays. About sixty-two million CT scans are given a year now in the U.S., contrasting to about three million per year in 1980. Brenner and Hall (2007) reported in the *New England Journal of Medicine* that a third of CT scans given today are unnecessary. The article also estimated that between one and two percent of all cancers are caused by CT scan radiation exposure.

In contrast, the maximum radiation a nuclear electricity generating plant will emit at the perimeter fence is about .05 mSv per year. A set of dental x-rays will render a dose of about .05-.1 mSv. Meanwhile, a CT scan will can render a dose of about 10 mSv—over a hundred times the dose of an x-ray.

A grand electromagnetic human self-experiment is unfolding. Unsuspecting humans and animals are the subjects of this experiment. The findings will be available in a decade or two from now.

Most researchers are quick to say gamma rays—from radon and other natural sources—produce significantly more radiation than do appliances. This might be true for someone with a minimal amount of electrical appliances who rarely visits the hospital and the dentist's office.

The question that persists is whether humankind's synthetic "nonionizing radiation" is as innocuous as is currently assumed.

Power Lines

The American Physical Society, an association of 43,000 physicists said in a 1995 National Policy (95.2) statement *"....no consistent significant link between cancer and power line fields...."* This statement was reaffirmed by the APS council in April of 2005.

Power lines emit electromagnetic radiation at ELF or *extra low frequency* levels. Power lines typically release about 50 hertz of pulsed radiation. As an electric current moves through a wire or appliance, magnetic fields move perpendicular with electricity in a cross pattern. Thus electricity fields form from the strength of the voltage while magnetic fields rise and break away from the electronic waveform's motion. While electricity voltage can shock us or burn the body, magnetic fields have more subtle yet lasting influences upon the body's natural biowave systems—such as brainwaves, biorhythms, hormone production, and so on.

While magnetic influences are difficult to perceive directly, it is apparent they may substantially interrupt our immune systems. Between 1970 and 2000 about fourteen international studies analyzed the potential link between power lines and cancer among children. Eight of those studies showed a link between cancer rates and power line proximity, with four associating power lines with leukemia.

One of the U.S. studies to show a positive link in between cancer took place in 1979 in Denver, led by Dr. Nancy Wertheimer and Ed Leeper (1979). This studied showed a more than double likelihood of cancer among children living within forty meters of a high-voltage line. Another Denver study published in 1988 (Savitz *et al.*) also found a 1.54x odds ratio (OR) positive link in all childhood cancers and high power lines. A Danish study (Olsen *et al.* 1993) also linked general cancer rates (1.5 OR) with power line proximity. A study done in Los Angeles (London *et al.* 1991) showed a 2.15 OR rate, a Swedish study (Feychting and Ahlbom 1992) showed a 3.8 OR risk and a Mexican (Fajardo-Gutierrez *et al.* 1993) study showed 2.63 OR increased rate of leukemia cancer rates among children with close proximity to high-voltage power lines. One Swedish study (Tomenius 1986) showed a 3.7 OR increased risk for central nervous system tumors among children living close to power lines. The Danish study mentioned above also showed a 5.6 OR increased potential of all cancers among children. The other positive link studies showed rates above 1x to 1.5 OR, which are not considered by mainstream science to be statistically significant.

Following the release of these studies, a number of governments took steps to warn housing developers of the potential risks of building close to high frequency power line hubs. In some municipalities across

Europe and the U.S., building departments have even taken steps to dissuade or ban developments close to larger power lines.

Adult cancer studies have yet to illustrate as large a correlation between power line proximity and cancer rates. Still a few have been significant enough to confirm the need for concern. While Werthheimer and Leeper's (1982) studies showed increased rates of all cancers, the 1.28 OR rate was not too significant. However a U.K. study (McDowall 1986) showed a SMR 215 increased rate of lung cancer and a SMR 143 increased risk (SMR 100 or less = no risk) of leukemia. Another study in the U.K (Youngson 1991) showed a statistically insignificant 1.29 OR rate for leukemia and lymphoma and Feychting and Ahlbom's (1992) Swedish study showed a 1.7 OR risk for leukemia subtypes. One significant study was Schreiber *et al.*'s (1993) study that showed a SMR=469 rate for Hodgkin's disease.

It must be noted that these studies are epidemiological. They are population studies where groups living in close-proximity to high frequency power lines are compared with groups living further away. The problems that can occur with these studies focusing on cancer are several. In cancer pathology, there can be a two to twenty year delay between exposure and cancer diagnosis. While some of the populations involved in these studies might have been living in a particular house for many years, most may have only lived there for a year or two at the most.

In addition, some of the studies limited the disease group population, restricting the usefulness of the information. Cancer is seen primarily in the elderly and middle-aged, where there may be a host of various different types of exposures. These would include smoking, alcohol consumption, job-related exposures, chemical toxins, and so on. For this reason these studies can be difficult to weigh against the costs of preventing exposure. The economic issues involving power lines are quite substantial. Relocating schools and families away from high-voltage lines or even relocating power lines comes with a substantial economic cost.

Nonetheless, this is increasingly becoming a problem for both homeowners and utility companies. For example in the mid-nineties the New Jersey Assembly enacted legislation requiring disclosure from home builders of vicinity transmission lines in excess of 240 kilovolts (kV). Other states have followed with real estate disclosure laws for power lines. Lawsuits have followed on power line proximity issues between schools, buyers, builders and utility companies.

One of the problems existing with some of the power line studies is the comparable limits of the distances between households and power lines. For example, is the effect of a transformer 40 meters away significantly different from one 50 meters away? Another difficulty with these epidemiological power line studies is that some of the studies measured utility wire codes (wire thickness) and distance, while other studies used spot physical measurements to determine exposure levels. In addition, there have been a variance of controls related to whether the child was born in the house or moved there recently.

With regard to the significance of the leukemia studies that showed positive link factors, we should consider the incidence of leukemia cancer among the childhood population, which is close to 1 in 10,000. A 2 or 3 OR among a group, unless the size of the groups are in the millions (most of the studies were significantly smaller—in the thousands), would relate to only a small handful of disease cases over the entire study population. If the study group size was five or ten million, then these numbers might be considered more reliable. As the increased rates have been smaller (rather than the 4 or 5 OR rate that appears in many study groups) then the size of the disease group is not considered to be a significant factor with which to judge the quality of the study. To this point, D'Arcy Holman, a professor at the University of Western Australia, calculated that the UK studies' worst projections might mean one extra childhood leukemia death in Western Australia every fifty years (Chapman 2001).

Occupational studies regarding exposure to EMR have shown unclear results with regard to leukemia and cancer (Kheifets *et al.* 2008). However, studies have pointed to the increased risk of amyotrophic lateral sclerosis (ALS) due to EMR exposure (Johansen 2004). Studies on electricians, electric utility line workers and other electrical workers have consistently showed higher rates of leukemia and central nervous system-related cancers. In a 2006 meta-study of fourteen studies by Garcia *et al.* (2008), Alzheimer's disease was associated with chronic occupational EMR exposure.

One of the difficulties with assessing the data on EMR effects is the sheer volume of studies of different types that has been published over the past twenty years. The breadth of variances between the studies of plants, animals, and human response to various degrees of radiation is substantial. Because of this huge base of studies, most researchers have been forced to rely upon various reviews by publications and government agencies to assess the implications of this large base of varying research. These groups have assessed and compared studies to figure out whether there is a correlation between study results, and whether they are significant. Government-sponsored reviews have included the United Kingdom's *National Radiological Protection Board*, the *Associated Universities of Oak Ridge*, the *French National Institute of Health and Medical Research;* councils in Denmark, Sweden, Australia and Canada together with U.S. agencies such as the *Environmental Protection Agency* and the *Department of Transportation.* In addition, the *U.S. National Council on Radiation Protection and Measurements* and the *US National Academy of Sciences* have also put together major reports on EMR research.

A number of respected journals have published reviews of EMR research as well. While some of these studies have found some epidemiological evidence notable, few found conclusive results, and some have presented skeptical views of any significant positive pathological correlation with non-ionizing EMR exposure. A group of these reviews were presented by Savitz (1993) in *Environmental Health Perspectives.* Here no interaction mechanism between EMR fields and biological organisms was found.

An electrical field is substantially different a magnetic field. An electrical field is generated when there is a charge differential between two terminating points, regardless of whether current runs between them. Thus an electric light bulb will still generate an electric field even when it is turned off. This electrical field allows alternating current to run between the two points when the switch is eventually turned on.

A magnetic field is created by a current flowing with electricity. The magnetic field will be emitted outward with perpendicular orientation to the electrical field. However, because magnetic fields have a particular polarity or direction, a current flowing in the opposite direction placed next to the current wire will cancel the magnetic field. Most power cords with double wires (hot and ground for a circuit loop) effectively cancel the magnetic field of the incoming current directly related to the distance between the wires. An increase in this separation increases the strength of the magnetic field. This occurs in power lines, where conductors are typically separated by poles and shields for fire protection.

For these reasons excessive magnetic fields are considered to have the greatest potential for harm. The level of potential harm are thought to be related directly to the distance from the generating source, the distance between other conductors, the size of the coils on the transformer (if any) and of course the amount of current flowing through the wire. It is generally accepted that the relative magnetic field strength halves with the amount of distance from the line. In other words, a line 100-foot away will have one-quarter of the magnetic field strength of a line 50-feet away.

Li *et al.,* (1997) after testing 407 residences in northern Taiwan ranging from 50 meters to 150 meters from high-voltage power lines, found that the magnetic fields at the houses ranged from .93 mG for 50 meters to between .51 and .55 milliGauss for residences under 149 meters, and .29 mG for residences beyond 149 meters.

This data is somewhat contradicted by a 1993 cohort study from the Netherlands that revealed magnetic field intensities, ranging from 1 to 11 milliGauss from two kilovolt power lines connecting to one transformer substation (Schreiber 1993).

Higher voltage wires are typically thought to be an issue because the voltage and speed is boosted to travel longer distances. With a high-speed voltage line comes an increase in magnetic field. Magnetic fields have been connected with decreased melatonin secretion (Brainard *et al.* 1999). A number of studies have linked lower melatonin levels with higher incidence of a number of types of cancers. It would thus seem probable that lower melatonin levels associated with high voltage, high speed power lines could well be a mechanism for cancer (Ravindra 2006).

In comparison, a typical house or office will range from .8 to 1 mG in magnetic fields. The magnetic field strength from a kitchen appliance at close range for a person working in the kitchen is significantly greater than the strength coming from power line 50-100 feet away. Stepping a few feet away from a microwave oven will dramatically reduce this field strength, while that same relative power line reduction will require a more significant change. A typical microwave oven might cause a field strength of 1000 mG, which can be reduced to a minimal 1 mG by stepping a few feet away. Moving ones house further away from a power line obviously requires a significant commitment to the reduction of magnetic field strength, and a few feet will not make a significant difference.

Epidemiological studies involving electrical appliances have been limited. They are more difficult because of the control parameters. Nonetheless, a few appliances have undergone controlled studies over the years. Electric blankets have undergone several studies. Some of these illustrated significantly increased risk factors for postmenopausal cancer (Vena *et al.* 1991), testicular cancer (Verreault 1990), and congenital defects (Dlugosz 1992). Although it was postulated that electric blankets were associated with breast cancer, study results have been equivocal. Again these studies are difficult to assess due to the population of disease occurrence and the control over other causative exposures.

Radiofrequencies

Radiofrequency waves range from about 3 hertz to 300 gigahertz. This means their waves travel from speeds of 3 cycles per second up to 3,000,000 cycles per second. *Extremely low frequency* (ELF=3-30 Hz) and *super low frequency* (SLF=30-300 Hz) broadcasting has primarily been used for submarine communications, as these wavelengths transmit well through the water. This is also the frequency range that sound travels. *Ultra low frequency* (ULF=300-3000 Hz) has primarily been used in mines, where the waves can penetrate the depths. Above these levels, *very low frequency* and *low frequency* (VLF and LF = 3-300 kHz) have been used by beacons, heart rate monitors, navigation and time signaling. *Medium frequency* (300-3000 kHz) radiowaves are typically used for AM broadcasts, while *high frequency* (HF = 3-30 MHz) is used primarily for shortwave and amateur radio broadcasting. *Very high frequency* (VHF = 30-300 MHz) waves are used for FM radio, television and aircraft communications while *ultra high frequency* (300-3000 MHz) waves are used for certain television ranges, but also cell phones, wireless LAN, GPS, Bluetooth and many two-way radios. While often considered outside the radio spectrum, *super high frequency* (SHF = 3-30 GHz) waves are used in microwave devices, some LAN wireless systems and radar. *Extremely high frequency* (EHF = 30-300 GHz) is used for long-range systems such as microwave radio and astronomy radio systems. The audio frequencies are primarily ELF through VLF brands, covering 20-20,000 Hz.

Note that radiofrequency wavelengths inversely vary to their frequency (for naturally occurring EMR such as light, the frequency will equal the speed of light divided by the wavelength), so while an ULF wave can be between 10,000 and 100,000 kilometers long, a UHF wave will range from one meter to ten millimeters in length, while an ELF wavelength will be between one millimeter and ten millimeters.

Radiofrequencies have been utilized by humans only for about the last seventy-five years. Early use was primarily for radio transmission, while the past few decades various communication and signaling systems have been developed to utilize radiofrequencies. Radiofrequencies are generated with alternating current fed through an antenna at particular speeds and wavelengths.

Studies on radiofrequency radiation proximity at work have also studied possible reproductive and cardiovascular effects. While many of the reports are inconclusive, there have been positive correlations between radiofrequency exposure and delayed conception (Larsen *et al.* 1991), spontaneous abortion (Quellet-Hellstrom and Steward 1993; Taskinen *et al.* 1990), stillbirth (Larsen *et al.* 1991), preterm birth after father exposure (Larsen *et al.* 1991), and birth defects (Larson 1991). However, many of these results have either not been replicated or remain uncorroborated. Three studies examined male military personnel exposure to microwaves and radar (Hjollund *et al.* 1997; Lancranjan *et al.* 1975; Weyandt *et al.* 1996). All three found sperm density reductions.

A number of animal studies have illustrated adverse health effects from radiowaves but doubt has been raised regarding the dose comparison with humans. In one study, GSM phone frequency radiowaves caused the cell death of about 2% of rat brains. Researchers hypothesized that the blood-brain barrier was being penetrated by the radiation (Salford 2003). This was correlated by three earlier studies that reported blood-brain barrier penetration with radiowave exposure (Shivers *et al.* 1987; Prato *et al.* 1990; Schirmacher *et al.* 2000). In the four years following the release of this latter study, several other studies on rats could not replicate the findings, nor could they establish a confirmation of the permeation of the blood-brain-barrier from radiofrequencies (Orendacova 2007; Finnie 2006; Franke 2005; Kuribayashi 2005; Franke 2005; Paulson 2004; Finnie 2004) Shivers and colleagues (Shivers *et al.* 1987; Prato *et al.* 1990) had previously examined the effect of magnetic resonance imaging upon the rat brain. They showed that the combined exposure to radiofrequencies with pulsed and static magnetic fields gave rise to a significant pinocytotic transport of albumin from the capillaries into the brain.

Rates of breast cancer, endometrial cancer, testicular cancer and lung cancer have been studied with close range radiofrequency radiation, primarily in occupational settings. Slightly positive correlations with endometrial cancer (Cantor *et al.* 1995) and breast cancer (Demers *et al.* 1991) were found. A potential link between testicular cancer and radiofrequency radiation from traffic radar guns, particularly among a small group of police officers (Davis and Mostofi 1993) was also established. Slightly increased ocular melanoma was established among occupational radiofrequency exposure (Holly *et al.* 1996) in another small group. French and Canadian utility workers were found to have an increased likelihood of lung cancer (Armstrong *et al.* 1994). However this couldn't be replicated in a U.S. study.

Cell phone tower radiofrequencies are popular concerns. The first cell phones communicated with analog frequencies of 450 or 900 megahertz, for example. By the 1990s, cell phones were using 1800 megahertz, and various modulation systems. Now the Universal Mobile Telecommunication System is adhered to, which uses 1900 to 2200 megahertz.

In 2000, over 80,000 cell tower base stations were in use in the United States. By 2006 this number was estimated at 175,000. CTIA, the *International Association for Wireless Telecommunications Industry,* estimates that by 2010 there will be about 260,000 towers. These base stations transmit radiowaves using around 100 watts of power. The range of GSM towers is about 40 kilometers, while the CDMA and iDEN technologies offer ranges of 50 to 70 kilometers. This obviously is relative to terrain. In a hilly area, the range can be a few kilometers.

In populated areas, cell base towers are placed from one to two miles apart, while in urban areas they can be as close together as a quarter of a mile. Some cell phone bases are mounted on primary towers, and some are built onto elevated structures such as buildings and hillsides.

A base cell tower antenna is comprised of a transmitter(s), a receiver(s)—often called transceivers—an electrical power source, and various digital signal processors. The circuits will utilize copper, fiber, or microwave connections. They may be connected to the network via T1, E1, T3 and/or Ethernet connections. They are typically strung together through base station controllers and radio network controllers, typically connected to a switched telephone network system. The radio network controller will connect to the SGSN network.

There has been scant research on the risks of radiofrequency waves from radio stations or television stations. The primary reason for this appears to be because most of these have been located outside of densely populated areas, on high towers enabling greater ranges. Cell towers have created more concern because of their close proximity and relatively lower heights.

Research has suggested that exposure from cell towers is reduced by a factor of one to one hundred times inside of a building, depending upon the building materials and style of the building. However, exposure also increases with height. Upper floors can have substantially greater exposure levels than lower floors (Schuz and Mann 2000). Whether this is a factor of pure height or whether the earth provides a buffering factor is not known.

Exposure levels in regions surrounding cell towers will range from .01 to .1% of ISNIRP (*International Commission on Non-Ionizing Radiation Protection*) permitted levels for general public exposure directly around the station, to .1 to 1% of ISNIRP permitted levels between 100 meters and 200 meters from the tower. Beyond the 200-meter level, the exposure returns to the .01 to .1% level and reduces as the range increases. It should be noted also that exposure levels from cell phone towers are not substantially greater than exposure levels of radiofrequencies (RF) emitted by radio broadcasting towers (Wood 2007). In one Australian study, the greatest level found was .2% (Henderson 2006).

In a 2006 randomized double-blind study performed at the *Institute of Pharmacology and Toxicology* at the University of Zurich (Regel *et al.*) in Switzerland, UMTS signals approximating the strength of a cell phone tower emission were tested on 117 healthy human subjects, 33 of which were self-reported as sensitive to cell towers with 84 reporting non-sensitivity. Physiological analyses included organ-specific tests, cognitive tests, and well-being questionnaires. Apparently significant negative physiological or cognitive results were not found, although there appeared to be a marginal effect on one of the cognitive tests for each of the two groups. Because the difference was slight, and each group (sensitive versus control) had different results, this effect was considered insignificant.

In 2006 the British medical journal *Lancet* (Rubin *et al.*) reported a study done at the King's College in London which tested 60 self-reported sensitive people and 60 control subjects with no reported sensitivities. Six different symptoms such as headaches were tracked, and subjects took questionnaires in an attempt to find whether the sensitive subjects could successfully judge whether a cell tower signal was on or off. While 60% of the sensitive subjects believed the tower signals were on when they were on, 63% believed the tower signals to be on when they were indeed off.

There have also been several international studies done on radiofrequency transmissions from masts. Tests in the United States, Britain, Australia and the Vatican City have shown no or low correlation between RF levels and health effects, rendering these studies for the most part, inconclusive. One study in the Netherlands using simulated mobile phone base station transmissions did conclude, however, that the UMTS-like spectrum of cell transmission might have an adverse affect upon the well-being of questionnaire respondents.

In July of 2007, an independent team of researchers (Eltiti *et al.*) from the University of Essex reported findings from a three-year double-blind study using a special laboratory to test potential cell phone tower effects. The study included 44 people who reported sensitivity to cell phone towers and 114 healthy people who had not. The study measured various physiological factors like skin conductance, blood pressure and heart rate while being exposed (or not) to 3G tower signals. During periods where the researcher and the subject knew the signals were on, sensitive people reported feeling worse, and their physiological factors were affected negatively. However when neither the subjects nor the researchers knew the cell tower signals were on during a series of tests, there was no difference between either the sensitive or non-sensitive subjects with regard to physiological factors. In fact, only two of the forty-four sensitive subjects were able to guess the cell tower signals being on correctly while five of the control subjects (non-sensitive) were able to guess correctly. Subjects who reported sensitivities to cell phone towers prior to the study reported negative symptoms more often regardless of whether the cell tower transmitters were on or off.

Phones

Typically, a digital cell phone operates at a power range of about .25 watts, while the newest digital phones might transmit as low as .09 watts. Analog phones were much higher power transmitters. The exposure level of a cell phone will depend greatly upon the way the phone is designed. The location of the antenna and the power supply/battery will typically govern the strength of the transmission to the dermal layers of the skin. The further away the antenna is from dermal contact (hand or ear), the less exposure.

The orientation of the power supply will also govern exposure. Some phones have shielding between the power supply and the antenna and earpiece. This is thought to reduce dermal exposure. Obviously the manner of carrying and holding the phone will vary the exposure.

There is another factor called *adaptive power.* When a cell phone is further away from a tower, or in a moving car, it will typically increase its internal transceiver power to send and receive signals. This increases the level of electromagnetic exposure as the phone is boosting power and transmissions. EMR cell phone exposure is thus typically less out of doors than indoors, because there is less interference from building materials out of doors. In addition, exposure to radiowaves is greatest on the side of the head the phone is most used and closest to where the antenna is located (Dimbylow and Mann 1999).

Radiofrequencies from handset use have been confirmed to heat the ear canal. In one controlled study of 30 individuals, 900 MHz and 1800 MHz phones against the ear for more than 35 minutes resulted in an increase of 1.2-1.3 degrees F (Tahvanainen *et al.* 2007). Other studies have confirmed this effect (Wood 2007). For this reason there has been a great concern regarding the potential for tumor development either in the brain or in the areas surrounding the ears—referred to as *acoustic neurinomas.*

Adverse effects of tissue temperature rise are not clear, but it is thought that the body's thermoregulation mechanisms may create an increased immune burden on the body. Lab studies have suggested a one-centigrade temperature rise at the tissue level will have immunosuppressive effects (Goldstein *et al.* 2003).

The *International Agency for Research into Cancer* has sponsored studies in thirteen countries to study the line between cell phone usage and cancer. So far Australia, Canada, Denmark, Finland, France, Germany, Israel, Italy, Japan, New Zealand, Norway, Sweden and Britain have participated. Through 2005 the research tracked 6,000 glioma and menigioma cases (brain tumors), 1000 acoustic neurinoma cases and 600 parotid gland cancers. Of these, the acoustic neurinoma results, primarily from Sweden, showed a significant link with handset use—for both cell phones and cordless phones. The German study also revealed a significant link between uveal melanoma and unspecified handset use. Other types of tumors had OR levels of around or just above 1 to 1.7 OR. The 2001 Swedish study on all brain tumors found a 2.4 OR link with ipsilateral cancer—more prevalent on the same side of primary handset use.

Again, we are faced with the fact that many of these associations are occurring at between 1 and 3 OR. A 2 or 3 level OR risk level creates questions in the minds of meta and review researchers. This, combined with the fact that the rates of these tumors are so small among the general population (10-15 per 100,000 per year for malignant brain tumors (Behin *et al.* 2003)) together with the typical ten year or more delay from exposure to diagnosis, gives some researchers a myriad of reasons to question the evidence that cell phone use is definitely linked with cancer.

Other researchers firmly disagree from the flip side perspective, stating that the weak evidence is actually enhanced by the cancer delay. Research from the Japanese nuclear victims of World War II has shown that many cancers arise ten to twenty years and more after the initial exposure. If we extrapolate this with cell phone use, we estimate that because cell phone use among the general population is still within this twenty year period, especially for many younger adults (who were barely using cell phones five years ago). This means we should expect to see higher cancer rates among heavy cell phone users within the next five to ten years from now.

One of the more dramatic releases on cell phone use emerged in 2003 from a study conducted by Dr. Michael Klieeisen at Spain's *Neuro Diagnostic Research Institute*. This study revealed from a CATEEN scanner linked to a brainwave activity imaging unit that radiowaves from cell phones could penetrate and interfere with the electrical activity of an eleven-year-old boy and a thirteen-year-old girl. Various hypotheses resulted from the release of this data. Among them, that radiowaves affect the moods, memory, and activities of children. Because brainwaves have been closely linked to moods, recollection, response time and other cognition skills, it is assumed that cell phone use has a disturbing effect upon cell phone users—particularly in children and adolescents.

In a 2004 study (Maier *et al.*), eleven volunteers' cognitive performance was tested with and without being exposed to electromagnetic fields similar to cell phones. Nine of the eleven (or 81.8%) showed reduced performance in cognition tests following exposure.

It should be noted that there is a tremendous market resistance to the information that cell phones and remote phones could be dangerous when used consistently. The cell phone industry is now a multi-billion dollar international business. The damage undeniable evidence of a health risk would have upon this industry is nothing short of monumental. It goes without saying that this would also have a significant impact upon the human lifestyle.

This effect may be effectively illustrated by the events reported by Dr. George Carlo and Martin Schram in their 2001 book *Cell Phones: Invisible Hazards in the Wireless Age.* Dr. Carlo was a well-respected epidemiologist/research scientist and pathologist. He was retained by the cell phone industry's chief lobbyist to study and comment on research regarding potential dangers of cell phone use. However, it was not expected that Dr. Carlo would speak out against cell phone use after examining the research data. In his book, Dr. Carlo describes the extraordinary efforts of the cell phone industry to discredit him. As Dr. Carlo began to announce negative cancer-related findings, his clients began to apply both political and financial pressure upon him.

We should however note that although brain cancer rates have increased substantially over the past three decades, brain cancer incidence increased until 1987, and has been slowing decreasing from that point (Deorah 2006). This statistic does not concur with a model of increasing brain cancer rates with increasing cell phone usage. Quite possibly, some of the environmental etiologies involved in brain cancer prior to 1987 have been somewhat mitigated. Perhaps some of the toxin exposure levels—such as the rampant use of DDT and toxic waste dumping in waterways—have been curtailed due to some of the EPA actions of the 1960s and 1970s—decreasing brain cancer rates in the years following. We also cite further controls on nuclear leaks. Epidemiologically, these could well be masking a slow rise in brain cancer levels due to cell phone use.

Cancer is not the only issue to consider with regard to cell phones. There have been a number of other disorders that have also been examined by researchers with respect to radiofrequency exposure. Heavy cell users commonly report a wide variety of negative symptoms. In a study of 300 individuals at Alexandria University in Egypt (Salama and Naga 2004), cell phone usage was positively correlated with complaints of headaches, earaches, sense of fatigue, sleep disturbance, concentration difficulty and burning-face sensation. The results showed that 68% of the study population used cell phones. All of the above health complaints were significantly higher among the cell phone users, and 72.5% of the cell phone users had health complaints. The frequency duration of cell phone usage was also extrapolated together with health complaints, and it was discovered that the higher the cell phone use, the greater the incidence of health complaints. While the burning-face sensation complaint correlated positively with call frequency per day, complaints of fatigue were significantly correlated positively to both call duration and the frequency of calls.

The warming of the ear, face and the scalp around our ear from cell phone is logically taking place as a result of frequency and waveform interference between our body's own natural waveforms and these synthetic waveforms. Should these electronically-driven waveforms interfere with the natural waveforms produced by our bodies—notably with the shorter waves of the brain and nerves and the weaker waveforms of cells, along with the molecular electromagnetic waveforms produced by our DNA—the very molecular structures of genetic information could gradually become damaged.

The effects of this interference should appear on a number of fronts. We should see lower cognition levels and brain fog as unnatural waveforms interfere with our brainwave mapping system. We should see body temperature interference within the basal cell network. We should see damage to the blood-brain barrier and damage to nerve and brain cells. These effects should release greater levels of oxidative reactive species from the imbalanced molecular structures—damaging cells and tissues. All of these effects have been documented in the research.

This waveform interference mechanism is illustrated by a recent study (Thaker and Jongnarangsin 2007) showing that a certain popular brand of MP3 player will interfere with the mechanisms of a pacemaker if held close to the chest for about five seconds. Appliance interference has been directly correlated with waveform interference. This is one reason why the U.S. Federal Communications Commission closely monitors and licenses bandwidths. When we consider that the body maintains various natural biowave "pacemaker" systems as it processes hormones, thermoregulation, cortisol, melatonin and the Krebs energy cycle to name a few, it is not difficult to connect the waveform interference of cell phones and other appliances with the disruption of these natural cycles.

Video Display Terminals

VDTs and televisions emit about 60 hertz of electromagnetic fields. Although a number of early studies suggested the potential of a health risk, many studies over the past few years have suggested that VDTs pose little if any health risks. The *National Academy of Sciences* reviewed a number of studies in 1999 and stated *"....the current body of evidence does not show that exposure to these fields presents a human health hazard..."* In 1994, the *American Medical Association* stated, *"no scientifically documented health risk has been associated with the usually occurring levels of electromagnetic fields..."* Their review included both epidemiological studies and various other direct studies of EMR effects.

Another report published in *Lancet,* the British medical journal, documented the largest childhood study comparing childhood leukemia and cancer rates and exposure to 50-hertz non-ionizing magnetic fields. No link was found.

The *National Radiological Protection Board* in 1994 confirmed that while existing conditions might be aggravated, their review of the research showed no link between skin diseases or cataract formation and VDT use. However, the chairman Sir Richard Doll did confirm that VDT use may aggravate conditions that have already formed.

In addition, a bevy of clinical research regarding pregnancy outcome for those working around or on computers has failed to show any links between miscarriage or birth defects and VDT use. The *National Radiological Protection Board* from Oxford UK confirmed this during a review of the research.

In 1998, the *International Commission on Non-Ionizing Radiation Protection* submitted low emission field guidelines. They suggested an upper limit of magnetic field exposure of 833 milligauss (mG). The electric field limit was set at 4,167 volts per meter (V/m).

Both VDTs and televisions are far below these exposure levels when measured individually.

Regardless of these reports, problems associated with vision, fatigue and headaches have been reported from VDT use. These problems have been attributed to such ergonomic issues as the potential for glare on the screen, lighting location with the position of the screen, the distance from the screen, and whether there are regular breaks from looking at the screen.

Other issues reported have been associated with static electricity generated through the keyboard and screen, posture problems, and repetitive injuries such as keyboarding without rest, which creates a risk of carpal tunnel and other motor difficulties.

As for television, there have been a host of efforts that have studied the effects of television on children and adults. Most of these have leaned towards its behavioral effects, but a few have reported significant effects on health. In 2007, Cronlein *et al.* found a significant link between television viewing and adolescent children insomnia. Thakkar *et al.* (2006) and Paavonen *et al.* (2006) found that watching violence on television increased insomnia and sleep disturbances among young children. Bickham and Rich (2006) showed that increased television viewing—especially violent TV—was associated negatively with friendships. Hammermeister *et al.* (2005) showed that viewers who watched two hours or less television per day had a more positive psychosocial health profile. Viner and Cole (2005) determined that early childhood television viewing was associated with people who had a higher body mass index later in life. Other studies have also correlated increased television viewing with childhood obesity (Robinson 2001).

Meanwhile Zimmerman and Christakis (2005) found that children who watched a significant amount of television before the age of three years (2.2 hours/day) scored lower on Peabody reading comprehension, memory and intelligence testing at ages six and seven. Hancox *et al.* (2005) found in New Zealand that increased television viewing was associated with higher dropout rates and lower rates of university attendance. Collins *et al.* (2004) found that watching sex on television increases sex activity at a younger age in children. Huesmann *et al.* (2003) found that watching violence on television increased violent behavior during adulthood. Vallani (2001) illustrates that studies after 1990 progressively show that increased television viewing is linked with violent behavior, aggression, and high-risk behavior such as smoking, drinking, and promiscuousness.

However Anderson *et al.* (2001) cleared up the fact that it is the content not the medium that creates these associations. 570 adolescents were studied since preschoolers, and their programming was monitored. Educational program watching was linked with higher grades, increased reading, greater creativity and less violent activities.

Radon

As research in the nineties focused on power lines, researchers illuminated the fact that electromagnetic fields can interact with various elements in the atmosphere, creating radon gas. A further potential danger was proposed for households not properly wired with copper and insulation. A lack of shielding will increase the potential interaction of household electricity with radon.

Radon 222 comes primarily from the nuclear decay of uranium. This natural process takes place within the earth. As this decay proceeds, radon gas is released, together with decay byproducts, called *radon daughters* or *radon progeny*. These particles are known carcinogenics. Should we breathe these particles, they can be caught in the lungs. Breathing radon gas brings the potential of it continuing to decay inside our bodies. This will effectively deposit the radioactive daughters inside our bodies.

The *National Council on Radiation Protection and Measurement* has developed a maximum safe dosage of radon to be 200 mrem per year.

The relationship between radon and outdoor power lines has not been clearly established, because in order to measure the interaction, an aerosol component (a pollutant of some sort typically) must accompany the electromagnetic field. Nonetheless, significant *radon daughters* have been measured (Henshaw *et al.* 1998) in power line fields.

The subsequent dose and tolerance of radon particles in the human body is also in question. In some research, heavy electromagnetic fields have been shown to penetrate with no more than about .0001 of the original field strength of radon emissions. Still this penetration effect alerted researchers to the fact that there might be a radon penetration into the lungs and basal tissues of the body (Fews *et al.* 1999).

The link between radon and lung cancer has become more evident in recent research. Lung cancer has been the most prevalent form of cancer worldwide since 1985, and has been responsible for more than one million deaths worldwide. The highest rates of lung cancer occurred in 2002 in North America and Northern or Eastern Europe. Although smoking is thought to be the primary etiology of lung cancer; uranium miners—who are exposed to increased levels of radon along with dust—experience higher rates of lung cancer (Tomasek *et al.* 2008). Epidemiological studies on radon-exposure and miners have also revealed that thousands of miners die per year of radon exposure (Field *et al.* 2006).

Research has illustrated that while living out of doors does not increase ones risk of lung cancer, unnatural living or working quarters without enough ventilation lead to a drawing in and encapsulation of radon radiation. A household with poor ventilation poses a higher risk of radon exposure than a well-ventilated house. This is exasperated by other electromagnetic radiation in the local environment. Research has illustrated that ventilation around electromagnetic current exposure is an absolute requirement because of a release of radon daughters into the immediate atmosphere (Karpin 2005).

Darby *et al.* (2005) reported in the *British Medical Journal* that a collaborative analysis of thirteen case studies of 7,148 lung cancer cases together with 14,208 control subjects that increased radon exposure is responsible for about 2% of European cancer deaths. Further research has revealed that most buildings, especially work environments that are full of various power lines and equipment, retain higher levels of radon. Radon levels are additionally increased with unventilated soils, higher air temperatures and higher atmospheric levels. Higher household radon levels are particularly associated with leaking and unventilated soils in the house. This research has caused legislation in many states in the U.S. requiring property sellers to disclose known radon issues.

The majority of our everyday radiation input comes from radon. Natural concentrations of radon are found in some granites, limestones and sandstones. Higher radon levels come from disturbed ground. Disturbing the normal landscape allows more permeability, allowing the release of the normally-contained daughters. Once a house is built upon the disturbed ground, the radon can come in through cracked foundations and spaces around piping and wiring. Because radon gas is pulled in through pressure changes within the house created by temperature gradients, it is important that the house is well ventilated. This is particularly significant during the nighttime and during cold weather, as the warmer temperatures inside with colder temperatures outside cause the most pressure differential—the *Bernoulli Effect*. Ventilation will not only allow the escape of indoor radon gas, but it will release some of this pressure, resulting in a lower draw of radon gas into the house.

Household radon levels tend to increase dramatically during the winter, and decrease substantially during the summer for these reasons. Radon levels also go up dramatically during the nighttime hours, as the outdoor temperature cools. This is when ventilation is most important.

Disturbed landscape ground can leak increased radon daughters also.

Adulterated Magnetism

Nature's magnetic fields surround us, and pose no threat. Many species utilize nature's magnetic fields to navigate migration and nesting.

Synthetic magnetic fields on the other hand, are dispersed with the distribution of unnatural alternating current. The proliferation of electricity and electrical appliances created by generating plants that convert nature's kinetic energy into alternating current has deluged our atmosphere with unnatural magnetism.

Most early research on the health effects of electrical appliances and wires focused on the electrical fields and ignored the magnetic fields given off by appliances. While most electrical fields are shielded by insulators within most appliances, magnetic fields can be more disruptive and insidious to the health of the body. This is because they can directly interfere with the body's internal biowaves. Normally, synchronic and harmonic biowaves—including brainwaves, nerve firings, and so on—travel with synchronicity throughout the body.

A magnetic field surrounding the body can induce an abnormal electrical current flow within the body. In a Swedish study (Wilen *et al.* 2004) of RF operators exposed to high levels of magnetic fields, currents were induced within the body at mean levels of 101 mA and maximum levels of one Amp. During this study, exposure levels correlated positively with the prevalence of fatigue, headaches, warm sensations in the hands, slower heart rates and more bradycardia episodes among the subjects.

In a study done by the *Fred Hutchinson Cancer Research Center* and the *Epidemiology Division of Public Health Services* in Seattle, Washington (Davis *et al.* 2001), 203 women aging from 20-74 years with no breast cancer history were studied between 1994 and 1996. Magnetic field and ambient light in the bedroom were measured for a 72-hour period during two seasons of the year. Urine samples were taken on three consecutive nights for each subject. After adjusting for hours of daylight, older age, higher body mass, alcohol use and medication use, those women with higher bedroom levels of magnetic fields had lower concentrations of *6-sulfatoxy-melatonin*. Thus it was concluded that increased levels of synthetic magnetic fields depress nocturnal melatonin.

While this illustrates how unnatural magnetism can significantly affect the body's biochemical rhythms, reduced melatonin also causes negative effects throughout the body. Over several decades since melatonin was discovered in 1958 by Dr. Aaron Lerner and his Yale colleagues, decreased melatonin levels have been linked to a variety of pathologies and immune function deficiencies.

A 3 milliGauss magnetic field at 60 Hertz will induce about one-billionth amp per square centimeter of the body. A magnetic field at 120 Hz frequency will have double the current effect the same field will have at 60 Hz. A typical American office building or home—filled with various electrical appliances—will contain magnetic fields at levels between .8 and 1 milliGauss. In a study done in a Canadian school by Akbar-Khanzadeh in 2000, workers, schoolteachers and administrative staff environments had magnetic field exposure levels ranging from .2 to 7.1 mG.

MilliGauss levels will be substantially higher in instrument-heavy environments, however. Hood *et al.* (2000) recorded the pilot's cockpits of a Boeing 767 with magnetic field levels of 6.7 milliGauss, while the Boeing 737 recorded at 12.7 mG of magnetic field strength. Nicholas *et al.* (1998) documented mean magnetic field strength of 17 mG among the cockpits of B737, B757, DC9 and L1011 planes. Meanwhile cabin measurements ranged from a high of 8 mG in the forward serving areas, 6 mG in the first class seats and 3 mG in the economy seats.

Rail maintenance workers experience magnetic field levels from 3 to 18 mG (Wenzl 1997). In a study published in the *Journal of the Canadian Dental Association* (Bohay *et al.* 1994), dental operating rooms with various ultrasonic scalars, amalgamators, and x-ray equipment revealed levels ranging from 1.2 to 2225 mG, with equipment distances from zero to thirty centimeters. Most of these magnetic field readings were accompanied by lower level radiation frequencies ranging from 25 hertz to 100 hertz (though the airline cockpits research recorded up to 800 hertz).

In a population study of 969 women in San Francisco, miscarriage levels positively correlated with higher magnetic field exposure. Li *et al.* (2002) concluded that levels in the region of 16 mG or higher had the greatest risk of miscarriage. While higher levels of magnetic fields have been shown not to significantly affect nervous system biowaves such as cardiac pacemakers (Graham *et al.* 2000), 12 milligauss magnetic fields operating from radiation frequencies of 60 hertz were shown to block the inhibition of human breast cancer cells by both melatonin and tamoxifen *in vitro.* While melatonin and tamoxifen have different mechanisms of retarding cancer growth, it was confirmed by Harland and Liburdy (1997) that synthetic magnetic fields prevented both of their immunity effects. When we consider that the magnetic fields blocked the immune activities of *both* substances, which work on different mechanisms, the affect of synthetic magnetic fields on the human body illustrates an *immune system magnetic interference* model.

This magnetic field interference model of electromagnetic exposure is further supported by research published in 2002 by Saunders and Jefferys. Brain tissue testing showed that even very low frequency electric and weak magnetic field exposure will induce electric fields and currents inside the body. These fields excited various nerve cells and retinal cells, inducing abnormal metabolic activity.

The immune system magnetic interference model mechanism is further confirmed by a study of magnetic and electric fields on neural cells by Blackman (1993). While magnetic fields stimulated abnormal neurite outgrowth between 22 and 40 mG, increased electric fields did not stimulate the same morphological change.

In contrast, the natural magnetic field strength of the earth ranges from about .2 gauss to .6 gauss (200-600 mG)—often also measured as .05 Tesla (1 Tesla=10,000 gauss). We should note, however, that the earth's natural magnetic fields are static. The field direction maintains consistency, pointing in the same direction over the short range. Over the longer range of time, natural magnetic fields may slowly adjust, as they balance with other activities occurring in nature. Synthetic magnetic fields, on the other hand, are typically oscillating fields. In other words, magnetic fields emitted from alternating current sources and appliances will rapidly switch direction.

To give some reference with nature's levels, an MRI magnet will range from one to three Tesla, or 10,000 to 30,000 gauss. This is equivalent to 10,000,000-30,000,000 mG.

Microwaves

Microwave ovens produce two different forms of radiation: High frequency radiowaves produce electromagnetic frequencies in the range of 2450 megahertz and magnetic fields at 60 hertz. The central question is whether this is enough bombardment to cause harm to the food. While some claims have been made that microwave ovens cause the food particle to spin and rotate, this statement has not been confirmed by scientific investigation. What we know is that the microwaves increase the waveform energy states of the molecules through thin microwave beams in much the same way fire increases waveform energy states. Whether this is accompanied by a spinning or rotation of the molecule is appears to be speculative and unsubstantiated.

Microwaves do create interesting molecular structure results, however: A well-cooked microwave dinner reveals dry and rubbery textures not seen in other forms of cooking. Is microwaved food healthy?

Dr. Becker (1985) reported various disorders such as cardiovascular difficulties, stress, headaches, dizziness, anxiety, irritability, insomnia, reproductive disorders, and cancer in the Soviet Union among microwave-exposed workers when the Soviets were developing radar during the 1950s. Though technically correct, it must be noted here that these are workers working amongst microwaves, not people eating microwaved dinners.

Dr. Becker also reported that research from Russia indicated nutritional reductions of sixty to ninety percent in microwave oven tests. Decreases in bioavailable vitamin Bs, vitamin C, vitamin E, minerals, and oil nutrients were observed. Alkaloids, glucosides, galactosides and nitrilosides—all phytonutrients—were found damaged by microwaving. Other proteins were degraded as well.

Research (Knize *et al.* 2007) at the University of California Lawrence Livermore Laboratory concluded that microwaves produced *heterocyclic aromatic amines* and *polycyclic aromatic hydrocarbons*. Both are suspected carcinogens. Frying forms primarily the polycyclic aromatic hydrocarbons.

Dr. Lita Lee reports in her 1989 *Microwaves and Microwave Ovens* that the *The Atlantis Rising Educational Center* in Oregon reported that a number of carcinogens form during microwaving in nearly all types of foods. Microwaving meats caused formation of the carcinogen *d-nitrosodiethanolamine*. Microwaving milk and grains converted amino acids into carcinogenic compounds. Thawing frozen fruit by microwave converted glucosides and galactyosides into carcinogenic chemicals. Short-term microwaving converted alkaloids from plant foods into carcinogenic compounds. Carcinogenic radicals formed from microwaving root vegetables according to this report.

In December of 1989, British Medical Journal *Lancet* reported that microwaves converted trans-amino acids to cis-isomers in baby formulas. Another amino acid, L-proline, converted to a d-isomer version. These isomeres have been classified as *neurotoxins* (toxic to the nerves) and *nephrotoxins* (toxic to the kidneys).

Swiss food scientist Dr. Hans Ulrich Hertel and Dr. Bernard Blanc of the Swiss Federal Institute of Technology reported in a 1991 paper that microwave food created cancerous effects within the bloodstream. The small study had eight volunteers consume either raw milk; conventionally-cooked milk, pasteurized milk; microwave-cooked milk; organic raw vegetables; conventionally-cooked vegetables; the same vegetables frozen and warmed in a microwave; or the same vegetables cooked in the microwave oven. Blood tests were taken before and after eating. Subjects who ate microwaved milk or vegetables had decreased hemoglobin levels, increased cholesterol levels and decreased lymphocyte levels. The increase in leucocytes concerned Dr. Hertel the most. Increased leukocyte levels in the bloodstream are generally connected with infection or tissue damage.

The controls in some of these studies may be in question, however. For example, in Dr. Hertel's study he was a participant, the group knew whether the food was microwaved or not, and the group members were predominantly macrobiotic. The Russian studies and the *Atlantis Rising* report statistics all come unconfirmed from secondary sources.

Various forms of cooking will also destroy nutrients and generate carcinogens—especially frying and barbequing. Overcooking in general destroys nutrients and can create a variety of radicals that can be cancer-forming if eaten in too large a quantity.

There are other dangers reported from microwaves. The leakage of various toxins from packaging during microwaving has been documented. A 1990 *Nutrition Action Newsletter* reported various toxins will leak onto microwaved foods from food containers. Suspected carcinogens benzene, toluene and xylene were among chemicals released into food. Also found was polyethylene terphtalate (PET). Various plasticizers are almost certainly to be included in this list, as they will quite easily out-gas when heated.

In addition, microwaving—unless done for extended periods—rarely completely sterilizes a food. This should be a warning for all those who pack leftovers into storage containers and assume a few minutes in the microwave will produce a sterile cooked food. This fact has been become obvious from the *Salmonella* outbreaks among those who took food home in doggie bags to microwave later.

Approaching this logically, it is apparent that nature did not design food to be cooked in microwaves.

This is evidenced by a simple experiment conducted in 2006. Marshall Dudley's granddaughter completed a science fair project that compared plant water feeding between stove-boiled filtered water and the same filtered water source microwaved. She started with sets of plants of identical species, age, and health. One of each set was fed filtered water boiled in a pan and cooled. She fed another the same filtered water, but microwaved until boiling and cooled. This 'watering study' went on for a period of nine days, and pictures of the plant sets (which sat together in identical potted condition) were taken each day. The simple assessment of each plant's health was clear by looking at the photographs:

Each day the plant watered with microwaved-water looked worse. It became increasingly withered and slumped over in obvious stress. By the ninth day, the microwave-watered plant had lost most of its leaves. Meanwhile the boiled-watered plant stood tall with crisp green leaves, growing healthier by the day.

EMR Harmonization

❖ How much EMR we can tolerate depends upon the type of EMR, its proximity, and how strong our body's defense systems are. These effects were studied intensely during the 1940s as the U.S. military measured the effects of atomic radiation in relationship with proximity to a nuclear explosion.

❖ We can first determine whether the EMR source is emitting ionizing or non-ionizing radiation. Ionizing radiation should be avoided or minimized to the greatest degree possible.

❖ Lead aprons can be worn whenever taking x-rays. CAT-scans can be questioned and avoided if unnecessary.

❖ Contact with Non-ionizing EMR appliances can be as brief as possible. We can turn the knob or dial then step away. At three feet, EMR from most electric appliances falls off rapidly. For televisions, this expands to about ten feet because of the effect cathode rays can have upon the eyes. We can keep a distance of at least five feet away from a working microwave oven.

❖ Laptops can be taken off our laps and hands can be pulled away as much as possible. We can use a peripheral keyboard.

❖ A TriField® EMR meter can be used to measure our workplace, home, phone, vehicle and other environment to determine our safe zones. Depending upon our sensitivity, levels over 5-6 mG cause concern, especially for any length of time.

❖ We can select housing that is at least 100 feet away from any high-frequency power line, and further for a large transformer.

❖ We can run our power cords through surge protectors and switch them off after using the appliance. This can reduce the amount of idling power moving through the appliance, and lower electric bills.

❖ We can limit our cell phone use to urgent or important calls, talk briefly, and use a headset or better yet a speaker. The Bluetooth headsets emit 2.4 GHz short-range radiofrequencies. Though non-ionizing, there is little research on the long-term effects of this technology. Wires can have their own effects, although a shielded, double-strand wire will emit little in the way of magnetic fields.

❖ We can judge whether our EMR dose is burdensome by assessing how we are feeling as we use these devices. Do we feel dizzy? Tired? Do we have brain fog? Does part of our body feel hot or feverish? These symptoms can lead to more serious health concerns later.

❖ There are a number of field canceling devices on the market. While lead and nickel aprons and garments have been shown to protect against EMR doses, the effects of the various biowave products on the market are controversial, and beyond the scope of this discussion.

❖ The stronger our body is, the more we will be able to resist the negative effects of the synthetic EMR of our environment. This means we can eat healthy foods, drink plenty of water, exercise enough and breath deeply to strengthen our body's biowaves.

❖ We can also get out into nature and get enough sun. Nature's biowaves can strengthen the body as discussed. Daily safe sun can help balance our EMR dosage with normalizing waveforms.

Chapter Eight
Vital Chemistry

The biochemicals natural to our environment undergo constant molecular recycling. They cycle through changes in electron-wave orientation, energy states, polarity, spin, angular momentum, and bonding valences, all while maintaining the laws of thermodynamics. While modern science suggests this process is accidental and chaotic in nature, practical observation indicates these processes are biologically driven by design and congruency. As we trace the cycles of specific biochemicals, we observe their journeys through different states, atmospheres, and local environments. A single biochemical may be eaten by one organism, excreted by another, inhaled by yet another and become the skin of yet another organism. At the same, an intricate chemical balance is kept through these state and molecular changes, keeping the movement consistent and orderly. In other words, the biochemical recycling maintains its nutritional nature on a consistent basis. It is not as though one day a food is nutritious and the next day poisonous to a particular organism.

The rules of engagement between these cyclic combinations are complex and vast. To interfere with any one part of the system, we have learned, will easily interrupt the balance of the rest of the cycle, often causing reverberations on an exponential basis. This is precisely what humans have unfortunately arrived at after two hundred years of industrial technological development.

An example is the delicate chemical pathway of nitrogen. Animals and humans consume nitrogen from the atmosphere and through nutrition—by breathing, eating plants, root nitrogen fixing, and so on. When an organism dies, *ammonifying bacteria* decompose the body and release ammonia into the soil. *Nitrosifying bacteria* oxidize the ammonia, converting it to nitrites, and *nitrifying bacteria* then oxidize the nitrites to soil nitrogen and ammonia ions. Plants utilize these to form amino acids and proteins. Nitrogen is also released into the air with *dentrifying* bacteria. As plant protein is eaten by animals and humans in plant food, these nitrogen amino acids become part of the proteins that make up our bodies. When our bodies die, the cycle begins again.

Throughout the nitrogen cycle, there is a precise balance of nitrogen in the atmosphere, the soils, and within each organism: Just enough to serve the combined purpose of all involved. Every species is balanced; exchanging and converting nitrogen to useable forms of nourishment as needed.

Enter chemical fertilizer. It is not enough to add natural forms of nitrogen to the soil such as compost. The beneficial addition of nitrogen chemical fertilizers into the soil has increased crop production for agri-business-based farms. While adding something already available seems innocent enough, the dumping of pure nitrogen without the nutrients necessary to engage the bacterial decomposition process ruins the entire cycle. Without the complex nutrients produced by the nitrogen-fixing process, the soil begins to erode and thin. The heavy load of unused nitrogen leaches through the soil, settling into the ground water. This nitrogen leaching creates a build-up of dangerous nitrates within the ground water. Nitrate build up has been poisoning ground water in agricultural areas throughout the world. In areas of heavy fertilizer use, undrinkable ground water is reaching increasingly deeper wells. Nitrate levels above about 50 parts per million can make a person sick. Higher levels have been known to be fatal. In addition, nitrogen-fertilizer-rich soils choke rivers and oceans with extra nitrogen, causing abnormal blooms of algae, cutting off oxygen supplies and leading to dead ocean zones as was mentioned earlier.

The use of nitrogen fertilizers illustrates how the precise rhythms of nature—the use and recycling of natural biochemistry—can so easily be disrupted.

Our Synthetic World

Over the past century, humankind has opened quite a Pandora's box of chemical manipulation. Public perception of the twentieth century chemical revolution—through the brilliant marketing efforts of chemical business giants—was that synthetic chemicals were beneficial for society. The chemical "miracle" made life easier and more productive.

As this grand synthetic experiment has unfolded, we have discovered many of these chemicals are not only toxic, but they now risk humankind's future existence. After only a few decades of massive synthetic chemical manufacturing, we are beginning to suffer the horrific price synthetic chemicals come with: We are faced with increasing epidemics of asthma, lung cancer, COPD, and other bronchial diseases. The water we drink has become toxic. Much of our drinking supplies are laced with DDT, PCB, nitrates, and hundreds of other dangerous toxins. Much of the non-organic food we eat is now to full of various synthetic residues. We are gradually discovering that agribusiness' use of chemical fertilizers and pesticides is slowly poisoning our bodies. The toxins are building up in our cells—mutating DNA and suffocating our immune systems.

Most of the furnishings we purchase now are filled with formalde-hydes, synthetic materials and other synthetic preservatives. Most office buildings and many houses still contain hazards like asbestos and other components that cause toxicity. Our entire environment is laced with synthetic chemistry. If the human race stopped chemical production to-day, we still would have done so much damage over the past fifty years that it will take centuries for the earth's detoxification systems to purify herself.

The core issue with synthetic chemicals is that they run contrary to the fragile balance existing among nature's biochemical recycling systems. As a result, they clog the arteries of our ecosystem. Today there are mountains of synthetic chemistry loading up our dumps, landfills, lakes, rivers, and oceans. These mountains are decomposing very slowly—outgassing and breaking down into potent toxins into the environment. *Time Magazine* reported on June 25, 2007 that Americans generated 1,643 points of trash per person in 2005. A mere 32% of it was recycled.

Much of this waste is plastic. The problem with plastic is reflective of its benefit—it lasts far longer than do natural materials. While a plastic bag might not tear and rip as fast as a paper bag as we walk from the grocery store, a plastic bag will have as much as a 500-year half-life—depending upon its material. That is a long time. Where happens to the bag while nature works biodegrade it? It clogs our soils and waters. For this reason our lands, waters, and bodies, are steadily becoming laced with polymers and plasticizers.

Plastics are made through reactions between monomers (small molecules) and plasticizers to create longer-chain molecules. Monomers are typically hydrocarbons such as petroleum. Combining ethane monomers and plasticizers forms polyethylene. Combining styrene monomers and plasticizers renders polystyrene. Combining vinyl chloride monomers and plasticizers results in polyvinyl chloride, or PVC. Combining propylene monomers and plasticizers gives us polypropylene.

Nature produces its own types of natural polymers such as rubber. In an attempt to improve upon nature, in 1855 the lab of Alexander Parkes mixed pyroxylin from cellulose with alcohol and camphor to form what is thought to be the first type of plastic. This clear, hard plastic was 'improved' by Dr. Leo Baekeland decades later with a polymer process using phenol and formaldehyde in the first decade of the twentieth century. "Bakelite" became a wildly successful product as it effectively replaced shellac and rubber as a general sheathing material. Because it was heat-resistant and moisture-proof, it quickly became the insulator of choice for engines, appliances, and electronics. Dr. Baekeland eventually sold his General Bakelite Company to Union Carbide in 1939 and retired a very wealthy man to Florida. His life was made easy through the 'miracle' of chemistry.

Nylon was an invention of DuPont researchers in the late 1930s. It was made initially with benzene from coal. The introduction of polypropylene as a synthetic rubber followed shortly thereafter. Polypropylene was an accidental discovery by a couple of researchers who were trying to convert natural gas for Phillips Petroleum. The American industrial complex gearing up for World War II focused its attention on this synthetic version due to a shortage of natural rubber. Thanks to synthetic rubber, each military person was estimated to have 32 pounds of rubber in clothing and equipment. A tank needed about a ton. We might consider that America's military might is at least partially due to its synthetic rubber making. Again, chemistry has seemingly made our lives easier.

The synthetic polymer revolution surged after the Second World War. The plastic revolution raged, as both consumers and manufacturers worked together to replace naturally derived goods to synthetic polymers.

One might argue that that combining earth-borne commodities like hydrocarbons cannot be so unnatural. After all, hydrocarbons are produced by the earth as part of her recycling process. However, the process of converting these hydrocarbon monomers into polymers requires various catalysts and *plasticizers* to complete. Plasticizers are used in plastic production to give the long polymer chain its flexibility. Without plasticizers inserted between the polymer chains, there could be no flexibility among the plastics. Without plasticizers, polymers are clear, hard substances: rock-like. The various gradations of flex added to polymer chains give the resulting plastic its particular usefulness and characteristic. A plasticizer will provide strength along with this flexibility, making the material difficult to tear or break.

Most plasticizers are *phthalates*. Phthalates are derived from phthalic acid, an aromatic ringed carbon molecule also referred to as dicarboxylic acid. Originally synthesized in 1836 through the oxidation of naphthalene tetrachloride, phthalic acid can also be synthesized from hydrocarbons and sulfuric acid with a mercury catalyst. Most aromatic carbon rings like the phenyl ring or the benzyl ring made using this process have proven to be hazardous to our environment and physical well-being. Note there are a number of aromatic carbon rings that are produced in nature as well. These do not come with the same hazards for some reason.

There are hundreds of different plasticizers now in use in humankind's production of different plastics. Most are either aromatic carbons or similarly hazardous compounds. While bound within hydrocarbon polymer chains they appear innocuous. However, as plastic polymers break down in the environment, these plasticizers are released. Our backyards, landfills and oceans—our entire environment for that matter—are silently being inundates with the release of these plasticizers.

We have discussed many of the specific plasticizer toxicity issues and the types of common plastics in our chapter on water. Here we add that researchers are becoming increasingly aware of the risk plasticizers present to our environment. Various studies confirm that plasticizer release has toxic effects upon various species, including humans.

Benzene for example, is the typical source of the phenyl plasticizer. Benzene has been classified as a volatile organic compound and a carcinogen by the *Natural Institutes of Health's* National Toxicology Program. Benzene is among the top twenty most used industrial chemicals. It is used to make adhesives, paint, pharmaceuticals, printed materials, photographic chemicals, synthetic rubber, dyes, detergents, paint and shockingly, even food processing equipment to name a few. Today benzene is found throughout our environment—in our air and water—and has been implicated in numerous cancers.

The problems of synthetic chemicals are pervasive. Estimates suggest some 80,000 chemicals have been approved for commercialization over the past fifty years. The 1976 *Toxic Substances Control Act of 1976* was set up to evaluate chemicals being introduced. Only about 65,000 have been reviewed. However only a small percentage of these chemicals have been carefully analyzed and reviewed as to their environmental and health effects. Hundreds of thousands of people die each year—and some say the number approaches one million just in the U.S.—from chemical pharmaceuticals—which undergo the greatest scrutiny—alone. Several hundred thousand die in the U.S. from cancer each year. Millions die worldwide from a disease suspected caused primarily by cellular mutation due to chemical toxicity.

The *Environmental Working Group's* Human Toxome Project has revealed some frightening statistics regarding the poisoning of our bodies by chemicals. In one study of nine adult participants, blood and urine contained 171 of the 214 toxic chemicals for which they were analyzed. These included industrial compounds and pollutants like alkylphenols, inorganic arsenic, organophosphates, phthalates, polychlorinated biphenyls (PCBs), volatile and semi-volatile organic compounds (Voss and Socks) and chlorinated dioxins and furans. In another study, the EWG found 287 of the 413 tested chemicals in the umbilical cord blood of ten mothers after giving birth. These included the chemical types mentioned above and more, including fifty different polychlorinated naphthalene compounds (EWC 2007).

A polychlorinated biphenyl is a grouping of chlorine atoms bonded together with biphenyl. Biphenyl is a molecule composed of two phenyl rings. It is an aromatic hydrocarbon occurring naturally in coal and petroleum. When synthetically combined with chlorine—another naturally occurring element—the result is highly toxic. PCB was banned in the early 1970s when biologists studied a population of dead seabirds and found they died of a toxic dose of PCBs. For more than forty years, PCBs have been used in paints, pesticides, paper, adhesives, flame-retardants, surgical implants, lubricating oil and electrical equipment. Referred innocently as "phenols" for many years, the PCB ban followed suspicion by a number of years. Massive PCB contamination in the Hudson River was found caused by local electrical manufacturing plants. Some two hundred miles was eventually designated a toxic *superfund site*. This woke us up to PCB toxicity. PCBs break down slowly and bio-accumulate in living organisms. When PCBs get into our waterways they build up in the smallest organisms and work their way up the food chain, eventually reaching humans. Today the ban on PCBs does not include many applications considered "closed," such as capacitors and vacuum pump fluids. This means there are still considerable PCBs in our buildings and electrical equipment. PCB poisoning can cause immediate liver damage. Symptoms can include fever, rashes, nausea, and more.

Biphenyl A is another sort of biphenyl, used in many types of containers, including baby bottles. BPA can easily leach into food or formula when the bottle is exposed to heat or sunlight. A 2000 Centers of Disease Control study found 75% of those tested had phthalates in their urine, and subsequent studies have found some 95% of the U.S. population has detectable levels of biphenyl A within body fluids. Biphenyls are considered endocrine system disruptors. Long-term effects as their residues build up in our cells, organs and tissue systems are largely unknown.

By some accounts there are nearly nine hundred different pesticides being used in the United States. Of those, at least thirty-seven contain organophosphates—one of our more toxic chemical combinations. Organophosphates kill insects through nervous system disruption. These neurotoxins are also toxic to humans' nervous systems. The nerve gases Serin and VX are organophosphates, for example. Organophosphates block cholinesterase—a key neuro-enzyme—from working properly within the body. With cholinesterase blocked, acetylcholine is not regulated. Unregulated acetylcholine causes an over-stimulation of nerve activity, resulting in nerve damage, paralysis, and muscle weakness.

Organophosphates are spreading through ground water, air and through dermal contact. They are exposing us through our breathing, touching, swimming, and drinking. Initial symptoms can include nausea, vomiting, shortness of breath, confusion, and muscle spasms. Some of the more common organophosphates include Malathion, Parathion, Diazinon, Phosmet, Clorpyrifos, Dursban and others. The EPA actually banned Diazinon and Dursban in a phase-out beginning in March of 2001, to last through December 2003. Curiously, both Diazinon and Dursban are still in use today. Phased bans like this theoretically take several years to allow companies to run out their inventories. Also since these bans were aimed at consumer products, organophosphates are still used profusely in commercial agriculture—our food production.

In a 2003 study done by the *Centers for Disease Control and Prevention*, thousands of people were tested for 116 chemicals. Thirty-four of these were pesticides such as organophosphates, organochlorines, and carbamates. Nineteen of the thirty-four were found in either the blood or urine.

The use of pesticides on agricultural land, playgrounds, parks, home lawns, and gardens throughout the United States is growing by a staggering amount. In 1964, approximately 233 million pounds of pesticide active ingredients were used. By 1982, this amount tripled to 612 million pounds. In 1999, the U.S. *Environmental Protection Agency* reported that some five *billion* pounds of these chemicals were used per year throughout America's crops, forests, parks, and lawns.

One of the fastest growing of these pesticides has been imidacloprid, a neonicotinoid. Introduced by Bayer in 1994, imidacloprid is used against aphids and similar insects on over 140 different crops. Touted as a chemical with a fairly short half-life of thirty days in water and twenty-seven days in anaerobic soil, imidacloprid's half-life is about 997 days in aerobic soil. While it has a lower immediate toxicity compared with hazards like DDT, imidacloprid's use is now widespread. It is rated by the EPA and WHO as *"moderately toxic"* in small doses. Larger doses can disrupt liver and thyroid function. While this pesticide does well at killing off increasingly resistant pests, it also can decimate bee populations.

A world without bees, as described in Rachel Carson's classic <u>Silent Spring</u>, would insure a destiny of hunger and destitution in human society. In France for example, some 500,000 registered hives were lost in the mid-1990s. Imidacloprid was implicated, and was subsequently banned for many crops in that country. Massive bee destruction has occurred in other regions of Europe also appear connected to imidacloprid use. A 2006-2007 loss of hives throughout Europe and the U.S.—referred to as *colony collapse disorder*—is now increasingly being connected to imidacloprid, although it also appears that other chemicals as well as possibly unnatural electromagnetic radiation work together to weaken the bees' immunity to viruses and other diseases. Imidacloprid and other chemicals can weaken the bee's immune system just as they weaken the human immune system—depending of course on the level of exposure.

Chlorinated dioxins are also pervasive in today's environment. Significant sources include cigarettes, pesticides, coal-burning factories, diesel exhaust, and sewage sludge. Dioxins are also byproducts of the manufacturing of a number of products, including many resins, glues, plastics, and chlorine-treated products. Dioxins also bio-accumulate in fatty tissues and can take years to fully degrade. Dioxins are known endocrine disruptors. They have also been linked to liver toxicity and birth defects.

Thanks to the human industrial complex, there are now thousands of *volatile organic compounds* in our environment. A VOC is classified as such if it has a relatively high vapor pressure, allowing it to vaporize quickly and enter the atmosphere. Gasoline, paint thinners, cleaning solvents, ketones, and aldehydes are a few of the chemicals considered sources of VOCs. Methane-forming VOCs like benzene and toluene are also carcinogens. VOCs are often used as preservatives for pressed wood and other building materials. As a result, many buildings contain VOCs locked within its building materials. Once soaked in or inborn with the fabrication, VOCs are trapped within the material, causing them to outgas over time. This outgassing process is speeded up when the building is demolished or taken apart. As the building materials are broken up, VOCs can be released at toxic exposure levels.

VOCs will form ozone as they interact with sunlight and heat. VOC poisoning symptoms include nausea; headaches; eye irritation; inflammation of the nose and throat; liver damage; brain fog; and neurotoxic brain damage. Using cleaning or painting solvents indoors is a common cause of VOC poisoning.

In a study by Janssen *et al.* (2004) and *The Collaborative on Health and the Environment,* some two hundred diseases were found to be attributable to exposure to industrial chemicals. The diseases listed are some of the most prevalent diseases of our society—cancers, cardiovascular disease, autoimmune diseases and so on. The researchers found that over 120 diseases have been specifically linked by research to exposure to specific industrial chemicals. For another thirty-three diseases, the evidence for linking to specific chemicals was considered "good." For the rest of the diseases, research indicated a definite link but the evidence was considered "limited" (Lean 2004).

Synthetic versus Natural

Nature's biochemicals have profoundly different results within the body and the environment. The assumption of modern chemistry is there is no inherent difference between a chemical synthesized in a lab or manufacturing facility from one made in nature by living processes. This assumption has led humankind to haphazardly create new synthetic chemical molecules with reckless abandon. Driven by economics and cost-reduction, the industrial chemical complex has assumed there is no environmental cost to synthetic chemical production.

The problem with this assumption is that nothing comes without a cost. Every type of result comes with a particular price—balanced precisely with the effort. This is the design of the physical world. As soon as we think we are getting away without working hard in a sustainable way, we will be surprised by the unforeseen consequences.

The over-riding assumption of modern chemistry is there is no overall governing design, order, or balance within the mechanisms of nature. This is the assumption of chaos. It is assumed nature is randomly unfolding from an accidental beginning. This assumption of a chaotic unfolding of events leads to a destructive and erroneous conclusion: There is no functional difference between the molecules nature produces and the molecules we force together. A molecule is a molecule, right?

The chemical industrial complex assumes all molecules are inherently the same. The classic definition of chemistry is that molecules are made of atomic "building blocks," which are simply linked together by covalent bonds. Regardless of how a molecule is formed, its molecular structure will determine its usefulness. This assumes a synthetic version of a particular chemical formula will function the same as its biological version.

Research and observation does not confirm these assumptions. This is illustrated with just a few examples:

When palm and coconut oils are cooled, they become hardened. This makes them good thickening agents for cooking and good for frying. In an attempt to match nature, in 1902 German Wilhelm Normann patented the first hydrogenation process, which was eventually purchased by Proctor and Gamble, leading to Crisco® oil and eventually margarine. When nutritionists convinced us that *all saturated fats are bad* in the sixties and seventies, margarine sales took off. Processors also found that frying oil had a better shelf life and was cheaper if cottonseed oil and soybean oil were *partially hydrogenated.* Because these oils do not normally harden at room temperature as does palm, coconut and lard, hydrogenation allowed processors to use the less expensive oils for frying, spreading and cooking.

Hydrogenation means to "saturate" hydrogen onto all of the available bonds of the central molecule. Whereas a natural substance might have a double bond between carbon and other atoms, hydrogen gas can be bubbled through the substance—using a catalyst to spark the reaction—to attach more hydrogen to the molecule. To saturate carbon bonds with hydrogen, catalyst is added, and soybean oil must undergo bubbling of hydrogen in a heated catalytic environment. This saturation synthetically changes the oils melting point, giving it more versatility at a lower cost.

We took real foods—oil extracted from soybeans or cottonseed—and synthetically converted it into what appeared to be the same molecular structure of another real food. The end result was unhealthy. After decades of use, health researchers began realizing that partially hydrogenated foods have damaging effects upon the cardiovascular system. While the saturated or partially saturated molecule was the same formula, the synthetic process of hydrogenation created an unusual molecular structure called a *trans-fat*. Trans-fat is now implicated in a various degenerative disorders, including atherosclerosis, liver disease, irritable bowel syndrome, and Alzheimer's disease among others. While the epidemic increase in cardiovascular disease has focused billions of dollars into research, trans-fats were altogether overlooked, simply because researchers assumed soybean oil—also considered polyunsaturated—was a safe and healthy food—and its saturated molecular formula was identical with saturated or partially-saturated fats.

Hydrogenating soybean oil was not the only form of trans-fat. Other polyunsaturates—such as canola oil and sunflower seed oil—were long considered healthy until they underwent chemical refining and overheating. Because food scientists assumed the molecular formula told the whole story regarding their action in nutrition, our society suffered decades of increased cardiovascular disease. There was no realization of other molecular structural issues to consider.

Eventually research has revealed that the polarity and spin orientation of the molecule was changed during the heating and/or chemical reaction with hydrogen gas. Nature normally orients these healthy oil molecules in *cis* formation: They are oriented so that the hydrogens are on the same side with the other molecular bonds. A *trans* configuration on the other hand, has hydrogens on the opposite side of the rest of the bonds. Depending upon the use of the molecule, this *cis* and *trans* bonding orientation allows the molecule to exert its specific polarity consistent with its natural environment. A *cis* molecule in an environment designed for *trans* structures would cause the same sort of mechanical deficiencies.

It became evident over time that the *trans* orientation of the partially hydrogenated soybean oil molecule is a toxin because it does not fit the environment of our bodies. Because the body's cell membranes and other cell parts are constructed with lipids or lipid derivatives such as phospholipids, the bonding orientation polarity of these lipids allow them to fit together like stacking chairs. This structure forms the semi-permeable cell membrane. Lipids stack together to form cell membranes because their bonding orbits are arranged in such a way that they form *weak hydrogen bonds* between each other. These weak hydrogen bonds have particular electromagnetic waveform characteristics, which resonate with bonds with similar waveform characteristics. The primary orientation for these stackable cell membrane lipids is the *cis* formation.

The *cis* and *trans* orientation issue is also evident in the case of resveratrol, a phytochemical constituent of more than seventy different plant species, including many fruits such as berries and grapes. Studies have shown natural resveratrol has many biological properties important to health. These include antioxidant, anti-bacterial, anti-viral, anti-fungal, liver-cleansing, mood-elevation, and amyloid-plaque-removal abilities. Resveratrol activates an enzyme-protein called sirtuin 1, which seems to promote DNA repair. These effects are exclusive to the natural form: *trans*-resveratrol. However, heat and light from pasteurization and other processing converts *trans*-resveratrol to its less effective form, *cis*-resveratrol.

Another example of natural versus synthetic molecular structure is vitamin E. While natural vitamin is d-alpha-tocopherol, the synthetic version is dl-alpha-tocopherol. While d-alpha-tocopherol has one isomer, dl-alpha-tocopherol has eight. One of those eight is similar to the one natural isomer. As a result, the natural version is more readily bioavailable, and is excreted at a lower rate.

In one study (Burton *et al.* 1998), subjects took two doses of vitamin E, one of natural vitamin E and one with synthetic vitamin E. The natural vitamin E levels in the blood stream for all subjects were at least twice as high as the synthetic versions. After twenty-three days, tissue levels were also significantly higher for the natural vitamin E group than the synthetic vitamin E group. These tests illustrated that natural form vitamin E is more readily absorbed and retained than the synthetic version.

In Traber *et al.* (1998), Oregon State University researchers discovered that humans excrete synthetic vitamin E at a three-fold rate of natural vitamin E. This ratio was confirmed in another test (Kiyose *et al.* 1997) showing that it took three times the quantity of synthetic vitamin E to reach the same levels achieved by natural vitamin E in seven women.

Gas chromatography has also revealed structural differences between synthetic vitamin C and natural forms of vitamin C. While many people supplement with isolated ascorbic acid, nature provides a completely different structure. Not only is the molecule itself slightly different, but natural vitamin C comes naturally chelated to other compounds. It is interlinked with various bioflavonoids, minerals, rutin, and other biochemicals, often referred to as *ascorbates*.

Synthetic vitamin D—referred to as vitamin D_2 or ergocalciferol—is also molecularly different from naturally produced cholecalciferol. In a study published in 2004 by Armas *et al.*, twenty healthy male subjects were given single doses of 50,000 IU of either vitamin D_2 or D_3. Their serum 25-hydroxyvitamin D (also 25-ODH) levels were measured over a twenty-nine day period. The measured 25-ODH levels were the same between the D_2 and D_3 in subjects for the first three days. The levels of 25-OHD fell dramatically after that for the D_2 subjects, yet continued to rise, and peaked at fourteen days after the initial dose for the natural D_3 subjects. Over the 29 days, the subjects taking D_2 had concentrations of less than a third of the levels of the D_3 subjects. Using a larger *area under the plasma curve* (AUC), this translated to the synthetic D_2's potency being about one-tenth of the potency of the natural D_3.

Research is increasingly indicating that certain isolated synthetic nutrients may even provide some detrimental affects, especially in cases of over-supplementation. Supplemented folic acid provides an example of this. The *British Journal of Nutrition* (2007) published a report revealing that while natural folates from green leafy vegetables and many other foods are metabolized within the intestinal tract, supplemental folic acid is processed in the liver. This means that folic acid can stress the liver. Livers burdened with other metabolizing processes may release unmetabolized folic acid into the blood. This, said Dr. Sian Astley of the U.K.'s *Institute of Food Research* in a *Statement of Concern*, *"...the liver becomes saturated and unmetabolized folic acid floats around the bloodstream... This can cause problems for people being treated for leukemia and arthritis, women being treated for ectopic pregnancies, men with a family history of bowel cancer, people with blocked arteries being treated with a stent and elderly people with poor vitamin B status. For women undergoing in-vitro fertilization, it can also increase the likelihood of conceiving multiple embryos, with all the associated risks for the mother and babies. It could take 20 years for any potential harmful effects of unmetabolized folic acid to become apparent."*

In a recent study at the Washington University School of Medicine (Nicola *et al.* 2007), 168 healthy postmenopausal white women were tested for supplemented calcium versus dietary sources of calcium. Women whose calcium intake was primarily from supplementation had lower bone mineral density levels than the women whose calcium intake was primarily from dietary sources. Those taking supplements plus good dietary sources had the highest totals.

Naturally occurring biochemicals are electromagnetically unique from synthesized chemicals. The resulting quanta of spin, angular momentum, and so on will significantly vary from even a synthesized molecule with the same chemical formula. When two different ions with different waveforms are brought together, the resulting compound typically displays different spectra than either did independently. The combination's waveforms are not necessarily a composite of the two. It is a new waveform combination resulting from the resonating interference of the combining elements together with its natural catalysts and environment.

Most biological combinations require a catalyst to complete. These catalysts may supply or borrow electron-waves in an oxidizing or reducing mechanism to assist in the molecular combination. They supply the needed electromagnetic energy to speed up the reaction. The natural catalyst thus intimately affects the structure and orientation of the resulting complex molecular combination—creating a unique *fingerprint* of sorts.

Nature's catalysts produce a variety of oxidizing and reducing mechanisms. These are balanced in a natural atmosphere, with a design that effectively works towards a balance of decomposition and recomposition among ions and molecules. Therefore, oxidized radicals form naturally to help break down the molecular combinations. If oxidized radicals did not exist, nature would be imbalanced. By design, nature's elements work harmonically to oxidize and reduce molecular structures in an orchestrated fashion.

A synthetically combined compound throws off the balance between oxidizers and reducers, simply by its forced oxidative or reductive addition to the system. When this balance is pushed off, nature's processes are thrown off. While the system is built to handle small imbalances with ease, an overload will require a larger response, just as our bodies must respond with a cold or other detoxification to throw off an overload of toxins.

Bombarding natural molecules with radiation will typically alter their orbital waveforms, boosting some into separation to accommodate the new radiation input. The resulting ion may then become a reactive species, requiring combination with other ions or molecules to rebalance its valence structure. The over-processing of foods creates this scenario.

When we process a food to extreme temperatures or infuse synthetics, reactive species can result. Once consumed, the reactive species attempt to rebalance their valences by utilizing molecules from the body's cell tissues. The prime example of this is atherosclerosis, where oxidative radicals tear apart artery molecules making up artery walls. The body then attempts to reduce the damage by implementing a cascade of plasmin and fibrin to repair and replace those molecules. This repair process tends to harden the artery walls, and even—if there is too much damage—sloughs off fibrin to create potential clots in other parts of the vascular system.

Nature's fingerprinting processes accompany various *scanning* processes such as our immune systems. The discovery of artery damage, chemical invaders or bacterial invaders is a result of a constant scanning of our body's biochemistry by our various immune cells. The earth's metabolism—technically the *ecosystem*—also performs these tasks, many unbeknownst to us. For example, when the earth's surface is damaged by humans, it begins to protect and heal itself with the production of irritating plants such as poison oak and poison ivy. It also leaks more radon. As a result, we only notice that poison oak and ivy is more prevalent in cleared or threatened areas. A study from Duke University (Mohan *et al.* 2007) confirmed that the oils in poison oak and ivy have worsened as carbon dioxide levels have risen. Again, this is an immune response to the earth's current infection: *humans*.

If we can accept that a molecule's unique waveform fingerprint is affected by its catalyst, we can logically accept the notion that a biochemical produced in a natural environment with natural catalysts should be different electromagnetically than one synthetically catalyzed.

The completion and evidence of our proof lies in the reality that a molecule created within a living system acts differently than the same molecule created synthetically in an artificial environment. We have described only a few examples above.

The more complex chemicals discussed earlier are only a small sampling of the 80,000 or more synthetic chemicals that humans have produced. Out of all of these, do we know of any that are healthy for living organisms? Can we recall any that are not rejected by the immune system as being foreign in one way or another—either as they enter or as they combine with other molecules? Even artificial hearts, which have undergone extensive investment from private industry to avoid rejection and immune response—are still rejected by the immune system. This is why there are so few artificial heart transplants as compared to human transplants (Eschenhagen and Zimmermann 2005).

While we have also found materials that can be inserted into fatty tissues such as pacemakers and pain drug injectors and seemingly not rejected, the body will often form a barrier around such objects to protect the body from them. This surrounding of invaders is also a form of scanning and rejection. Researchers have discovered this effect among pacemakers. Pacemakers and wire leads are subject to gradual immune rejection, and thus need careful consideration as to the insulation materials used. A breakdown of their chemistry can lead to poisoning and other disastrous results as reported by Bruck and Mueller (1988).

Simply using an understanding of the complex voltage potentials regulated and gated at the cell membrane—such as those within ion channels—we can see how synthetic chemicals can damage a cell. Chemicals with different voltage potentials have differing waveform characteristics and thus disrupt the voltage parameters of the channel. For example, some alkaloid-based toxins have a tendency to bind to and block some ion channels within the cell membrane. This causes the cell an insufficient flow of ions. The voltage on either side of the channel is thus altered. Other alkaloid toxins may repeatedly activate ion channels, forcing the gateway to stay open, overloading the cell's voltage potentials. Other synthetic chemicals may otherwise modulate ion channels. These sorts of modulations change the natural voltage potentials and change the natural electromagnetic waveform transmissions within and around the cell.

Even minute amounts of a synthetic chemical have the ability to disrupt channels within the cellular network. Because neurotransmitters and hormones activate these gateways with waveform broadcasting mechanisms, their molecular components are sometimes only required in the parts per trillion dilution range. As we have found in carcinogenic research over the past two decades, some synthetics may become toxic to the body—interfering and disrupting ion and endocrine channels—in the parts per billion range. At this level, their waveforms can easily overwhelm the balance existing within the gateway messaging system of these channels. For this reason alone, we can see how synthetic chemical production is like playing with fire: Disrupting the sensitive balance of natural waveform conductance. Untangling and restoring that balance, on the other hand, can sometimes be quite a gargantuan task for the body.

Even with trillions of dollars of research, chemical manufacturers cannot seem to synthesize safe chemicals. As one chemical's toxicity is discovered, another one is launched in its place: Perhaps with different or delayed toxic effects. While there might be improvements in dilution or half-life to reduce toxic effects, almost all synthetics will accumulate within living organisms in one respect or another—forcing destructive mutations in later generations. We must therefore ask ourselves: With all the financial exposure of the chemical industry in potential toxicity claims, how come they cannot seem to synthesize non-toxic synthetic chemicals?

In a 1998 report published in the *Journal of the American Medical Association* (Lazarou *et al.* 1998), approximately 2,300,000 people either end up being hospitalized, permanently disabled, or fatally injured as a result of pharmaceutical use every year. That is over 2.2 million people annually with *reported* injury from pharmaceuticals. This study, done at the University of Toronto, also showed that approximately 106,000 people die each year from taking correctly prescribed pharmaceuticals approved by the FDA. This does not include the number of deaths resulting by overdose or by addiction to these same drugs. The U.S. FDA was sent 258,000 adverse drug events in 1999. Harvard researcher and associate professor of medicine Dr. David Bates told the Los Angeles Times in 2001 *"…these numbers translate to 36 million adverse drug events per year"* (Rappoport 2006). The plausibility of this number is confirmed in another study published in the *Journal of the American Medical Association* in 1995 (Bates *et al.*). This revealed that over a sixth month period, 12% of 4031 adult hospital admissions had either a confirmed adverse drug event or a potentially adverse drug event. If we extrapolate this rate using the population of 300 million Americans, we would arrive at the same 36 million Rappoport calculated.

We might compare these horrific figures to the same incidence factor among herbal medication usage. According to the FDA, a total of 184 deaths and 2621 adverse reactions resulted from consumer use of herbal supplements over a five-year period. Most of these deaths were associated with weight-loss formulas involving obesity and cardiac events. Still, this is an average of 37 deaths and 524 adverse reactions per year. This means that herb deaths are .00037 of drug deaths or .037%. The adverse drug reaction relationship is 524/36000000 = .0000145 or .00145% of adverse drug events are herb-related. This report also estimated from several studies that herbal use in the U.S. ranges from 27% to 36% of the population during that period (Hirshon and Barrueto 2006)—as widespread if not more than pharmaceutical use. Furthermore, while pharmaceutical use is regulated and prescribed specifically by doctors, most herbal supplementation is self-medicated. We can thus conclude that while herbal use is as widespread as pharmaceutical use among the general population; the number of adverse events and deaths among herbal use is practically nil, compared to the millions of adverse events of pharmaceuticals.

For thousands of years, traditional doctors and scientists have gradually assembled and documented the particular botanicals associated with particular ailments. One of the earliest advanced records of herbal medicine is the *Pen Ts'ao*, written some 4500 years ago by a Chinese herbalist. The *Pen Ts'ao* recorded 366 different plant medicines and their specific uses. Ayurvedic texts—some even older—also document the use of hundreds of botanicals, as do the documents and lineage of the Greeks, the

Romans, the Polynesians, the New Zealand Mauris, the Aborigines, the North American Indians, the Indonesians, the Mayans, the Egyptians, the Arabs, and the Northern Europeans. Over the centuries, traditional yet scientifically established treatises on herbal medicine have accumulated from so many societies and traditions—all confirming tremendous healing activity among botanicals. Plants from various parts of the world were analyzed and tested for their usefulness by different cultures. Many cultures used the same plants, while some plants grew more prolifically in some areas and their usefulness was focused upon by that culture. In all cases, we find all of these cultures found tremendous use in botanical medicine.

All of these traditions together assemble a grand study, complete in its double-blindedness and substantial in its population. In this study were billions of participants of different cultures, many unknown to the others. Each culture developed and used various botanical medicines, many unaware that other cultures were using some of the same botanicals or botanical families, yet experiencing the same results. These different cultures—some ancient and some a few hundred years old—all found botanicals with primarily the same curative properties.

It is only over the past century or two that humankind's quest for ownership and profits drove humans to isolate and attempt to synthetically reproduce the constituents of these natural medicines. The primary fact that modern societies have given patent rights to those who develop an isolated synthetic chemical has promoted an increasing concept of molecular ownership. This chemical ownership policy has spawned several generations of chemical manufacturing businesses who have exploited and damaged our environment in their exclusive attempts to gain wealth.

Yet a number of sources estimate at least 60% of pharmaceutical drugs are founded upon one or more naturally occurring biochemicals, many from botanicals. Some sources have gone on to say that at least 30% of currently prescribed medicines are synthetic analogues of botanical constituents.

However, we need to carefully distinguish between an herb and a pharmaceutical. While pharmaceuticals are isolated constituents with typically have one mechanism of action, most botanicals have multiple—some even hundreds—of pharmacological constituents. For example, according to the research of James Duke, PhD of the U.S. *Department of Agriculture* and Norman Farnsworth, PhD, a research professor at the University of Illinois, ginger has at least 477 active constituents (Schulick 1996). As we compare single-constituent pharmaceuticals, this fact is simply astonishing. While nature produces active chemical substances that act in a healing manner, she also produces active multiple constituents that buffer and balance each other.

Separately one or two of ginger's active constituents might produce side effects. The other constituents present in the whole plant balance the various constituents, however. As a direct result of these many constituents in ginger that make it one of the most active and effective medicinal botanicals for so many ailments. Many herbal medicine experts consider it one of the best herbal anti-inflammatory substances. Ginger has been shown to suppress the expression of both cyclooxygenase-1 and cyclooxygenase-2, two enzymes that stimulate inflammatory responses. Ginger also slows leukotriene production through the blocking of the 5-lipooxygenase enzyme (Grzanna *et al.* 2005). Unlike the parade of side effect-ridden COX inhibiting non-steroidal anti-inflammatory drugs introduced by the pharmaceutical industry over the past two decades, ginger provides a level of safety and a myriad of nutritional benefits. We might add that these pharmaceuticals—some of which have been pulled from the market for serious cardiovascular problems—have required billions of dollars of commercial research by the pharmaceutical industry and vast amounts of taxpayer investment in government oversight. If even a small amount of this wasted effort was diverted to the development of better ginger delivery systems, we could have saved the planet and our population from millions of tons of toxic synthetics.

On March 10, 2008, a report was released by the *American Press National Investigation Team* that pharmaceutical drugs taken for pain, high cholesterol, asthma, epilepsy, infection, heart problems and mood disorders had contaminated the water supplies of over forty-one million Americans. In a finding that confirmed a 2002 report by the U.S. Geological Survey, pharmaceutical drugs are infecting the country's rivers and streams nationally. These drug residues slip through the filtration systems of municipalities, as they were not designed for treatment for these pollutants. The source of this pharmaceutical pollution appears to be primarily the flushing of medications down the toilet as those medications expire or are otherwise thrown out. This is of course compounded by the waste streams of pharmaceutical manufacturers.

One report recently stated, for example, that a quite large and well-known pharmaceutical manufacturer generates some 2,400,000 pounds of hazardous waste every year, which escapes into the surrounding countryside and its waters. The report also stated this same manufacturing facility caused genetic pollution among a variety of crops near its manufacturing facility.

According to Schulick, through 1996, the pharmaceutical industry has launched at least two hundred pharmaceuticals and spent over $70 billion aimed at reproducing the eicanosoidal-inhibitory effects ginger and other anti-inflammatory botanicals. Not only does ginger act thoroughly and without side effects, but it is also a flavorful ingredient in food and grows profusely in natural environments. How much better can nature get?

The issue modern medicine has with herbal medicine is the speed in which its therapeutic effects can be seen. This is also its benefit, however. While pharmaceuticals are fast acting, they also create imbalances within the body that require the body to work harder in other ways to detoxify the drug. Medicinal botanicals have no toxicity, but they work slower and more gradual. This forces the patient to be disciplined, and well, patient. It is because natural herbal medicines are complex and balanced that they tend to act more slowly. They also produce a result that is long-lasting, working with the body to strengthen its immunity and ability to resist the problem in the future.

Pharmaceutical eiconasoid manipulation with NSAID pharmaceuticals has, quite frankly, been a disaster in almost every respect. Each has some negative side effects. Some alter the stomach lining, causing ulcers. Some damage or overburden the liver. Some have been linked to hypertension. Some have been linked to thin skin and bone weakness. Some have been known to weaken immunity. Ginger does none of this. Rather than *cause* negative side effects, ginger and other botanicals in the same category—such as curcumin, feverfew, holy basil, oregano, rosemary and baikal skullcap—modulate eicanosoids, reduce fever and inflammation while having other healthy side effects (LaValle, 2001). According to various sources, ginger also stimulates immune function; increases healthy appetite; reduces nausea; protects against ulcers; increases liver vitality and circulation; calms nerves; stimulates endocrine function; encourages bone healing; increases lung and gum health; and neutralizes oxidative radicals. Ginger is antiseptic, anesthetic, and antimetic—among so many other characteristics and affects (Schulick 1996). Laboratory studies have also shown ginger to be protective against cancer (Shukla and Singh 2007).

There are a number of other excellent botanical treatments for pain to consider before resorting to the array of increasingly toxic pharmaceutical painkillers. Botanicals such as *cayenne pepper, white willow tree bark* and *meadowsweet* are good examples. The latter two botanicals contain a constituent called *salicin*. This of course is the natural version of the synthetic *acetyl-salicylic acid,* known by its expired patented trademark name of *aspirin*. Acetyl-salicylic acid and its non-acetylated *salicylic acid* come with a side effect of gastro-intestinal upset, sometimes leading to ulceration and stomach bleeding. Other documented side effects of aspirin include *Reye's Syndrome*, liver damage, nephritis, central nervous system issues such as tinnitus, dizziness, vertigo, and the risk of uncontrolled bleeding due to the thinning of the blood.

Far from being natural, today's aspirin is usually manufactured from phenol, which can be derived from coal or isolated from other components. The phenol is treated with sodium and then carbon dioxide under pressure, rendering salicylate. After acidification into salicylic acid, it is acetylated with acetic anhydride to yield the final product. The environment surrounding each of these reactions produces significant magnetic influence to yield the synthetic result. Though it successfully relieves pain, it also creates imbalances. We know that aspirin has been shown to disrupt the cyclooxygenation (COX) process, which oxygenates fatty acids utilizing an enzyme called cyclooxygenase. The natural process that produces prostaglandins and prostanoids such as thromboxane—which inhibits the inflammatory immune response of the body—is thus blocked. When the body's immune system senses an invader, this cascading inflammatory process helping to rid the body of the invader is interrupted. Aspirin inhibits this natural inflammatory response without the detoxifying biochemicals contained within the natural versions.

This means the fever the body produces as part of its detoxification process to rid the body of a toxin is interrupted without an outlet. Yes, the headache—a signal indicating a metabolic problem exists—may be squelched. The joint pain—indicating there is not enough water in the system or an internal infection to detoxify—may be eliminated. While most of us might consider these effects as positive reasons to take aspirin, without a correction of the cause of this inflammatory signal, the problem will accumulate. The headache will come back. The joint pain will remain. Pain is a signal for us to make a correction to remove the cause.

Aspirin's disruptive mechanisms are illustrated in its effective depletion of a number of nutrients from the body. These include iron, potassium, folic acid, and vitamin C. According to Kauffman (2000), death rates among populations of aspirin users are significantly higher than non-aspirin populations. Conservative estimates state 7600 deaths and 76,000 hospitalizations occur within the U.S. from NSAIDS like aspirin (also including Motrin®, Aleve® and Celebrex®).

White willow bark and meadowsweet—by nature's design—contain salicin buffered by a variety of constituents to balance these side effects. Whole botanical salicin plants do not have a history of these side effects. Their molecular arrangements are synergistically oriented with a balance of constituents, which resonate with and stimulate the natural rhythms of our bodies. While botanical pain relievers have the ability to gradually modulate the eicanosoid response (to decrease the headache or pain), they also stimulate healing mechanisms to solve the root of the problem. Cayenne contains the alkaloid *capsacin*—known to reduce the amount of *substance P* in nerves, thereby reducing pain transmission, for example. Cayenne also has other healing effects, which help to mitigate the actual cause of the pain though. *Capiscium frutescens,* as it is known in botanical terms, contains not only capsaicin but also various carotenoids, flavonoids, volatile oils and vitamins A and C. These constituents work together with harmonic synergy. As a result, cayenne pepper is antibacterial and antiviral; increases circulation; increases sweating; stimulates appetite; increases liver and heart function; stimulates the immune system and increases metabolic function.

Meadowsweet, or *Filipendula ulmaria,* has three types of salicylates; opiraein, salicin and gaultherine. These three act conjunctively to reduce fever, and reduce inflammation for such issues like headaches or rheumatism. Whole meadowsweet also contains mucilage and tannins, which buffer the salicylates' potential negative effects within the gastrointestinal system. The tannins in meadowsweet also have therapeutic effect upon the digestive tract. For this reason, meadowsweet has been used successfully in cases of digestive disorders, including diarrhea and colitis. Meadowsweet also has antiseptic diuretic action, making it useful for detoxifying excess uric acid. It increases detoxification through sweating while still reducing fever through its salicylate action.

The contrast between the action of synthesized aspirin versus its natural ancestors clearly illustrates the significant advantage nature has over synthetic chemicals. When these points are overlaid against the tremendous focus humankind has given to the effort of substituting traditional natural remedies with synthetic *patentable* chemicals, it says something else: We are missing a significant reality. There is a process occurring in living organisms we simply cannot replicate. This is the harmonic of nature.

Another prime example illustrating the difference between biomolecules produced within the cyclical environment and those produced synthetically is the controversial *hormone replacement therapy* (or HRT). Synthetically produced estrogen and progesterone (or progesterone extracted from horse urine) has become increasingly controversial following several research projects. A study of nurses reported in 1985 that women taking estrogen had a 50% less risk of heart disease (Stampfer *et al.* 1985). This study reverberated through the 80s and 90s among prescribing physicians who suggested to their women patients that estrogen was not only helpful for menopause difficulties, but had a wonderful side effect of being good for the heart.

Eventually, this helped provoke the Women's Health Initiative, a 15-year study launched to study cardiovascular disease, cancer and osteoporosis prevention strategies in women aged 50-79. One of the major focuses of the study was hormone replacement therapy. Diet and other factors were also examined in this study of 161,000 women. While the study was to run until 2005, The National Institutes of Health—who oversaw the study—suddenly halted the study. Apparently, the hormone replacement group showed a higher risk of invasive breast cancer, as well as increases in strokes, embolisms, and heart disease. Several conflicting reports followed. A 2003 New England Journal of Medicine documented that HRT women who took estrogen within 10 years of menopause actually experienced a *reduced* risk of heart disease by 11% (Manson *et al.* 2003). The NIH published another report in 2007 (Rossouw *et al.*) illustrating that additional statistical analysis of the same study revealed different risk levels depending upon age and years from menopause HRT was initiated.

This type of statistical analysis of research data has been a source of increasing confusion over the years, as data can be looked at one way once, and then recalculated using different assumptions and comparisons. This took place in the WHI studies. Re-interpretations of the study results have continued to be made, despite the size of that study. The age variances and years from menopause are just two of many variables that could be considered in this study. Other variables might also have been considered. Overall, it appeared that women who took estrogen around the time of menopause and perimenopause did not experience the same risk factors as those who took HRT later in life—such as over 70. In 2007, the Journal of the American Medical Association report noted above (Rossouw *et al.*) calculated that women who took estrogen between ages fifty and sixty had a 7% reduction in heart attack risk, but had a 13% increase of stroke, and then—curiously—a 30% lower risk of death of any cause. Meanwhile women over 60 and under 70 had a mere 2% lower risk of heart disease and a whopping 50% greater risk of stroke, and then a 5% greater risk of death of any cause. Finally, women between 70 and 80 years old had a 26% greater risk of heart attack, a 21% greater risk of stroke and a 14% greater risk of death by any cause.

The pharmaceutical industry concluded these results should allow peri- or post-menopausal between 50 and 59 years of age to take HRT without a significant risk. This age group did not seem to have the negative heart health risk that older or younger women had with HRT. What about the 13% increase in strokes? Stroke must not be a major concern to these manufacturers or physicians.

At the end of the day, the results have proved confusing for many doctors, let alone their women patients. The real question might be why are doctors still prescribing estrogen at all after greater risk factors were discovered in any age group?

Prempro® was the primary agent used in the WHI study. Prempro® is a combination of Premarin® and Provera®. Premarin® is a hormone extracted from horses. This hormone is called *equilin*. Provera® is a synthetic version of progestin, patented in the 1940s. While the progesterone naturally produced in the body is active throughout the body's cell network, Provera®'s progestin-progesterone primarily affects the uterus. This of course creates further imbalances among estrogen within the body, requiring more HRT manipulation.

Certainly, the body naturally produces its own (endogenous) estrogen and progesterone complex. While we might summarize that a healthy body produces its own estrogen and progesterone, the body's estrogen is actually made up of a balance of different molecules. *Estrone, estradiol,* and *estriol* comprise the bulk of the body's estrogen. However, a healthy body will also alternate dietary source phyto-estrogenic compounds to provide a buffered balance between estrogen receptors and ligands. The balance of estrogenic molecules resonates with botanical phytoestrogens as the body ages and begins to produce less. This is part of the harmonic between nature and the physical body.

Estrogens and progesterone are primarily manufactured in the ovaries, and perfectly timed in conjunction with an environment of nature's other waveform cycles. Because these are natural biochemicals produced by the rhythmic body, when unnatural or synthetic estrogens and/or progesterones are consumed, the body's natural rhythms become confused. This creates a number of subtle metabolic dysfunctions as the body seeks to rebalance itself. For example, some HRT medicines contain higher proportions of estradiol. Interestingly, most breast cancer patients also test higher in estradiol and lower in estriol.

The types of dysfunctions created (heart disease, strokes, and cancer among others) by hormone replacement therapy are created primarily because we simply do not understand all of the cyclical harmonics at work within the body. When we attempt to replace a cycling biomolecule normally produced within the body or consumed naturally with a synthetic or analogue version, negative results are to be expected. The introduced molecules are simply foreign to the body and its cycling processes.

There are a number of ways to naturally balance hormone flow within the body. These must by synergistically applied, however. The 'shotgun method' comes with various risks. There are a number of botanicals that contain isoflavones and lignans, which work as naturally weak phytoestrogens as mentioned above. These include beans, nuts, a number of grains and certain roots. There are also various herbs such as black cohosh and chasteberry, which have been used successfully for thousands of years to balance the female's natural rhythms when they get seriously out of whack. These happen to not have the confusing array of side effects documented by HRT.

There are various other strategies to add, to help achieve the natural rhythms of hormones and metabolism. These include adequate exercise, sleep, diet, hydration, sun, color exposure, and cycle timing. In addition, exposure to various toxins including plasticizers, VOCs, smog, fragrances, and other chemicals can dramatically alter our hormone levels. These are termed *hormone disruptors*, because they can stimulate over- or underproduction of certain hormones, throwing off the balance of the entire hormone cycle. Eliminating or reducing toxic exposure is necessary if we want to regain our natural biochemical balance.

Unless the ovaries or uterus have been removed, the female body has an innate ability to produce the right mix of hormones given the right natural stimulation and environment. This is our challenge: To instigate the right natural mix stimulated by the body. There are also a number of alternative therapies to add to the above, such as homeopathic compounds, and in the worst case, bioidentical therapies can be considered. A number of physicians and pharmacists are now applying bioidentical therapies to their practice.

As cancer researchers have probed for various anti-cancer prevention strategies, the overwhelming evidence points to natural foods—especially those from plant sources—as most assuredly chemo-protective. In other words, those chemicals produced within botanical living organisms prevent cancer. They prevent DNA mutation. They stimulate the immune system. They augment the process of detoxification. Foods noted for anti-cancer benefits include *apples, asparagus, barley, basil, beans, beets, various berries, broccoli, Brussels sprouts, buckwheat, bulgur, cabbage, cantaloupe, carrots, cauliflower, celery, cherries, chili peppers, corn, fiber, flaxseed, garlic, grapefruit, grape juice, mangoes, milk, mushrooms, nuts, oats, olive oil, oranges, pears, peas, pectin, prunes, raisins, rhubarb, rice, saponins, soy, various spices, wheat,* and many others (Yeager 1998). New anti-cancer biochemicals are continuously being found among nature's foods, herbs, and fungi.

For example, research indicates the intake of selenium—a mineral contained in nuts, grains and other foods—appears to protect the cell from the mutation of the p53 gene (Fischer *et al.* 2007). The p53 gene is now referred to as the *tumor suppressor gene*.

We are hard pressed to find edible botanicals that do not contribute to the immune system in one way or another, in contrast to pharmaceuticals, where we are hard pressed to find safe synthetic medications.

As researchers have investigated the causes of cancer, overwhelmingly the biggest culprits appear to be synthetic chemicals or our generally pervasive chemical-laced environment. Air and water pollution, synthetic building materials, cleaning agents, chemical food additives, and many others have been implicated as absolute carcinogens. Unnatural elements are carcinogens because their waveforms inflict, disrupt, and damage cells throughout the body, notably the vital cells of the artery walls, the stomach, the intestines, the heart, and the brain—creating a basis for genetic mutation while switching off the protective p53 gene.

It is evident that plant-based natural biomolecules are for the most part chemoprotective, while synthetic substances are mostly carcinogenic.

What does this tell us about the difference between the two types of chemistry—synthetic versus natural? While natural biochemicals are produced within conscious living organisms within a natural environment, synthetic chemicals are created in sealed laboratories and factories using extraordinary heat and unnatural catalyst combinations. While natural biochemicals work harmonically within nature's balance, nature's exposure to synthetics must be limited to minimize disease and detoxification.

Living Biochemistry

The processes of the natural environment are driven by living organisms. This component is illustrated by the fact that specific plants help cure specific ailments. Is this not a curious mechanism? What connects these particular botanicals to particular ailments? Why does ginger treat digestive ailments while eucalyptus treats respiratory ailments? Why does cayenne stimulate circulation while skullcap slows metabolism and nerve firing to help us relax and sleep?

The connection between the botanical species and the disease is more clearly understood from the perspective of consciousness. This issue was discussed in great detail by early twentieth century Dr. Edward Bach, a well-educated British physician. Dr. Bach described that at its very fundamental aspect, a disease is an expression of a particular deficiency or malignment between the inner self and its physical body. He taught that this deficiency created the need for learning particular lessons. In other words, particular diseases are associated with particular lessons to learn.

This concept was also shared by a German physician now considered the father of homeopathy, Dr. Samuel Hahnemann a century earlier. Though Ayurvedic and Greek physicians also utilized homeopathy, Dr. Hahnemann perfected the techniques and discovered hundreds of different natural remedies. Dr. Hahnemann also introduced the concept of *miasms*. A miasm, Dr. Hahnemann proposed, was a particular deficiency of the inner being, expressed through physical tendencies and disease models.

The ability of plants to produce reflective medicinal effects yields a rather complex yet revealing discussion. As the thousands of years of traditional research have been compiled, it has become evident that while certain botanicals have particular medicinal effects, there also appears to be a connection between the medicinal effect and the physical appearance of the botanical. This link between botanicals' physical shape and medicinal benefit is illustrated in the North American Indian medical *Doctrine of Signatures*. This doctrine stated that each medicinal herb reveals its therapeutic properties through a *sign*. This sign shows a visual relationship between plants and ailments. This might be the shape of the root, the shape of the leaf, the color of the plant, or another aspect. For example, milkweed and plants with milky fluids are linked with ailments of the breasts, while roots like ginger—twisted and contorted like the digestive tract—is linked with stomach and intestinal health. It is also interesting that the root is where plant nutrient absorption takes place, while plant milky fluids are often a byproduct of their procreative activities.

The healing practice of the North American Indians, like so many indigenous medicines, also (independently) utilized the famed Hippocratic theory (and later homeopathic concept) of the *law of similars,* or *"like cures like."* Many medicinal botanicals reflect their properties by mimicking some of the same symptoms they treat. As these ancient pretexts have wound their way into modern medicine practice, there has been confirmation by research for much of these basic tenets. Today we find the same law of similars being applied in the form of vaccination, allergy treatment, and digestive issues.

As Ayurvedic medicine began to utilize thousands of years earlier, Dr. Hahnemann further documented the effective use of the law of similars with thousands of *provings*. A proving is the clinical establishment of the therapeutic effects of a natural element by observing the effects upon a healthy person. If the element created symptoms, that element was considered curative for precisely those symptoms in diluted doses. Elements used in *provings* over the last 250 years of clinical homeopathy have included botanicals, rock minerals, venoms, and animal secretions. In all, over 2000 homeopathic treatment elements have been *proved.*

Dr. Bach extended the curative nature of homeopathics beyond physical ailments, by disclosing and proving the curative effects of various flower essences for psychological and mood disorders. In addition, many centuries before, Ayurveda elaborated on the curative effects of gemstones for healing not only physical ailments but specific mood and psychological issues.

As we examine again the usefulness of botanical biochemicals for therapeutic purpose, we cannot escape connecting the living nature of these botanicals with their therapeutic properties. Just as our weight, skin complexion; even hair color and eye color our consciousness, we may consider each plant's nature a reflection of that plant's inner self. If we consider that each plant is also housing an individual inner self, just as the higher forms like animals and humans portray an individual self within. When we assume that our physical body reflects our conscious intentions, we can then cross-reference our ailments with the strengths expressed by the consciousness of certain medicinal plants. In other words, their expression perfectly juxtaposes with our expression of deficiency. This understanding was arrived at by Dr. Bach.

Because plants are stationary, they must protect themselves with their various biochemicals. This means the biochemicals they produce interact with environmental threats in the same way those biochemicals will interact with environmental threats within the human body. The consistency of nature's rhythms allows for a consistency of therapeutic biochemistry.

Flower essence biochemistry provides the ideal expression of plant consciousness as it interacts with its environment and other plants. While the constituents of plant roots and plant leaves will each have different effects when we apply them medicinally, the essence of the flower produces yet another effect. Furthermore, each type of flower produces a distinct effect. From biology we learn the flower of a plant is its expression of its sexual activity. A flower will display color and delicate beauty, just as a female human looking to attract a man might wear a colorful dress with delicate high heels. The flower appropriately expresses the plant's personality and moods on during its pollinating season. This contrasts with the photosynthetic activities of the plants' leaves, or the nutrient-absorptive root activities. As Dr. Bach illustrated and evidenced by the billions of people who have benefited from Dr. Bach's flower remedy discoveries, these expressive flower essences happen to help us with our expressive areas of emotions and moods.

Though research on flower remedies overall appears scant, a famous flower essence blend called rescue remedy has undergone a variety of controlled studies with humans and animals (Muhlack *et al.* 2006). Made from essences of the flowers of the cherry plum, clematis, impatiens, rockrose, and the star of Bethlehem, this remedy quickly treats mood-related issues of anxiety, trauma, and emergency-related stress. The efficacy of the rescue remedy has not only been used successfully by humans over the years, but on a variety of animals, including pets and horses. In both animal and human use following a traumatic experience, a few drops of rescue remedy consumed directly, or in a glass or basin of water has repeatedly been shown to bring about a state of relaxation without any side effects. For this reason, rescue remedy is a favorite among many who are feeling nervous before traumatic events such as driving tests, public speeches, or following accidents and other traumas. While Dr. Bach's research focused on thirty-seven flowers growing throughout England, further research over the years has revealed hundreds of other flower essences with a variety of effects.

This discussion illustrates how consciousness is expressed through the biochemistry of our environment. We can see this directly as we review the thousands of studies linking moods and emotions with the various neurotransmitters and hormones produced by the body. We have observed that a conscious emotion of fear will stimulate the body to produce adrenaline, acetylcholine and other 'fight or fight' biochemicals. We have observed how the conscious feeling of love or joy stimulates the production of oxytocin and dopamine. These different chemicals are produced by the body in response to the self's conscious feelings and emotions. Connecting consciousness with the production of biochemistry is not such a far-fetched concept after all.

As we ponder the fundamental differences in consciousness between the intention to respect and work with nature and the intention to manipulate nature, we can each individually examine our own attitudes regarding this discussion. What, we may ask, is the difference between harvesting a natural herb or a natural fiber from the fields and the production of synthetic chemicals for the same ultimate purpose?

The difference is our consciousness. Synthetic chemical production expresses a consciousness of *arrogance*. While we are aware of many of the nuances associated with the delicate nature of the nature's biochemistry, we ignore this delicate balance while we seek to accomplish our goals. When arrogance is combined with greed, we surge with self-confidence. This self-confidence directs us to believe we can duplicate nature's processes better with our own technologies.

Humankind's recent need to create this massive industrial chemical complex—with its infrastructure of chemical manufacturers and large-scale distribution centers around the globe—is built upon a falsehood:

That falsehood is that we can do better. That falsehood assumes that something is missing from nature. That there is no design in place. That there is no meaning and reason for our existence. As a result, we assume we can do what we want without any consequences.

We can see this particular virus of arrogance deeply entrenched when we hear of many scientists' views of the future and how technology and synthetic chemicals will bring about greater luxuries and happiness. Meanwhile, there is no net increase in happiness or ease existing today as a result of our modern technologies. We are only experiencing an increase in complexities and problems. For every chemical we manufacture and release into the environment, we are faced with the problem of how to remove or neutralize that chemical to avoid our own extinction.

It is not hard to rediscover the natural biochemical options that can provide immediate balance. Over the past decade, as a society we are re-discovering natural alternatives to just about every synthetic chemical. The environment gives us natural antibiotics, natural pesticides, natural cleaning agents, natural pharmaceuticals, natural health and beauty agents and of course natural foods. Some of us have found we can clean with natural soaps, disinfect with vinegars and citrus, repel insects with various oils and soaps, and fertilize with compost, to name a few. Some of us have recently discovered that the same creativity that went into making synthetic chemicals can also be applied in working intelligently with nature's own biochemistry. These realizations are still a tiny drop in a large ocean of destructive forces at work in our pursuit of a synthetic chemical environment. A mass change will require a big dose of humility: An admission that our environment has been designed by a greater Intelligence for a purpose greater than our self-gratification. Will a mass environmental disaster be required to force this humility?

Harmonization: Global Detoxification

❖ Quite frankly and sadly, it is too late for some things. We cannot turn around this super-tanker of synthetic pollution easily because there is just too much momentum and the iceberg is looming. About the only thing we can do at this point is understand this society's arrogance will result in a series of traumatic lessons, and take our own personal steps to harmonize ourselves with the planet and get clear on our own values. The earth will be able to cure herself in her lifetime—after she throws off this infectious virus of humankind bent for tox-icity. She will surely throw off this toxicity just as we might recover from a bad case of influenza or a cold. The global warming problem is not unlike a fever to throw off a viral or bacterial infection.

❖ Our infectious society must undergo a necessary environmental cleansing as a result of our toxic production. We can see this effect taking place throughout the planet today. The planet is running a fever, with gyrations of temperatures, rainfall, storms, fires, earthquakes and rising seas. Again, the increased level of natural disasters we are experiencing is part of the planet's process of cleansing, and we are the infection to be cleansed. How bad the cleansing will be depends upon what we do from this moment forward. Either we begin to detoxify our environment by choice or the planet will detoxify us. It may not all happen this year, or even this decade. Gradually, the earth will increasingly become uninhabitable for humans unless and until the human race learns to become probiotic (or at least symbiotic) instead of pathogenic.

❖ Detoxing the planet from our chemicals will detoxify our bodies at the same time. By removing our pollutants from the atmosphere, waters and soils, the planet will begin to slowly revitalize as we gain cleaner air to breathe, cleaner waters to drink and better soils to grow food with.

❖ We can act personally in many ways to become increasingly aligned with the design of the planet. This will require some adjustment. A chemical-cleansing process can be begun immediately and consistently. We can begin the withdraw from synthetic chemical use in our foods, cleaners, furniture, and houses.

❖ This will take teamwork. We will have to work together as consumers and manufacturers to make this change. It will require a monumental economic change. It may take years, but we must resolve and intend to begin making a complete withdrawal now.

❖ We are reminded of Rachel Carson's approach to the spraying of pesticides: *"Spray to the limit of your capacity..."* meaning spray as little as possible. As we weigh the benefits versus the costs of either producing chemicals or succumbing to massive infestations of insects, we need to have another plan. Today our massive chemicals are disrupting our environment with synthetic toxicity in so many ways. We must immediately begin using natural-based solutions to achieve the same results: There are many natural methods for reducing insects and infections. These include vinegars, sulphur, minerals, oils, and so many other deterrents. Organic farming is not only viable, but many organic farms produce better yields with more nutrients than their synthetic neighbors growing the same commodities. Many organic crops have proven to be more disease resistant and more drought tolerant than their conventional cousins. Organic plants have stronger immune systems, giving them more resistance to funguses and other diseases. This harmonically proves true for the human body as well. Organic foods are healthier and better for our immune systems.

❖ This will be a tough one, but we must cut our plastic production. It is nearly impossible in today's society to rid our homes from plastic. As far as plastic-wrapped foods, the only thing we can do for now is try to purchase more fresh foods and demand our manufacturers use recyclable containers. As consumers, we can gradually force this change. In the meantime, we can reuse the plastics we have now and try to stop using new ones. We have enough plastic now.

❖ As far as home furnishings, we can try to purchase natural fibers and natural woods as much as possible, while avoiding pressed woods with formaldehyde to the greatest extent. As for preserving our forests, New Zealand has illustrated how forests can be sustainably farmed. It simply takes discipline and determination.

❖ Our food can be purchased from farmer's markets, health food stores and farm stands. These markets provide fresh food with minimal packaging, and employ the least amount of carbon for distribution purposes. Buying from farmer's markets or local *community sustainable associations* also supports sustainable forms of local food production.

❖ So many goods are available used. We are a throw-away society. Buying used items recycles our goods and prevents more landfill trash. By buying used furniture, cars and used appliances to the greatest extent, we can reduce our burden on the planet.

❖ Using geothermal, wind and solar systems, we can reduce coal-, nuclear- and petroleum-based electricity. Progress towards implementing electric, fuel-cell, or solar-powered automobiles can eliminate carbon-based travel. Cellulosic ethanol also provides a good fuel source during the interim.

❖ There are so many options today for living in harmony with the planet. Our houses, clothes, transportation, personal care, and energy production can all be changed if we want to. If we beat the earth's detoxification of us by detoxing ourselves of our synthetic chemical addiction, we will be saving our race from extinction.

Chapter Nine

Dynamic Color

It was 1666 when Sir Isaac Newton first projected the sun's rays onto a wall after passing them through a prism and a narrow slit. As he contemplated the amazing rainbow of colors on the wall, he considered the cause. Did these come from the light or the prism? Rene Descartes had tried to explain it as refracted light—the colors were created by the refraction angle. Newton provided the answer to this debate as he then passed the light coming from one prism through another prism, which changed the color rays back to the original single white ray. Upon passing through yet a third prism, the light again resumed the color spectrum. This clarified to Newton that the refraction explanation could not provide the solution. If so, the second prism would yield yet more colors rather than reverting back to white light. Light, Newton proposed, must actually contain these colors. The concept of the *electromagnetic spectrum* was born.

The confluence of spectra driven in the electromagnetic is codified as light. Light in turn is captured within the umbrella of an all-encompassing *white light*. The white light of course has never been proven to exist, although Georg Cantor, a German mathematician at the turn of the nineteenth century and inventor of the *set theory*, spent many years attempting to prove the *continuum hypothesis*. The continuum hypothesis related finite sets to infinite sets. Extended into the plane of spectra, this continuum would connect visible light to an all-encompassing white light.

The white light has been discussed for thousands of years, first documented in the ancient *Vedas* of the Asian continent. Here the white light was discussed as the *Brahman effulgence,* or emanation from the Supreme, and all material components were derived from it. The white light has since been documented within many other spiritual texts as a vehicle of transcendence.

The concept of the white light containing many other spectra of material densities and waveforms is quite similar to the notion of visible light rays containing the various spectra, visible as colors as they refract and reflect through our environment.

Color, or *chromatics,* is the translation of particular energy waveforms by the cones of the eye. This means that colors have a number of characteristics besides what our minds perceive as color. William Snow, M.D. documented that blind people can perceive color without the use of eyesight. He explained that the *"radiant light, heat and color are capable of setting up responsive vibrations in animal tissue, inducing responses relative to their intensity…. their wavelengths and frequencies."*

We have discovered through research that other rays of the electromagnetic spectrum are capable of unique physical effects. Consider cosmic rays and gamma rays, which can penetrate tissue, causing various organ and cell damage. Consider x-rays, with their potential for radiation damage with too much exposure. Consider ultraviolet rays with their potential to damage skin cells. Consider radio and television waves with the ability to carry information through buildings and other physical obstructions. The visible spectrum contains waveforms within the same range of spectra. The colors of the visible spectrum certainly influence physical structure just as do these other waveforms. However, their effects are generally more subtle and less damaging. Similarly, each color has distinct effects.

Color has been used therapeutically for thousands of years with overwhelming success. It has been an important element of Ayurvedic, Chinese and Egyptian medicinal therapies. Goethe's 1810 book, _Theory of Colours_ related color with Hippocratic medicine. He described the four basic colors intertwining with the four basic humours of the physical body within a circular wheel, which he coined the _"Temperamental Rose."_ Goethe tested subjects and moods, describing character associations with colors, stating that, _"Every colour produces a corresponding influence on the mind."_ Light and color therapy has been used amongst psychologists thereafter. Colors were used therapeutically in European asylums. Painted walls with violet or blue brought about a calming effect for anxious patients; while red, yellow and orange brought about increased activity among depressed patients.

The Ayurvedic science utilized color therapy using gems for thousands of years. Utilizing some of these principles, during the first part of the century, an Indian Colonel and self-described metaphysician and psychologist named Dinshah Ghadiali wrote and lectured famously about color therapy. In 1920, he invented a machine called the _Spectro-Chrome._ Equipped with a 1000-watt light bulb, the device had five sliding glass color plates that could be mixed and matched to create up to twelve colors. Ghadiali's instruction manual for the device—_The Spectro-Chrome Metry Encyclopaedia_—documented various therapeutic case histories of the machine's use. While Ghadiali was subsequently dubbed a quack and his machine described as a fraud by the FDA and others in the medical establishment, some 10,000 of his devices were sold and used by a wide range of healthcare providers for many decades. There were also several other similar _chromo-therapy_ devices commercialized in that era. Ghadiali's was simply one of the best known. Ghadiali was said to have been influenced by Edwin Babbitt's _The Principles of Light and Color_ (1878). Babbitt's book proposed that everyone has a distinct energy color. Babbit suggested illness is at least partially caused by disturbing our unique color balance. Healing, he proposed, could be hastened by re-establishing ones color balance. A schoolteacher, Mr. Babbit also invented a popular device for this purpose, called the *Chromolume*.

In 1946, Ghadiali was tried and eventually convicted of fraud. The FDA put the theory of color therapy on trial along with Ghadiali himself, and color therapy was functionally discredited in western medicine. Ironically, in that same year a Swiss psychologist named Dr. Max Luscher designed a well-received study using colors to assess personality characteristics along with a risk assessment of potential disorder trends. Developed for psychiatrists and physicians, the test indicated patients with higher risk factors for ailments of cardiac, cerebral, or gastro-intestinal origin, depending upon the types of colors the subject selected. The test became a standard among psychologists, as it proved clinically useful.

Indeed, marketers and advertisers—who have been successfully using color in their marketing campaigns and packaging—have drawn upon a wealth of practical and measured experience relating colors with purchase decisions. It is for this reason we see fast food restaurants advertising in yellows and oranges—hunger colors. Banks on the other hand will pick blues and grays with some reds showing stability and professionalism. Healthy food brands will choose greens and browns to appeal to the healthy ideals of some consumers. Some marketing research has indicated green by far the most appealing color to food consumers. For this reason, we see lots of greens among labels on our supermarket shelves.

The use of color in the practice of psychotherapy has remained somewhat consistent over the past century despite the FDA's case against Ghadiali. Over the last decade or two controlled research has increasingly confirmed color's therapeutic effects. Today color therapy systems like *Colorpuncture* (Peter Mandel) and *Chromo-pressure* (Charles McWilliams) are emerging, combining color therapy with other established therapeutic methods (Cocilovo 1999).

There is a volume of research now confirming the usefulness of color therapies. In a study done by Lund Institute of Technology researchers (Kuller *et al.* 2006) on 988 subjects in indoor work environments, it was concluded that brighter colors and lighting created higher moods among workers in four different countries.

In 1992, poor reading children were studied by researchers in the psychology department of the University of New Orleans (Williams *et al.*). Color overlay intervention increased reading comprehension in about 80% of the children.

Hypertension has been successfully treated with color therapy (Kniazeva *et al.* 2006). Preterm jaundiced infants have been successfully treated with blue and turquoise fluorescent lighting. Significant reductions in plasma bilirubin have been achieved with color therapy (Ebbesen *et al.* 2003). Ultraviolet B light color has been associated with a lowering of blood pressure (Krause *et al.* 1998).

The visible part of the electromagnetic spectrum ranges in wavelength from 380 nanometers to 740 nanometers. Within this range, the color red has the longest wavelength, ranging from 625 to 740 nanometers. Meanwhile, orange ranges from around 590 to 625 nanometers; yellow ranges from around 565 to 590 nm; green from 500 to 565 nanometers; blue from 450 to 485; and purple from about 310 to 380 nanometers. Of course, as the wavelength increases, the speed or frequency of each waveform cycle decreases. Red ranges from a frequency of 405 to 480 terahertz (10^{12} hertz) while on the other side of the visual spectrum, purple ranges from 790 to 840 terahertz.

The color spectrum is called *continuous* because there is no cessation or absolute break between colors. Rather, the true color of a range is apparent in the middle while the edges of the color transitions into the next color, rendering a mixing of the two colors. For example, an indigo transition between purple and blue may be apparent from some observers, and cyan should appear during the green and blue transition.

Through a variant mixing of these central colors, the mind perceives thousands if not millions of colors. Some researchers have documented up to 10 million colors as potentially distinguishable by humans.

The eyes are equipped with two visual receptors: rods and cones. The cones of the eyes are the primary receptors of bright light and the color spectrum. Humans typically have three kinds of cone cells, each with a different pigment: one is sensitive to the short violet color waves; one is sensitive to the medium green waves; and one is sensitive to the longer yellow wavelengths. Cones are not very sensitive to light, yet they pick up colors better in brighter light than in darker light. The rods are primarily light-sensitive cells, and thus they are useful for night vision. In fact, rods will usually only begin functioning during weaker light, distinguishing primarily black and white images.

The perception of distinct colors takes place through a contrasting process between the three types of cone cells and a *bleaching* out of others. Each different cone type has specific photoreceptors oriented to receive particular wavelengths of light. Each particular wavelength will stimulate a specific type of cone over another. A blending of multiple images from each type of cone it transmitted through the optic nerve and brain cells, providing a pallet view of various colors on the mind's screen.

The conversion of light into the neural pulses takes place through a transduction process. Like many other sensory receptors in the body, cone photoreceptor pigments become polarized by particular waveforms. When light of a particular waveform strikes a receptor, a depolarization takes place. The depolarized photoreceptor acts as a gateway, opening micro-channels through which sodium ions travel. The sodium ion movement stimulates the release of glutamate. Glutamate in turn adjusts the polarity of the neuron membranes, blocking or accessing further ion movement between receptors. This causes the bleaching effect between color determinants. The cone pigment itself is a protein called *iodopsin*, which resonates with a molecule called *retinal*—a molecule derived from vitamin A. When light hits the pigment, retinal's molecular bonding structure changes from a *cis* configuration to a *trans* configuration. If we translate this oscillation to a static image, we might imagine it being similar to the wing of an airplane being bent downwards towards the ground. This *trans* configuration closes the ion channel through a protein messenger called *transducin*. When the ion channel is blocked, the neural signaling impulses are sent through the optic nerve with a negative feedback of calcium ions.

Rather than initiating the flow of current, the stimulation of light onto the cones *shuts off* the regular flow of ions through the membrane. The concept of the *dark flow*—where a steady stream of ions flow between these cells and the nerves is shut down when stimulated by light— was reported in 1970 by Hagins *et al.*, from rod pigments surgically removed from the eyes of rats. This dark flow halting process was evidenced through the measurement of tiny voltages and currents among the photoreceptors.

What we are discussing then is a steady rhythmic current moving between the optic neurons, only to be intruded upon with the reception of light through retinal cells. When our eyes are closed or we're walking in pitch darkness, the rhythms flow. When light hits the cones a preponderance of retinal pigments stimulate particular waveforms. These intercept and shut down the ion channels. This interception process of shutting off the ion channels is called *hyperpolarization* (Nakatani and Yau 1988).

The photocurrents running between the retinal cells and the brain have specific waveform frequencies. These are classified as alpha waves, beta waves, or otherwise (Breton and Montzka 1991). This classification of waveforms is defined by the orientation of the brain wave oscillation frequencies that pulse through the brain. The interaction between these *visual reception waves* and the other oscillating rhythms circulating around the brain's neurons integrates visual perception into the mind's web, enabling a reflective picture to be observed by the self.

Note that this image reflected onto the mindscreen is not the actual image. It is a composite of colors, expectations, and a process of filling in the blanks. This filling process creates a unique visual experience for each person. Though we may compare and confirm we are all seeing some of the basic transmissions, each of us brings together a slightly different impression. This differential allows each of us an interpretive element, infusing our individual goals and objectives into the scene. This is why many people can watch the same show and notice different things. Perception is not an automated process. It is an act of consciousness.

As further evidence of this, in 1999 the *Proceedings of the National Academy of Sciences* published a study out of Stanford University firmly establishing that speaking the names of colors invokes the same brainwave response in subjects as seeing those colors would (Suppes *et al.* 1999). Upon hearing a word describing a color, the brain triggers a translation into a mental image. The perception by the conscious self makes no distinction between the source of the input.

Harmonization: Color Therapeutics

❖ Research has continued to connect the visual perception of color and brainwave response to our moods and behavior. Brain imaging has indicated that color stimulates corresponding brainwave patterns, which have been linked with particular moods and behaviors. The ancient science of Ayurveda correlates colors with particular energy states and subsequently different chakras and energy centers around the body. The mechanism for this subtle electromagnetic bridge is explained using wave resonance and interference. If we were to hit a piano key in a room full of pianos, the other pianos would begin to vibrate in the same chord. With color resonance, we can associate particular waveforms with other oscillations occurring within the body. After all, colors are part of the electromagnetic wave spectrum. As these waveforms connect with our body in some respect, they stimulate internal waveform responses just as touching a hot burner stimulates an immediate nervous response to pull our hand away.

❖ Red increases energy. Its longer wavelengths tend to stimulate higher frequency beta waves in the brain, vibrating at more than thirteen cycles per second with wavelengths of 630-700 nanometers. Its longer wavelength is responsible for the redness of the sunset. The longer wavelengths scatter less than the blues and violet waves as they interact with the atmosphere. Red tends to stimulate the body's autonomic systems, increasing heart rate and blood pressure. Red stimulates physical stamina, circulation, hostility, violence, competition, and jealousy. It also is considered stimulating to sexual activity. For this reason, it seems to make sense that we find passionate people or very active people wearing reds and driving red cars. Red cars also tend to receive more speeding tickets. Whether this is because red car drivers drive faster or red cars are more noticeable is debatable. We might also consider those who choose red cars tend to want to be more flashy and noticeable. Red is also typically attributed to the planet mars, known for its connection to war, passion, and the struggle for survival. Its waveform is said to vibrate at the plane of the first 'root' energy center chakra in Ayurveda. Ayurveda further associates red with the health of the anus and coccygeal area, including hips and feet. While red can be stimulating and aggressive, it can also bring about a sort of grounding, especially when its color is tinged with yellow and/or orange. This color combination of course leads to the color of brown, the basic color of earth. Also combining with the brighter orange brings about elements of passion and sexual activity. For this reason, a bright red dress or red roses has been used to stimulate passion between the opposite sexes. Because it stimulates greater energy and stamina, red is considered helpful for completing projects requiring great amounts of physical energy and focus.

❖ Orange tends to stimulate high alpha brainwaves, which oscillate between ten cycles per second and thirteen cycles per second at wavelengths of 590 to 630 nanometers. Orange has many of the stimulatory effects of red, but without some of the intensity and passion. Orange is therefore warming and anti-congestive. It suggests enthusiasm, creativity, and inquisitiveness. Orange is often accompanied by sincerity, thoughtfulness, and health. Orange resonates with the second energy center chakra according to Ayurveda. Thus orange is associated with the sacral area—the back and lower spine—and the lower abdominal area. Orange stimulates reproductive activity—as opposed to the passion of red. It also stimulates appetite and the movement of the colon. Orange resonates with aspects of family and parenting. Orange is also associated with wisdom and enlightenment. Activities that resonate the most with orange include family relationships, friendships and group organizations.

❖ Yellow stimulates lower alpha brainwaves, eight or nine cycles per second with wavelengths of 560 to 590 nanometers. Because this color resonates with the third chakra energy center located around the solar plexus, yellow resonates with spontaneity, compassion, memory, learning, and appetite. It is physiologically associated with the healing capacities of the stomach and upper intestines. Yellow apparently also resonates with the adrenal glands. Thus yellow may be a trigger for stress. Activities associated with yellow include memorization, study, and focus. For these reasons, yellow is also considered draining. Because it can stimulate the adrenals, it can also stress the body and mind. It also reflects light with a greater intensity, so it can be exhausting on the eyes and mind after some time. Yellow can be cheerful, but too much of it can be fatiguing. Yellow rooms seem to cause more anxiety. Babies cry more and couples argue more in yellow rooms.

❖ Green stimulates brainwaves in the higher theta region, about six to seven cycles per second with a wavelength of 490 to 560 nanometers. Green is calming and balancing. It stimulates growth, love and a sense of security. Green resonates with the fourth heart energy center. It is connected to devotion and giving. Green is also a soothing color, stimulating healing response, particularly among ailments related to the cardiovascular system—and considered beneficial to the heart, blood pressure, and congestion. Green is also stimulates the immune system, especially the T-cells regulated by the thymus gland. As a result, green is considered invigorating to the immune system. This is reflective of the fact that most green foods are immunostimulating and detoxifying in nature. The green frequencies of light tend to suppress the body's endogenous melatonin levels. Melatonin is crucial for the smooth operations of the immune system, aiding the process of relaxation and sleep—both of which are integral in the body's rejuvenating mechanisms. These mechanisms are also part of the body's rhythm cycles, translating energy to movement. As the flow of melatonin wanes in the early morning hours, the adrenal glands begin to secrete an increasing flow of corticosteroids like cortisone. A healthy rotation of this cycle allows melatonin levels to increase substantially at night to help us relax and get good sleep. Green radiation is also associated with problem-solving, negotiation, resolution, gardening and cooking. In Ayurveda, green stimulates the heart chakra energy center. The heart chakra connects our body's vital systems to our conscious emotions. As we express ourselves through our relationships with others, the cardiovascular system reflects those expressions. From this comes many such expressions such as *"you broke my heart."* Increasingly doctors are beginning to connect clinical cardiovascular disorders with distress among relationships and family.

❖ Blue stimulates lower theta waves in the five to six hertz area at wavelengths of 450 to 490 nanometers. Blue is cooling and calming. It slows metabolism. The rhythm of blue is gentle and holistic. Blue is associated with creativity and communication on both a spiritual and physical level. Like green, blue wavelengths help lower melatonin as the body increases adrenal hormones in the early part of the day. Blue activity is associated with relaxation, playing music and cleanliness. For this reason blue is associated with purification. Rightfully, the ocean and sky—both key elements of the earth's detoxification systems—are blue. Blue is also considered a color of stability and conservatism. Corporate executives and government administrators often choose blue for this reason. Blue is also a very good color to use around children, as it is calming and relaxing. Blue resonates with our fifth energy center chakra located in the region of the throat. Therefore blue is associated with the lungs, breathing, sound and the functions of the thyroid. The thyroid is the key regulating organ in core body temperature and the rate of cellular metabolism.

❖ Indigo stimulates low delta waves around one cycle per second and wavelengths of 400 to 450 nanometers. It resonates with the sixth energy center chakra located between the eyebrows. It is also considered the color of the 'third eye.' Indigo is therefore associated with clarity, intuition and intelligence. It is a color linked with decision-making and meditative thinking. The sinuses, vision, and the immune system are therefore stimulated by and resonate with indigo. Activities most associated with indigo would be highly intellectual activity, humanitarian behavior, medical research, and philosophical contemplation. It is a color is often associated with exploring the reason for existence.

❖ Violet stimulates higher delta waves from two to four hertz. These waves vibrate at a faster frequency than indigo, primarily because they are rhythmically more stimulating. Violet waves are associated with the seventh energy center, the *crown* chakra. Thus violet is associated with the raising of our consciousness and our personal spiritual quest to new levels. It is also associated with brain circulation, spinal fluid movement, and joint fluid condition. Violet activities are associated with deep meditation and inspiration, prayer, and spiritual insight.

❖ Many associate the color purple—a color somewhat blending indigo and violet—with luxury, royalty and nobility. It is considered a favorite of nobility, expressing feelings of superiority and dignity.

❖ Colors with lower frequency and longer waveforms such as red, orange, and yellow tend to stimulate physical activity and resonate with instinctual activities such as sexuality, reproduction, survival, digestion, and thermal energy production. Meanwhile the higher frequency, shorter wavelength color waveforms of green, blue, violet, and indigo tend to stimulate brainwaves associated with thoughtfulness, problem solving, and intuition.

❖ Research indicates blended colors have yet their own unique effects. Pink, for example, has been associated with sedative and muscle-relaxing effects. Behavioral therapists like Alexander Schauss, Ph.D. have reported that pink colors create a tranquilizing effect, preventing or slowing anger and anxiety. Interestingly, Dr. Schauss also reported this same effect among colorblind patients.

❖ Light blue color seems to bring about better mental performance. This relates to not only the calming effects of blue, but the softness of the color, which seems to effect intellectual cognition.

❖ Black is a shade and not a color. Still, black appears to influence seriousness and even depression. Former prisoners of the former Soviet Union have reported that the KGB utilized black cells to induce depression in their interrogations. Black is of course the color most associated with death, as it is worn at funerals. Other black clothing events (such as "black tie affairs") are associated with seriousness and the need to establish dominance and order. On a brighter note, black is also considered elegant and clear. Black clothing or automobiles tend to reflect an intent to be conservative and uncomplicated. At the same time, black can also express an intent to be bold and overstated. Black can express both simply because black can have both an abrupt and muted waveform—depending upon the consciousness behind its use. When choosing colors to wear or otherwise surround our physical environment with, we might consider the goals we intend to achieve. The colors with the longest wavelengths such as red, yellow, and orange can be used when we need speed, energy, stamina, and immediate responses. When we need to be 'upbeat' about an activity or event, brighter colors will lift our mood and behavior. These brighter colors are also useful to fight depression and sluggishness. If we need to boost our enthusiasm, these brighter colors are significantly useful.

❖ The cooler colors of violet, indigo, blue, and green will taper down and balance our energy levels. These colors will provide a relaxing meditative environment. We can let the stresses and the traumas slide right over us, as we surround ourselves with these colors. Certainly the greens and the blues are quite easy to be surrounded by. All we have to do is take a walk in a natural environment with lots of green trees and of course blue sky. This environment will stimulate greater intuition, problem solving, intelligence, and even possibly spiritual insight.

❖ A significant amount of research has confirmed that the same pigments providing the color in foods also give those foods their nutritional and therapeutic value. For example, curcumin—a color pigment giving turmeric its yellow color—has been shown to enhance the immune system. Curcumin has been identified as stimulating the production of T cells, B cells, macrophages, neutrophils, and natural killer cells. Curcumin also works to downgrade proinflammatory cytokines (Jagetia and Aggarwal 2007).

❖ There are so many other examples of the healthy effects of color pigments in our foods. Color pigment phytonutrients exist in so many natural fruits and vegetables, each with their particular benefits:

❖ Red foods like tomatoes, watermelon, apples, strawberries, red raspberries, and pink grapefruit contain nutrients like lycopene and astaxanthins. These biochemical pigments have been associated with cancer prevention, healthy lungs, cardiovascular health, and prostrate health.

❖ Orange foods like squash, carrots, oranges, papaya, and mango contain alpha- and beta-carotenes as well as cryptoxanthins. These components convert and support various sensory cells and enzyme processes, provide strong antioxidant activity and increase metabolism. For example, vitamin A is a frequent component in orange foods, which promotes metabolism and nerve transduction, not to mention healthy vision.

❖ Yellow foods like yellow pears, bananas, corn, and summer squash contain limonenes, luteins, carotenes, and zeaxanthins, pigments that guard against cancer growth, assist in detoxification, provide antioxidants, inhibit atherosclerosis, and assist in cellular metabolism.

❖ Green foods such as wheat grass, mustard, onions, leafy greens, cruciferous vegetables, and spirulina contain zeaxanthins, sulforaphanes, isoiocyanates, allyl sulfides, amino acids, and various vitamins like K and C, which assist with liver function, DNA repair, cell metabolism, cancer prevention among others.

❖ Purple foods like grapes, cherries, cabbage, eggplant, and various berries contain ellagic acid, anthocyanins, pomeratrol, pycnogenol, and other polyphenols. These elements work conjunctively to inhibit bacteria growth, increase detoxification, prevent cancer growth, and balance hormones.

❖ Brown foods like whole grains, nuts, and legumes contain good levels of amino acids, fatty acids, vitamin E, and various isoflavones such as lignans and phytoestrogens. They thus help modulate and regulate hormones, increase immune function, build proteins, cell membranes and provide fiber. They also provide grounding and a feeling of fullness to our meals.

❖ The world around us gives us many natural colors to help balance our various physical rhythms. At daybreak, we can see the coming sun with a red-orange-amber hue, stimulating our physical energy. Through the later morning, the yellows of the sun's radiation help keep our focus and energy. The glowing mid-day sun lifts our mood and energy levels, so it is important to journey outside for at least a few minutes during the middle of the day. At the same time, we must limit our sun exposure in the middle of the day to avoid fatigue and potential skin damage.

❖ A source of soothing and healing energy prevails amongst the greens of the plants and trees of nature. For this reason, we can be around or surround our living environments with plants and nature in general. If we cannot live in a rural area with lots of nature's greens around us, then we can bring nature into our homes.

❖ The blueness of the sky can also provide a calming, feeling, while increasing our imagination and creativity. Looking up at the sky periodically throughout the day can be extremely soothing to the mind and physical body. The blue sky also increases problem solving, so we can look up to the sky as we are considering a solution to an important problem.

❖ Near the end of the day—at around dusk—the sky can have an array of colors, reflecting the rays of the sun reflecting through more of the atmosphere. When clouds are in the sky, the sunset can turn purple, violet, and even indigo. These colors influence spiritual insight and thoughts of higher consciousness as we ponder the meaning of our existence. We may also experience these meditative feelings as darkness begins to fall and the sky becomes increasingly violet and indigo.

❖ Engaging this beautiful natural pallet of color within our physical lives can instantly vanish with one simple and innocent act: putting on a pair of sunglasses. Sunglasses can modulate a significant part of nature's visible spectrum. While this may be very convenient and even good for the eyes when we are driving on a highway with significant glare reflecting off other cars, there is a negative side affect as well: Sunglasses refract light rays and thus bring to the eyes unnatural and unreal colors. While some sunglasses may appear to brighten the colors around us, others remove light and colors and thus dampen all color. Artificially brightening and changing the color pallet around us renders an ungrounded, almost eerie mood. While it might produce an immediate elation as we put the sunglasses on, this eventually turns to a strange mood as our brainwaves attempt to coordinate to these odd colors.

❖ Darker sunglasses can bring about other, darker physiological and psychological effects. For one, the darkening of light depresses the pineal gland, which suppresses serotonin and dopamine secretion. These two important neuro-messengers are critical for the balance of our moods and behavior. Depressing these secretions can result in feeling fatigued, depressed, and lethargic. Dark glasses can unnaturally raise melatonin levels. Increased melatonin stimulates sleepiness and lethargy. In the middle of the day—particularly while driving—this is not such a good idea. Light is extremely important to the rhythmic health of our bodies and minds.

❖ Worse are eyeglasses that automatically darken with increased light. Of course they are convenient, and we can appreciate the advances in *glass colorimetry* and *polarization* techniques. However, the problem with wearing automatically-darkening eyeglasses is that not only are we robbed of much of the light needed to stimulate important hormones and neurotransmitters and cheated of the benefit of natural colors, but we are also entraining our bodies, nervous system and endocrine glands to a world of dampened color and reduced light. This entrainment might be compared to sending a person to live in a darkened cave. Over time, the body will adapt itself to this depraved condition of being without the body's important rhythmic components: light and color. Research has illustrated that when a person is entrained to an indoor darkened habitat, the risk of depression grows substantially. Along with depression comes the risk of various other diseases such as fibromyalgia, back pain, digestive difficulties, decreased circulation and so on. While some research links these disorders to the body not getting enough vitamin D, others link these to the suppression of serotonin and dopamine. Still others have associated these disorders with simply not having enough of nature's colors. We would suggest that perhaps all of the above are applicable. The fact is; a generous amount of outdoor living resolves most of these for most people.

❖ The colors of our rooms, our houses, our cars, and our clothing can all be chosen with all of the above elements in mind. If we are going to a negotiation where we want to communicate calmness, we might want to wear dark blue. Should we want to convey and increase energy among our associates we might wear some orange. Should we be seeking a balanced approach in our relationships we might consider wearing green.

❖ The walls of our rooms can affect cognition. Studies have shown that color photographs with natural scenery are remembered more than black and white photos, though unnaturally colored photos are remembered no better than black and white photos. Therefore, to enhance learning and memory, we can pick natural colors and images to surround ourselves with as much natural visual environment as possible. A natural visual environment might include, for example, a large picture window to a natural setting, an array of indoor plants, and natural fiber and wood furniture. A selection of nature's greens, blues, oranges and yellows can provide our rhythmic body with a comfortable, balanced, and positive harmonic environment.

Chapter Ten

Coherent Sound

We might stop for a moment and listen to all the sounds surrounding us. If we are fortunate, we will be hearing birds, crickets, and maybe some wind or even sounds of the ocean. If we are in a house, we may hear the ticking of a clock and the hum of the refrigerator. We may also hear other sounds of our mechanical society. We may hear cars, trucks, and other machinery such as lawn mowers or chain saws. All of these sounds taken together form a layer surrounding us. This layer may be invisible, but it is nonetheless a thick layer of pressurized rhythm. This layer is composed of more than just the sounds we hear: The sound layer is an evolving web, containing an accumulation of sounds from the past and present.

The rhythmic layer of sound envelopes us just as thoroughly as flesh and blood surrounds us. Sound is not simply something we hear. It is a realm of waveforms with particular frequencies and amplitudes swirling with harmonic resonance. Researchers have proposed that this envelope together forms white noise—a conglomeration of all frequencies at equal power. This is analogous to the concept of white light, containing a variety of spectra. As a theoretical model, white noise is curiously proposed as random or chaotic. This lies inconsistent with the fact that through the modulation of sound we can communicate specific information. If the noise envelope were random or chaotic, any modulation would necessarily be dowsed by the overwhelming sounds of chaos. Instead, "background" sound is quite acquiescent. White noise is thus coherent rather than chaotic.

The Anatomy of Sound

As we survey sound outside of the environmental sounds of nature, we find sounds arising from consciousness. Thus sound transmits intention. If we include the assumption of a living planet, the sounds of wind, rain and the crack of an earthquake are also sounds of consciousness.

Each species has slightly different sound reception mechanisms. The body of each species is designed to pick up a slightly different waveform range from the sound layer. The human ear is tuned to frequencies ranging from twenty to 20,000 hertz or cycles per second, and is especially sensitive to the range between 1000 and 4000 cycles per second: The higher the frequency, the higher the pitch of the sound. The lower-pitched, softer sounds are the lower-frequencies. The high-pitched frequency of a dog whistle will not be picked up by human ears, for example. The dog's ears receive and recognize sounds in lower frequencies and higher frequencies than humans. Dogs' ears are also supported by muscle receptors that allow the dog to tilt and rotate towards the source of the sound to isolate the sound's location. Meanwhile, various marine mammals can hear frequencies in sonar or *echolocation* waves, which conduct through the water in much slower frequencies and longer wavelengths—around two to five cycles per second. Dolphins, for example, are believed to receive sound through their lower jaws. This form of tympanic membrane then transmits the vibrations through to the middle ear, which converts the pulses to neural signals.

Thirty cycles per second will barely produce musical sounds. The base organ notes will vibrate at fifty hertz, while a base human voice will beat at 100 to 200 cycles per second. Higher notes from the human voice can go as high as about 8,000 hertz, which is almost at the peak of what the human ear can pick up.

In the human body, sound waves are mechanically transduced through a series of stages before conversion into neuronal electromagnetic pulses. Once brought into the ear canal, the tympanic membrane (eardrum) begins oscillating at the same beat as the sound vibration. This oscillates the ear bones of the malleus, incus and stapes, and the resulting pulses are transferred to the liquid portion of the inner ear. The liquid labyrinth of the inner ear conducts the pulses into the motion of tiny cochlear hairs. These hairs conduct their movement to the auditory and *vestibulocochlear* nerve as electromagnetic signals. They then pulse through the ion channels of the vestibular nerve through a relay system of the *cochlear nucleus* and *interior colliculus* to the thalamus. The thalamus acts as a conversion and relay station for the waveforms, sorting and broadcasting the pulsed information through to the *auditory cortex*. At the auditory cortex, the pulses are integrated throughout the neural network as nerve gateway impulses. As these information waves enter the limbic system, they are 'switched up' to high-velocity gamma waves, where they are reflected through the neural web and onto the mind's screen.

An electrophysiological encoding takes place as sound waves are converted to mental perception. This process has been called *cortical auditory evoked potentials*. Researchers such as Kerr *et al.* (2006) using EEG spectra modeling believe it takes place within specialized neuron populations within the corticothalamic pathways.

A deaf person can also interpret sound vibration physically. A deaf person may indeed feel the shaking of the floor as their parents walk into the room. These vibrations can be translated into information just as the spoken word might be translated by a hearing person. With training, a deaf person can tell what kind of emotion another has by touching or simply sensing the seismic vibrations of their walking. Once these vibrations intersect with the rhythmic body, they can be converted to cortical perception just as sound waves or visual waves might.

Prior to its entry into the ear canal, sound is a longitudinal wave. These waveforms simply undulate through air molecules without actually moving this medium. We might compare this to the movement of tension through a spring. The spring does not go anywhere, but the tension travels through the spring as it undulates. This is often referred to as a transfer of motion through particulate matter without the actual particles moving. While symbolically correct, this would not be consistent with the waveform view of reality. While we might picture molecules floating around the air like tiny balloons, each of these molecules are combinations of waveforms themselves, as we have covered earlier. Each molecule is a stable combination of resonating waveforms, balanced by so many waveforms interacting through the environment.

This means that longitudinal sound waves move through this matrix of resonating waveforms by interaction. The specific method of interaction takes place with resonance and interference. As the wave pulse connects with an air molecular combination of waveforms, it is transduced through the molecule and its localized electromagnetic environment. The pulse is then handed off so to speak to neighboring molecules and their localized environment (which we might call a *microenvironment*). This pass-through effect creates a channel through which the existing waveform structure pulses, allowing a particular waveform to be conducted from one molecule to another. This conductance can be measured through pressure gradients, because the waveform interference modulates the density of waveforms within each microenvironment. (This microenvironment is somewhat symbolic because the localized environment around each molecule is continuous in a homogenous atmosphere. Still this local environment is important. For example, an air molecule bubbling within a liquid environment will not conduct sound in the same way that an air molecule might within the atmosphere of air. Its electromagnetic surroundings are completely different.) This modulation of air pressure through resonating wave interaction is enough to physically vibrate the eardrum.

We can see a similar effect as we watch a boat moving through the water. The boat's movement creates waves, which interact and ripple through the existing ripples on the water. The new ripples interfere with the existing ripples, and a new waveform is created from this interference pattern. If the existing ripples are small as on a lake, the boat's movement might create a bow wave. If the existing ripples are rather large—like large ocean swells—then the boat will be surrounded by larger waves and the boat's waves will create a different resulting waveform pattern—noticeably different from the bow wave created on the lake. The wave signature of the boat's movement is specific in each case, but the resulting effect on the water's surface will be different in each circumstance. At the end of the day, the specific waveform created by the boat will not only be specific to the type of boat and its speed, but to the existing waveforms in the water. Even shortly after the boat is gone, we will still be able to probably identify the size and speed of the boat by looking at the waves it created—as they will have changed the existing water surface. If the boat was a large supertanker, we would see very long, slow waves with large amplitudes. If the boat was a ski boat, the resulting waves would be shorter, faster and smaller.

This illustrates the movement of sound in many respects. Just as the boat is identified by its effect upon the existing water surface and existing waves, the information within sound is determined through its effect upon the air surface molecules and waveforms already existing within the air. The motion existing within the air environment will often determine the intensity of the sound. On a very windy day, our voice may be severely dampened, when compared to a still day, for example. If there are a lot of other sound waves in the air—say a train passes by—our voices will also be muted and even possibly transfigured a bit. We may have to shout loudly to communicate in these conditions. Just as the wave motion within the water affects the resulting waveforms created by the boat, sound is carried through interference with the waves moving through air rather than through particles of air. Again, because molecules are gathered envelopes of waveforms, interfering waveforms can be conducted through them.

Sound may be further interfered with a solid or fluid surface. The slower-moving waveforms of the solid and the fluid will subtly be affected by the sound as well. Sound cannot be conducted through solid mediums very well because latticed solid waveforms are slow and restricted. Sound waves simply do not provoke much interference in this long slow wave environment. Just as a small boat has little impact on the large waves of an ocean storm, sound has little influence during an interaction with the longer and slower waveforms of solids and fluids. However, sound does conduct through solids to some degree, especially sound waves with very low or high frequency and high amplitudes. In most cases, solids or liquids will severely dampen sound waves, just as a boat's bow wave will be overwhelmed by stormy seas.

Sound is not an electromagnetic waveform in the conventional sense. It is caused by a pulsed vibration due to the interaction of physical structures. The classic example is the vibration of vocal cords or the cricket's leg rubbing. Other sounds are further generated as objects interact. Wind and water may also create sound. Here again, it is the interaction of molecular matter that creates this vibration. Again all molecules are made up of electromagnetic waves. The interaction of these waves creates the physical vibration.

This is illustrated by the magnetic component involved in hearing. Research has illustrated that there is a magnetic field when a magnet is fixed to the eardrum of a deaf person. Here sound pulses are stimulated by the mechanical effects of a *magnetomotor*, simulating the sense of hearing (Matutinovic and Galic 1982). The fact that sound exerts pressure as well as has both an electronic and magnetic waveform illustrates the universality molecular waveform interaction taking place in sound transmission.

As with any waveform input, balanced sound vibration is critical to the well-being of the rhythmic body. Research on sound and stress has revealed that practically any mechanical sound over the 90 decibels level stresses the body and mind. This type of stress can be just as harmful and dangerous as any other toxic burden or stressor. Why?

Sound vibration affects not only the tympanic membrane, ear bones, cilia and neural network, but also the entire body, the mindscreen, and the inner self. Every tactile skin cell has the ability to respond to the minutest of sound vibrations. Our tactile nerves are "hearing" sounds throughout our body. It is only through the ear's sensitive transducing system that the information held within sound is translated. While the rest of the body receives these vibrations and their rhythmic characteristics as well, the information is not necessarily translated. Nonetheless, sound does affect the entire rhythmic body though sometimes only subtly. We can experience this effect more consciously in a rock concert, a jet plane engine, or the rumbling of an earthquake.

We have all heard about the effects of loud sound vibration on one's hearing, especially over time. Automobile traffic noise levels are often sited as major irritant, and in 1991 the *New South Wales Australian Environmental Protection Agency* cited that undesirable levels of outdoor traffic noise levels were between 55 and 65 dB(A), and *unacceptable levels* were greater than 65 dB(A). The AEPA noted that at these levels sleep and amenity were disrupted. Furthermore, clear behavior patterns in these high traffic areas demonstrated negative mental health effects resulting from *unacceptable sound levels*. In the Sydney area, the New South Wales EPA cited that an estimated 1.5 million people were exposed to chronic *undesirable levels*, while about 350,000 were exposed to chronic *unacceptable levels*.

One pervading outdoor sound source extensively researched has been the noise of mass transits—specifically subway systems. One study revealed an average noise level of 89 +/- db(A) (decibel-A weighting), while a maximum level of 112 db(A) was measured inside and around subway platforms. Meanwhile typical jet engine noise will often top 150 dB(A).

According to the *World Health Organization*, an estimated 278 million people worldwide have moderate to profound hearing loss in both ears. Interestingly, the WHO also reports that 80% of these hearing-impaired live in second or third world countries, and estimates that half of deafness and hearing impairment is avoidable. Prevention not only includes avoiding chronic loud noise, but by preventing prolonged exposure. These statistics indicate that in poor countries, sheltering from loud noises—especially in urban areas—is deficient.

The decibel system of measurement is a logarithmic scale designed to measure the electrical sound power per unit of area. Thus the unit called *decibel* is one-tenth (*deci*) of one *bel*, which is a unit of transmission named for Alexander Graham Bell—famous as the inventor of the telephone. The decibel scale for human hearing begins at 0 db, the point which sound is first heard by the ear. When measuring decibels, a filter is placed on the equipment to compensate for the dampening of sound frequencies by the human ear. The filtering system has three types of filters; A, B and C. It is the dB(A) filtering system that is typically used for sound measurement because it accommodates the dampening of the lower frequencies—as the human ear does.

The recommended exposure guidelines given by the *Occupational Health and Safety Association* defines a 90 dB(A) exposure for more than eight hours as hazardous to physical health and hearing. Consistent noise above these levels has been known to cause sleep disturbance and communication interference; have negative effects on coordination and performance, damage social behavior; and cause hearing loss. A number of studies have cited that sustained levels of 90-95(A) dB will cause hearing loss and otherwise damage hearing, and 125(A) dB levels will cause painful hearing.

140 dB(A) is considered the top painful threshold of human hearing, though 125 dB(A) may be the most that many of us can handle without pain. While a whisper ranges from 15-25 dB(A), a typical conversation is 55 to 65 dB. A washing machine is about 75 dB(A), while close-in city traffic can get up to 85 dB(A). A lawn mower might run up to 95 to 110 dB(A), while an average household hums at about 40 dB(A). A power saw will easily push out 110 dB(A) levels, while a pneumatic riveter will easily

create 125 dB(A) sound levels. The sound of a jet engine can easily top 150 dB(A).

Interestingly, we have a much greater tolerance for decibel ranges in music than for machinery noise. The typical concert piano ranges from 60-100 dB(A), while most classical instruments range from 85 dB(A) to over 100 dB(A). The trombone, clarinet, cello and oboe for example, can each reach 110 dB(A). Meanwhile a symphony might peak out at over 130 decibels, while a rock concert may easily reach 150 decibels. A walkman headset at mid-volume range will push some 94 dB(A), while a headset and digital music player will easily reach the 105 decibel level, often peaking at 110 decibels. However, harmonic distortion in headsets can add 10 dB(A) to these levels, so the equivalent of 120 dB has been measured with personal digital music players.

This all means our tolerance of harmonic or chromatic sounds is much greater than our tolerance of machinery-generated noises like automobiles, airplanes, or jackhammers. Why is this? As we look more closely into the difference between the two types of sounds, we find that both sounds have a particular beat and repetitious rhythm. However, musical sound vibration has a markedly different tempo and beat than does mechanical sound. One is irritating and stressful, while the other is enlightening and relaxing. Mechanical sounds, made of repetitious bursts of forceful pounding, are *monotonous*. Although music also has repetition, it is not monotonous. Note the breakdown of the word "monotonous:" *mono*, meaning "single," and *tonous*, meaning "tone." In other words, mechanical sounds are not only repetitious, but they have a singe tone range.

The prime distinction between music and mechanical noise is the variance of tone. For this reason, we often refer to music as *tunes*. Music is multi-tonal. While the beat and tempo of music is based upon a harmonic—meaning a particular rhythm is repeated or multiplied—in music we find a changing of notes. While a machine noise might also establish a harmonic with its repeating beat, there is no variance to the beat, and there is little or no tonal change. Furthermore, whatever tonal changes there are in machinery noise, these changes are also repetitious—they are considered monotonous because there is no variance.

The central reason for the variances in tone and tempo within music is that music is informational and communicative. Obviously, there is a difference between the two types of noise with respect to their sources. To find the central differences between the two types of noises, we must look deeper into the ultimate purpose for each.

As for the machine, the noise is simply a byproduct to its practical purpose. Specific machinery noise still reflects the type of machine; how it is being used; and even its purpose. A machine that is being utilized to break up the earth, or speed down the road unnecessarily fast will naturally sound different from a machine being used to express an emotion such as an electronic organ. When we consider not the machine but the machine's purpose, we can understand better the sound that is communicated and why one sound has negative effects and another sound has positive effects upon us. When we hear the roar of the highway, we are hearing not just the conscious purpose of earning a living or taking our kids to school—a quite innocent objective. We are also hearing the consciousness of our society as a whole in having no patience. We are hearing the consciousness of the designers and manufacturers of the automobile, who put little thought into the noise created or the natural resources used by the gas turbine engine. Even though the hydrogen fuel cell engine was invented in 1838 by Christian Schonein and the first prototype was put together in 1843 by Sir William Grove, automobile manufacturers have continued to push the gas turbine despite a diminishing supply of oil and the destruction of our atmosphere. Though fuel engines are not as noisy as gas turbines, solar-electric cars are also technologically available. The EV automobiles of the late nineties and early 2000s were apparently confiscated (Pain C. *Who Stole the Electric Car,* 2006) by the manufacturers despite mass protests of their owners. The problem again has been that people want to go faster—and longer. So again, the noise we hear on the freeways is the sound of an inpatient society.

When we hear the sounds of nature's birds and crickets, we are also hearing noises from conscious living beings. Though these sounds also relate to survival, these noises do not attempt to circumvent the natural flow within the rhythmic world. These sounds work *within* the flow of movement and consciousness, rather than outside of nature's harmonic flow. This is why the birds' songs and the crickets' chirping are calming and soothing while the roaring of the highway is stressful.

It is not as if birds and crickets are making joyous songs. They are either calling their mates or marking their territory. Still they make these sounds with repetition, yet not monotonously. This is because their sounds are harmonious with the rest of nature's sounds.

Harmony can be heard among singers in a choir or musical instruments in a symphony. On the other hand, should we hear a fighting crowd shouting in anger or greed; we would not consider that noise harmonious. That noise would be considered stressful. This is because the consciousness behind the noises was different. Humans can utilize their voices to stir up discord and create violence, or humans can utilize their voices for the purposes of entertaining, caring, and giving. A group of caring and giving voices is typically harmonious, while a group of angry, hostile voices is typically displeasing and irregular. This is again related to the consciousness behind the source of the sound.

With any particular sound, we might ask what its purpose is. Why is the sound being made?

As is all energy, sound vibration is inseparable from consciousness. Within each sound wave resides the print of a conscious source. Conversely, sound coming from a positive, uplifting consciousness has a better effect than sound coming from a hostile, angry consciousness. This is confirmed by the fact that humans—as well as plants and animals—respond similarly to similar kinds of music. Australian psychologist Manfried Klein illustrated this effect when he conducted research testing hand-muscle responses to music. He found that regardless of the language and cultural background—whether Japanese, Australian aboriginals, Americans or otherwise—the emotional response to the same music passages as reflected by hand-muscle tension were identical. It did not matter that the music had not been heard before. The emotional response was the same among those familiar with the music and those who were not (Ackerman 1990).

Every sound wave carries the consciousness of its origin. The conscious source of a sound is primarily its direct speaker, such as in a bird's chirping, a dog's barking or a person's spoken voice. At the same time, however, a sound's source goes back to the teacher or family of that animal or human. For this reason, barking dogs may convey the feelings of care, fear, or territorial marking of the dog, as well as the fear of invasion from the owner. For this reason, we are often irritated by the sound of a dog barking, because it is to some degree transducing the fear and greed of its owner. At the same time, the howling of wolves or coyotes in the wilderness do not provoke such a negative response. The howling of wolves or coyotes in the distant night may be just as soothing as birds chirping or owls hooting, because these sounds are harmonizing with the flow of the conscious natural world.

These natural sounds of animals are not unlike our singing. While natural animal and bird calls are communications between animals for the purpose of territory or calling their mates, our singing is also a form of conscious communication. Researchers have found that when we sing, our pupils will dilate and our endorphin levels will rise. When we hear music or singing, some of the same effects are experienced, depending of course upon the music. We also know that certain music enhances activity and invokes various emotions and feelings. This most of us know from experience. We have also known since World War II that even a comatose person responds emotionally to music. This is because again, sound waves carry consciousness—connecting one conscious self to another conscious self or group.

As we illustrated above with freeway noise, sound may also be an indirect result of consciousness. The rumbling of a jet overhead or the sounds of a lawnmower are examples of indirectly conscious sounds. They may not directly communicate, but they nonetheless carry the conscious intentions of their users, manufacturers, and inventors. For example, a neighbor's lawn mower sounds, which may be heard some distance away, will communicate the neighbor's intent to make his environment attractive as rapidly as possible. A more abrasive example might be the jackhammer or earthmover. These sounds indirectly reflect the intention of the users to impatiently grind up and move rocks to build against the flow of nature. If the consciousness reflected an intention to preserve the sounds and flows of nature, the jackhammer would be replaced by picks and shovels. As an interesting byproduct, the workers who performed the work would also not have to worry about weight loss diets.

If it was merely the decibel levels that cause stress and disorder among humans, the mechanical noise of urban areas should have the same negative effects—as pointed out from the Australian government research noted above—as a symphony orchestra or the crashing of ocean waves at the same decibel level. This is apparently not the case.

This was confirmed in 2006 by the research of Garcia-Lazaro *et al.* at the Oxford University's Laboratory of Physiology. The researchers determined that natural sounds exhibit primarily $1/f$ *spectra*—which have primarily gradual and gentle fluctuations in pitch and loudness. The researchers also found that human subjects tend to prefer melodies with $1/f$ distributions more than $1/f0$ (slower) or $1/f2$ (faster) distributions of fluctuations in loudness and pitch. The researchers then tested the sound fluctuations on the auditory cortex, and found that the auditory cortex responded more positively to the $1/f$ distributions. The researchers concluded from this research that the *"auditory cortex is indeed tuned to the $1/f$ dynamics commonly found in the statistical distributions of natural soundscapes."*

This is because natural sounds are also stemming from consciousness. The sound of a freeway and the sounds of ocean surf have about the same loudness and even sound somewhat alike. Yet they have completely different physical and mental effects upon us. Ocean waves breaking upon the beach are soothing and relaxing while freeway noises are stressful, irritating, and disturbing to our sleep. This is because the planet and the environment are both reflecting a larger consciousness—one that is peaceful, soothing and harmonic.

Once sound vibration of any sort enters the rhythmic body, it interferes with the existing rhythmic elements of the body. This interference creates information. The type of information created by the sound vibration depends upon the source of the vibration and the condition of the rhythmic receiver. Should the sending source of the vibration be congruent with the receiving source, the resulting information waveforms will resonate with the receiver. This resonation creates the positive characteristics relating to the sound. This resonation effect is also why the Oxford researchers noted above found the auditory cortex responded positively to the 1/f sounds of nature.

This effect is typically considered a mechanical issue, as if the human body is simply a machine run by a computer program. On the contrary, the source of the individual consciousness of the body relates to a distinct personality replete with choice and intelligence.

We also contend that all rhythmic energy has a conscious source. Should the conscious source of the rhythms have positive intentions, the effects of those rhythms create positive, resonating information as they interfere with the receiving consciousness-contained body. When we consider beautiful, melodious music originally composed by a person who wanted to portray or reflect the inner beauty he or she felt as they composed the music, the resulting musical composition will create a positive resonance with the rhythmic receiver—the listener. On the other hand, should the composer have more derogatory motive as to the reason for the music creation, the resulting music should not have the same effect.

Even if the composer is feeling pain, sorrow, or even honest anger when they compose the music, the expression of that inner pain—that of sincere consciousness attempting to communicate—will resonate with the listener. Often music critics will describe a musician who lacks the "talent" to resonate their music. What is this "talent?" This type of talent might be compared to the "talent" of an abstract painter. In both, while a painting or music expression may not be technically accurate, it may translate the soulful expression of the artist. This "soulful expression" is what resonates with the listener or art collector—not the technical precision of the art or music

We might notice that different types of people are attracted to different types of music. This is simply because their consciousness resonates most closely to the consciousness of the composers to whom they listen. This may seem obvious. Because music is known as an expression, and the musician typically expresses his or her consciousness through their music, the music is a carrier of that consciousness. Listeners will also be attracted to those types of music that most closely matches their mood and emotions. This element—mood and emotion—is also a reflection of our consciousness.

This is illustrated by how audience and musicians in a concert tend to dress very similarly. Conservatively seated and well-dressed symphony orchestra musicians will usually be attended by a well-dressed, conservatively seated, and calm audience. A rock concert on the other hand, will be performed by exuberant, standing casually dressed musicians attended by an excited, standing, loud, and casually dressed audience. Meanwhile a grunge-rock concert will reveal both audience members and musicians wearing gothic or dark clothing and hairstyles. The audience will mimic the musicians, and vice versa. This is because they are resonating on various levels.

Sound vibrations are certainly complex. There are so many types of sound vibrations—each distinct in waveform and information. When we hear the sound of a dog or porpoise for example, we might think these sounds are fairly crude and uninformative. However, within these seemingly simple barks or squeaks come nuances and intonations reflecting specific emotions and moods, originating from the consciousness of the self within that particular body. As we analyze the various animal sounds with electroencephalographs for example, we find they are amazingly varied. No two are exactly alike. They are—like all the physical expressions of each and every one of us—unique down to fingerprints and DNA.

Regardless of the channel of entry, sound vibration is transduced through a series of brainwaves and onto the mindscreen. This transmission happens rather quickly, depending upon the medium of the channel of course. For example, sound is transmitted over four times faster through water (about 1435 meters per second in fresh water) than through air (about 330 meters per second).

Once the sound vibrations are reflected by the neural network onto the screen of the mind, the inner self can observe and interact with them through the utility of intelligence. In the case of conversation, the mental images are blended with existing neuronal waveforms, and their interference patterns are analyzed for utilization by the intelligence tool of the inner self. The intelligence sorts out which pieces of information are to be stored for later use, and which pieces of information are not important. Those considered not important are pushed aside to allow focus on those considered important. The sorting of the intelligence is based upon criteria established by the inner self—the real *you* and *I* within the temporary physical cage of waveforms. With this, the mind can be loaded with a pre-sorting program to filter sound impulses before the neural network reflects the images. This allows us to dismiss things we do not want to hear even before they are imaged. This gives rise to the expression: *"We hear what we want to hear."*

Sometimes, however, the inner self may not be prepared for some sound images. A sudden loud noise, profanity, an obnoxious noise like a screeching or scratching, or an abusive command, might indeed shock the sensitive inner self, viewing the mindscreen. While the self-programming may filter and damper this, the self may also react adversely to the consciousness these sound waves reflect, leading to a type of shock injury.

Acoustic shock injury, or ASI, is the pathological term used for a person hearing and being disturbed by a sudden or loud sound. This response typically stems from unexpected sound waves at an unexpected moment. While ASI victims usually describe the sudden noise as loud, often it is not. The perception of loudness comes from the unexpected shock to the sound. Sometimes ASI will occur while using a telephone headset. While the phones and their headsets typically limit decibel levels, the frequency of the sound combined with a particular statement can cause this injury. Some reports show sound in the range of 4 kilohertz has the greatest potential of ASI. Symptoms such as pain around the ear, psychological issues such as *phonophobia, vertigo,* and *tinnitus* have been observed as a result from this injury.

Many attribute ASI to a failure or weakness of the *medial olivocochlear efferents,* which apparently modulate the waveform intensity elements of the sound transmission as it travels through the neural network. This, we propose, is part of the self-programmed neuronal screening system. They work on decibel level and meaning. When the filters are not functioning properly, they cannot damper the waveform appropriately. In extreme cases, ASI can even cause a longer-term trauma, disturbing sleep and causing chronic stress for some time.

The Resonance Principle

In 1863, John Alexander Newlands, a British chemist, was the first to construct a periodic table of elements arranged to atomic mass. Contrasting Mendleev's atomic number table, Newlands' table revealed that every eighth element apparently had amazingly similar properties. This led Newlands to propose the *law of octaves,* which also occurred to him due to early musical training. Newlands could not help recognize the numerical parallel occurring between the elemental spacing and the harmonic that occurred in music theory.

A musical octave—also referred to as the *perfect octave*—is the interval occurring between notes with a doubling or halving of frequency between them. In other words, the successive note octave has a frequency double

f   *2f*   same note, but
           diff. freq.
|_____|
 octave   TOTAL HARMONIC

that of the previous note. Another perfect octave will occur when the frequency is doubled again. A note at an octave above the previous is also written as that same note, simply because the human ear hears both as essentially the same sound even though the two sounds actually vibrate at different frequencies. This is referred to as *equivalency.*

The perfect octave is precisely double the previous note, while a *diminished* or *augmented* octave would have a slight variance from doubling, into a flat or sharp note.

Absolute harmonic is accomplished when sound frequencies are whole integer multiples of some particular frequency. The first four harmonics of a 200-hertz frequency are 400 hertz, 600 hertz, 800 hertz, and 1000 hertz, for example. The whole integer multiple of a sound will harmonize with the first sound simply because its frequency is reflective. By reflective we mean the successive sound mirrors the waveform of the first sound. This reflective waveform is typically referred to as the *fundamental frequency* in a harmonic sequence. If we look at the concept of harmonic from a broader perspective, we can understand that each harmonic is actually a reflective fraction of an even greater harmonic—a more expansive fundamental frequency.

The pitch of a sound is related to its frequency, yet frequency is not the only characteristic of pitch. The pitch of a sound is more precisely its perceived frequency. While the frequency is a two-dimensional measurement related to cycles per second, a sound's pitch may incorporate a variety of *overtones* into the total sound. These may include changes in amplitude, tempo, and intonation. These qualities give the waveform informative and instrumental variance. A sound may be *pitched in* such a way to appear very much like a note of a particular frequency, yet the sound waveform may not have that precise frequency. The pure note *A,* for example, should have a frequency of 440 hertz. Most concert tuning forks are set to the A-440 frequency for this reason. In music writing, the pitch is consistent with a tonal step from a fundamental frequency note, though musicians can subtly raise the pitch of a particular note or composition with intelligent instrumentation.

When notes move into the flat or sharp designation, the pitch or frequency adjustment moves into the *enharmonic genus,* which is based on the Greek *tetra chord.* The tetra chord concept is that notes can be tuned in intervals of *perfect fourths.* The four-stringed lyre was the basis for this early concept, but as the tetra chord concept expanded into other instruments, the *diatonic* and *chromatic* interval systems were added to the enharmonic system.

Pitch variances are measured in *tones.* A shift to flat or sharp may become a variance in *semitones, quartertones, duotones,* or even *microtones.* These shifts may also be represented as fractions. For example, a *ditone*—or third major tone—may be 16/13 of a full note. The octave concept expressed in tetra chords would thus be a whole tone plus two tetra cords.

The *chromatic scale* is a common scale used in music to denote the rise through a series of related notes. The chromatic scale is usually based upon the C note, but the B note or others may be used as fundamental pitches as well. There are typically twelve total pitches in a tempered chromatic scale. Each of these pitches is a half step or semitone step from the prior pitch.

As music math has further developed, the perfect fourth led way to the disjunctive *perfect fifth.* Also called the *diapente,* this is a music interval providing harmonious latitude with surrounding tones. On the piano keyboard, perfect fifths are separated by exactly seven keys. The perfect fifth also provides the root of the major and minor chords and their extensions. While the *just fifth* provides a 3:2 ratio, the *perfect fifth* has seven semitones, two less than the just fifth.

When notes harmonize, they resonate together. A tuning fork set to A-440 will resonate at the 440 hertz frequency, transferring this frequency through the air until interfered with. A concert tuning fork will typically be tuned to the violin's third string. As the violin is tuned, the tuning fork and the violin's third string will resonate together. This occurs because of a facility within the violin's construction to allow it to become an *acoustic resonator.* An acoustic resonator is a point on an instrument or body that carries the vibration of a note for a period of time. In other words, it vibrates at the same frequency. On a violin, for example, the string, the bridge, and the body of the violin all facilitate this resonating system. When the tuning fork is struck and the A note resonates through the concert hall, a violin tuned to the *A* note will resonate with the tuning fork, forming a harmonic to tune by.

If another *A*-note tuning fork is held nearby the struck tuning fork, the second tuning fork will also begin to vibrate to the same note and frequency. The second tuning fork will become *entrained* to the first tuning fork. These two tuning forks are resonating together, with the second having been entrained to the first.

Resonators and entrainers appear throughout nature. A canyon resonates with the sounds of the wind, birds, moving trees, and animals. The canyon would be comparable to the body or bout of a guitar or violin, creating coherent resonance and structure for those sound vibrations. An ocean beach or bay is a resonator of the waves marching in from distant storms. Nautilus shells and conch shells provide an interesting resonating mechanism. Most of us have "heard the ocean" by putting the ear up to one of these shell. The spiraling echo chamber provides an entrained resonating cavity for the rhythmic sounds of wind and water. Garcia-Lazaro *et al.*'s 1/f spectra result indicates that a resonance between nature's sounds also occurs within the auditory cortex.

The relevance of music with respect to the rhythmic body is critical. The rhythmic body is an entrained resonating cavity, but it is also a generator of rhythms. In other words, a living organism will seek to entrain its environment, while also becoming entrained by the environment. This makes the living body very much like a multi-dimensional violin. Waveforms from others and the environment entrain each living organism, which each organism contributes to the entrainment of the environment with waveforms created from the conscious inner self of each. Thus, there is a constant balance between the affect of the environment on the body and the effect of the conscious organism on the environment. This mutual entrainment creates the illusion of physical existence.

We might compare this activity to a stage play. In order to create a full scene, all of the actors must be performing together in sync within the visible section of the stage. They must perform when the curtain is up. They must have the stage lights on so the audience can see them. The sound system must be on. The set must be positioned behind the actors properly. Furthermore, the actors must all know their lines and must coordinate their lines to synchronize with the elements of the stage—the props. This creates a harmonic between the actors and the stage, which creates the illusion of reality. Without the set, the sound, the lights, the curtains up, and actors knowing their lines, the performance of the play would not appear credible enough to be entertaining to the audience.

The depth of sound vibration—with its multiplicity of pressurized air and electromagnetic qualities—enables it to transfer messages of consciousness. The conscious intensions of the inner self are expressed through the vocal cords into sound. In order to resonate within the context of this world, these vocal vibrations must have coherence. In order to be understood, these sounds must interfere coherently among the various waveforms of our immediate environment.

As an example, a Chinese person might walk into a party in America without knowing English. As he begins to speak to those around him, he quickly finds out that no one speaks Chinese. No one understands him. In order to communicate, he must make hand gestures to describe what he means to say. This is because the Chinese language was not coherent to these American people.

Here we use the word *coherent* to describe both understanding and the gap between a person's intention to communicate and the listener's inability to understand those communications. In other words, the intention to communicate does not connect (or interfere) with the other person's intention to understand, simply because they speak different languages. Now should another Chinese person be at the party, there can be communication between the Chinese speakers. This creates a coherency, as the intention to communicate connects with the intention to understand. Still, however, the two Chinese people may not have anything in common. This would mean while they have coherency, there is no resonance. There is no common chord between the two.

The body and the natural environment also have this sort of coherency and resonance. The body needs a natural environment of air, sunshine, water, and natural food for sustenance. Nature provides these elements. So there is a relationship of coherency between the body and its environment. Now should the body *be in tune* with the natural environment, it will be resonating with nature.

Without this resonance, the body will begin to become dysfunctional. On a gross basis, we might observe this with breathing and the heart beat. When our body's breathing is normal, there is a timed release of oxygen into the bloodstream. The heart paces the output of this blood throughout the body with the timed pumping of the valves. Now should we hyperventilate due to a lack of good oxygen levels; our rate of breathing will increase. This in turn pushes oxygen levels within the blood higher. The heart may also beat faster, which will push this highly oxygenated blood throughout the body. This blood is not good for the brain. The body may become dizzy and light-headed, leading to fainting or worse. While the air has a lack of oxygen, the body begins to hyperventilate to accommodate for the situation, leading to over-oxygenated blood.

The more subtle rhythms of the body will also reflect a resonating condition in much the same way. When the body is in resonance with nature, the various rhythms of the body also resonate together, creating a symphony of oscillating energy playing within the body. Every rhythmic plane of matter—solid, gas, liquid, thermal and electromagnetic—has particular types of oscillations. In order to achieve this coherent resonance, all must resonate together much as a symphony's instruments must be in the same key and must be playing the same song. When the orchestra is resonating with the conductor and the song the conductor wants them to play, everything falls into a harmonic. The body's various instruments must be orchestrated in the same way, playing the resonating song of nature, and tuned to the Grand Conductor.

Dr. Jacques Benveniste and his associates illustrated this directly when they found that metabolic biochemicals like hormones and neurotransmitters actually oscillate at unique signature frequencies. Furthermore, some of these frequencies were digitally recorded just as one might record a song. As we have described earlier, these recorded frequencies could be played back in the presence of organs, stimulating effects identical to those stimulated by the physical biomolecules themselves.

Furthermore, Professor Popp's weak electromagnetic biophoton measurements among cells revealed each cell vibrates at a particular frequency. Simple observation confirms that particular organ systems vibrate at particular frequencies: our heartbeats at one rhythm, our breathing rates at another, our digestive tracts at yet another rhythm and so on. We see how these rhythms have to coordinate together. It is not a stretch to assume these various body rhythms must also harmonize with the *micro-rhythmic* frequencies of cellular vibrations, nervous system neurotransmitters, and thermoregulation hormones. Dr. Popp's work went further, and confirmed these frequencies with specialized equipment—again as we mentioned earlier.

We can also observe how the *macro-rhythms* of our tissues and organs—growth, aging and so on—must harmonize with the microrhythms of our cells. Hormones cycle differently as we age in almost every aspect. We can see changes in growth hormones, sex hormones, and energy hormones as our bodies age. This in itself is a larger, macro-rhythm, which resonates with the microrhythms of our cells, leading eventually to that natural part of the rhythm, death. As we have observed clinically for thousands of years, each person has a unique macro-rhythm. We all age slightly differently, and we all die at unique times in our lives. Like fingerprints, no body is identical. The particular waveforms and their interference patterns of each individual body is unique. This would also logically mean the micro-rhythms of our cells must also be unique.

This very concept was proposed by Vera Stanley Alder in 1938: *"Every object and every person has a key-note—in other words the sum of their vibrations—responds to one particular note or chord of the musical scale. If a person's note or chord is sounded gently and melodiously, it has a healing and constructive influence upon him. If it is sounded loudly, harshly and continuously it has a correspondingly destructive influence, making that person ill and unhappy."*

Though we might question whether this proposal has scientific basis—as many have done over the years—a plethora of clinical observation seems to confirm its reality. Data from instrumentation including EEG, EKG, otoacoustical instruments, and various biofeedback devices have confirmed in different respects that different parts of the body are oscillating at different frequencies—and these different oscillations coordinate together. As we have examined brain waves, sound waves, sensual nerve input, visual spectrum and neurotransmitter frequencies, we arrive at the conclusion that the functions of the body are completely integrated with these various waveforms transmitting throughout the body.

Furthermore, the whole body appears to vibrate a central theme. We might compare this to listening to a crowd within a large stadium on television. As we are further removed from the crowd sitting in our living rooms, we can hear an overall hum of the stadium crowd. Although each individual audience member may be screaming or clapping uniquely, the noise of the entire stadium crowd may be represented by a single note or sound. Most of us have imitated or heard someone imitate the sound of an entire crowd with uncanny accuracy with one tonal note.

The notion of a body full of sound vibration has been knocked around for a number of years. As it tends to be communicated from a perspective of mysticism, it has hardly been given much serious thought by scientists. However, when we combine the research noted above with the perspective of effective clinical application, we discover that we can draw upon research provided by a respected branch of medicine called *sound therapy*.

Sound Therapies

There are several areas of emphasis among the elements of sound therapeutics.

Voice analysis therapy, for example, assumes the sound vibrations of the voice represent the body's harmonic frequency as a whole. The therapist will usually begin by recording and analyzing the subject's voice with spectroscopy. As we have mentioned, spectroscopy is a breakdown of the various frequencies resident in a particular group of atoms, or, in this case, sounds. Because our voice is a collective vibration of individual vocal cords, a variety of individual frequencies come together to create the total voice tone. The spectral analysis of the voice tone is charted against an octave and color chart, tying the voiceprint to the frequency gradations of musical notes and their corresponding color frequencies. There are a number of different schools of thought and methodologies that are used therapeutically on the basis of the voiceprint. Some utilize music, and others develop specific sound recordings for listening by the subject. Some simply utilize the voiceprint for diagnostic purposes.

One type of sound therapy was developed in the 1940s by Frenchman Dr. Alfred Tomatis, an ENT (ear, nose, throat specialist). Much of Dr. Tomatis' work formed the basis for the principles of *Audio-Psycho-Phonology*. The *Tomatis Method*—a process of attempting to improve ones learning capacity through the introduction of altered music and sounds from nature—eventually came into use from this work. His methods have been clinically applied for persons with learning disabilities and attention deficit issues. Dr. Tomatis' theories have been expanded into a larger format over the years. Today, practitioners of the Tomatis Method utilize special music to increase concentration, focus, and calm.

Music therapy has a long tradition. Pythagoras and his students explored the relationships between music, the universe, and health quite extensively. His *Music of the Spheres* treatise, handed down by Pythagoras' students, illustrated the harmony existing within the universe; and how this harmony related to music.

Music therapy has been documented through the centuries as a credible form of healing. The philosophy of music has also been found prominently in the writings of Aristotle and Plato. Homer prescribed music to counteract mental anguish, and Asclepiades of Bithynia is said to have prescribed Phrygian music for sciatica and other illnesses. Democritus prescribed various flute melodies, and Pythagoras is said to have clinically applied music for nervousness. The respected Roman physician Galen applied music to his healing repertoire. Among other therapies, Galen prescribed a *"medical bath"* inclusive of flute song for nerve pain. The famous sixteenth century Swiss physician Paracelsus was a strong believer in sound therapy as well. His recommendations included not only herbal remedies, but colors and sound to achieve health.

Both Aristotle and Plato also suggested music therapy was useful for the healing of society and people in general. Aristotle went as far as to describe a good man as a type of musician. Said Plato in *The Republic*: *"Musical training is a more potent instrument than any other, because rhythm and harmony find their way into the inward places of the soul, on which they mightily fasten, imparting grace..."* (III, S401e).

Modern western medicine rediscovered music therapy during the twentieth century, when it was found that music helped injured world war veterans heal faster. This led to musicians being hired by hospitals, which in turn led to a need for further research and training. Since that time, music therapy has become a credentialed, evidence-based healing modality in the United States. The first college degree program for music therapists was established in 1944 at Michigan State University. In 1950, the National Association of Music Therapy came into existence, which later became the American Music Therapy Association. In the 1980s, the Certification Board for Music Therapists began to certify music therapists. Today this organization claims over 4,000 board certified music therapists.

The clinical research documenting the usefulness for music therapy is impressive. It has been identified as a valid complementary and alternative medicine by the National Institutes of Health and many hospitals other organizations. Music therapy has been used with success in cases of cancer, schizophrenia, dementia, cardiovascular diseases, somatic issues, anxiety, pain, post-surgery recovery, eating disorders, depression, multiple sclerosis, deafness tinnitus, and a host of other physical, emotional, and psychological issues. Music therapy during prenatal care and during labor has been shown to be a valuable addition to other therapies like breathing exercises and specific exercises. In hospitals, music therapy is often used to help alleviate pain and increase post-surgical healing. It is often used in conjunction with anesthesia and pain medication. Music therapy has been shown to be dramatically effective in elevating mood, alleviating depression, inducing sleep and in general, decreasing hospital stays (Hillecke *et al.* 2005). Most of these effects have been observed in rigorous clinical conditions. One review of the research, published in the *Annals of the New York Academy of Sciences in 2005* (Thaut 2005) has called for *"a paradigm shift to move music therapy from an adjunct modality to a central treatment modality in rehabilitation and therapy."* As a result of the credibility such statements have rendered, music therapy is now covered under many states' Medicaid programs, and many insurance companies around the U.S. are starting to cover music therapy sessions.

There are a number of measurable physiological responses now shown by a growing library of research on music therapy. In one example, Wachiuli *et al.* (2007) found in a controlled study on 40 volunteers that subjects who engaged in recreational music-making using drumming protocol had better moods, lower stress-related cytokine interleukin-10 levels, and higher natural killer cell activity compared with the control group. These markers of course, translate to better immune system response, suggesting greater disease prevention.

Music therapists use a variety of strategies in clinical applications. After an assessment of the patient's physical and psychological issues, the therapist will typically design a personalized program for each patient. A music therapist might suggest a particular type of music or set of songs for the patient to hear. They might also design a musical session that includes playing instruments if the patient is so inclined. Sometimes a more gifted therapist will sing selected songs for the patient during sessions. Music therapists often compose specific songs to deal with specific patients, disorders, and situations. This type of music therapy seems to be especially useful for children's disorders. Children tend to respond very positively to original, imaginative songs it seems. Older patients, such as those with dementia or and those not responding well to other treatment in retirement homes, will often respond well to music therapy. Classical music or the traditional music they grew up with is primarily utilized, although original songs are also sometimes very useful for older patients (Schmitt *et al.* 2007).

One can self-integrate therapeutic music into ones life easily and quickly. The research has offered us a number of tools when it comes to music therapy. Most researchers agree it is a combination of the harmonies and the tempos of music that provides its soothing effect. While we might assume the slow, harmonious music provides this effect, music with drumming beats also appears to provide therapeutic results (Bittman *et al.* 2001). EEG testing has shown many relaxation tapes do not help as much as native music, Celtic music, and even certain rock and roll music can. Evidently, rhythmic beats provide a balancing effect between the left and right brainwaves. Flutes have also shown to be particularly therapeutic. Research has also shown that live performances seem to provide more therapeutic results than do recordings. The right mix of therapeutic music appears to increase deep breathing, lowers the heart rate, balances thermoregulation, and increases production of serotonin.

Playing an instrument or percussion appears to provide a special benefit. Playing music promotes not only the hearing of music, but allows an expression of creative rhythm. Most people can find at least one instrument they can play, even if it is just a makeshift drum of some sort. Playing music while working appears to lowers stress and increase productivity. Music with slower rhythms—slower than one's heart rate—has also been shown to provide relaxation. Music with faster beats—faster than one's heart rate—tends to increase energy and stimulate activity. Familiar music from the past seems to soothe anxiety and depression, especially among the elderly.

Rhythmic breathing with music will provide an additional therapeutic effect.

Supersound

The connection between music, cognition, and mental performance was extensively studied during the 1960s by Dr. Georgi Lozanov. Dr. Lozanov's early research investigated people with extraordinary memory and metal capability. This area of study became known as *supermemory*. Dr. Lozanov's research to find the causes for supermemory expanded into various realms, including ancient communities known for their lengthy historical memories such as the New Zealand Maoris (who could recite an entire tribe history of over one thousand years) or the Vedic Brahmins, who could recite thousands of ancient scriptural texts from memory.

The study of the yogic systems of *pranayama* and breathing led Dr. Lozanov towards what we might refer to as the rhythmic nature of memory. His trials studied learning and memorization with timed recitation. Among his subjects, he found if information was repeated either every eight seconds or every twelve seconds, remembrance was higher than repetition at other rhythms or randomly. This led to comparisons between combinations of beat (timing) and intonation. Tones were isolated and commingled, including normal declaration tones, quieter soft whisper tones, and loud commanding tones. A variance of tones combined with a predictable beat appeared to increase memory capacity the most. This led to an investigation of various forms of music.

As Dr. Lozanov's research progressed, he meticulously explored the relationship between music and learning. He soon discovered particular beats and melodies had greater effects upon the body than did others. In particular, he found music playing around sixty beats per minute—close to the average human resting pulse—substantially relaxed the body. This also had the effect of calming and synchronizing breathing rates. More importantly, music at this tempo consistently increased memory retention, recall, and learning. As Dr. Lozanov's research continued, it became apparent that Baroque-style music as composed by sixteenth, seventeenth, and eighteenth century composers such as Handel, Bach and Vivaldi had the greatest positive effects upon learning and memorization skills. These music forms seem to relax the body and focus the mind more than other types of music.

It is not simply the 60 beats-per-minute (or 3600 Herz) frequency rate that encourages this high cognition state—enabling better memorization and physical relaxation. It is the 4/4 or 4/3 tempo along with the various tonal and pitch variances. While numerous beats and intonations have been tested in this research, few have the effect of increasing learning and relaxation as music with these particular parameters (Ostrander 1979).

We also know that different effects are obtained through different beats and different music forms. As most of us have experienced personally, different music forms bring about different thought patterns and moods, with a degree of variance between people. While a fast-paced rock song might increase our mental and physical intensity, a slow-paced acoustic folk song will bring about a more relaxed, contemplative mood. This effect can also be observed in the lyrics of popular songs across the music styles. A hard rock song might proclaim *"I wanna rock all night long,"* while a folksy acoustic song might pledge *"the answer is blowing with the wind"* (Dylan 1963).

In a 2006 volume of the *Annals of the New York Academy of Sciences*, contributors documented research illustrating that music training increased the amount of grey matter in the right hemisphere of the brain. Increases in the auditory cortex were proportional to the amount of music listened to. The *Annals* also documented research confirming once again that classical music enhances cognition and learning. In addition, it was found that classical music increased immunity and wellness of listeners. More importantly, the findings indicated various other types of music—particularly types of music considered personally appealing by the listener—also had many of these same effects (Avanzini, *et al.* 2006).

Geneticist Dr. Susumu Ohno of California's *Breckman Research Institute* underwrote this effect within the realm of DNA. Dr. Ohno proposed that our gene expression rhythmically resonates with Fourier interference patterns. These interference patterns provide a repetitive sequencing apparently common among all life forms (Holmquist 2000).

Plants undoubtedly discern and respond to the rhythmic effects of music. Researcher Dorothy Retallack, author of *The Sound of Music and Plants* (1973), published controlled research whereupon plants preferred certain types of music over others. Ms. Retallack discovered plants growing in rooms with Baroque-like music (such as Bach or Ravi Shankar) had robust growth with generous foliage and root systems. Meanwhile, plants grown in rooms inundated with rock music shriveled, died, or were generally weak. Her trials concluded that plants were impartial to country music, but positively responded and grew more with Baroque style music and jazz.

Prior to this research, Dr. Jagadis Chandra Bose documented extensive experiments with plants in his 1902 book *Responses in the Living and Non-Living*. Dr. Bose utilized a sound transmission instrument called the *Crescograph*, which measured the responsiveness of plants. After numerous trials, Dr. Bose proposed that plants had a type of nervous system. Though controversial for many years, research confirmed many of his findings since. For example, research in the early nineties by Wildon *et al.* (1992), confirmed pulsed electrical activity transmission within tomato seedlings.

Dr. T.C.N. Singh of a Madras music college continued the research of Dr. Bose, as he investigated crop yields using traditional Indian music. Between 1960 and 1963, Dr. Singh was able to increase crop yields by 25% to 60% when the fields were exposed to particular types of (Baroque-like) music.

Eugene Canby's Bach violin sonatas played to a plot of wheat resulted in heavier, larger wheat stalks with 66% greater yields (Newton 1971).

Canadian researchers Mary Measures and Pearl Weinberger at the University of Ottawa tested spring and winter wheat germination on 5,000 cycles per second (83.3 BPM) and 10,000 cycles per second (166 BPM). While both frequencies significantly stimulated germination, germination was elevated with 5,000 cycles per second (Weinberger & Measures 1968). Note again that 83 beats per minute is closer to the 60 beats per minute rate suggested in Dr. Lozanov's research on humans.

A few years' later, horticulturist and dentist Dr. George Milstein worked with a commercial sound engineer to test various frequencies on plant growth. Dr. Milstein determined the best growth rate to be 3,000 cycles per second (50 beats per minute). He later produced a record using what he found were the optimal sounds and tempos most encouraging to plant growth. He recorded this music onto a well-received album recording called *Music to Grow Plants By* (Conely 1971).

In 1972 Dan Carlson, a student at the University of Minnesota, also conducted research into stimulating plant growth with sound. His research led him to a discovery that the tiny pores on the leaf's surface that absorb nutrients—called *stomata*—appear to open further around an environment of sounds with particular frequencies. The sound frequencies causing the greatest stomata opening ranged from 3,000 cycles to 5,000 cycles per second. As he investigated the frequency results further, Carlson discovered the sounds were remarkably similar to the sounds of common morning songbirds. The songs of swallows, martins, and warblers provided the closest match. The songs of these birds appeared to stimulate the opening of the stomata the greatest.

To take advantage of this effect, Mr. Carson developed a nutrient plant spray, which is sprayed onto the leaf during the sound exposure. The effect is apparently most evident during sunrise. This of course is the time when songbirds are naturally more actively singing. It is also the time of the day that plants naturally absorb the bulk of their nutrients. Ken Taylor, a college chemistry teacher and owner of Windmill Point Farms, is one of Mr. Carlson's spray-sound programs. According to Spillane (1991), Mr. Taylor averaged 400 pounds of carrots per 40-foot row—an incredible harvest using the Mr. Carlson's techniques. In 1985, *Acres USA* reported impressive results with Mr. Carlson's program during tests on alfalfa and corn crops. Mr. Carlson also currently holds the Guinness World Record for indoor growth of the *Purple Passion* plant. Normally growing to about 18 inches, Mr. Carlson's plant grew to an amazing 1,300 feet.

Both Baroque music and Indian music with frequencies of 3000-5000 hertz also stimulate plant growth. We can thus connect the waveforms of these forms of music with the morning songs of birds. From our discussion of Dr. Lozanov's and Avanzini's research, we also know that the 3000-4000 cycles per second Baroque-style music increases learning among humans.

Most of us have experienced the calming and rejuvenating effect of listening to morning birds singing outside our window. Our own experiences and observations clearly indicate that natural melodies can induce a calming effect, together with a higher level of alertness and mental activity. Biofeedback research confirms that these moods are connected to a preponderance of alpha brainwaves.

We all recognize that some music can be irritating and aggravating to some, while appealing to others. This effect is most likely related to the resonance between the consciousness of musician and the listener. The musician's consciousness simply may or may not resonate with the listener's. In the case of Baroque-style music or classical Indian music, these musical styles resonate most closely with nature. They are bound to resonate with most organisms as a result. Their waveform characters coherently interfere with the internal rhythms of living organisms.

This certainly is not experienced universally. An active adolescent with an abundance of energy might actually be irritated by the slower beat and soothing sounds of classical, jazz or folk songs. At the same time, the teenager might find comfort in heavy metal music, especially while that teenager is rebellious or torn with conflict and angry about being forced to adapt to an adult world. These conflicting issues will likely be the same ones expressed by the musicians of the music the teenager finds comfort in. Furthermore, commercial musicians and their labels tend to record the music other teenagers resonate with because they are in the business of selling music. In many cases, these musicians will become commercially successful in the first place because they reflect those same conflicting issues with their musical expression. Following their commercial success, they may continue the same formula, despite the possibility of maturing beyond their early rebelliousness.

As the adolescent grows older, the intensity of these conflicting issues begins to mellow. They may begin to work within the adult system, often compromising their more altruistic principles in the name of survival. As this takes place, the music that resonates closest may become alternative folk music, for example. A change from the rapid grinding electric beats to a slower and smoother acoustic pace may reflect the change from angry intoxicated aggression to a sober outlook.

Through observation we know that as a person grows older, their preferred listening music pace tends to slow. This evolution also parallels with the adult's arising attraction for sounds resonating within nature. With this trend comes a greater sense of inner wisdom. Assuming we are focused upon and growing from the lessons life brings us, there will most likely be a growing attraction to nature's rhythms. As the body ages and the conflicts of society mean less to us than finding inner peace. These efforts correspond with an attraction to hear the sounds that deliver a higher sense of awareness and spiritual identity.

As we correlate the orchestration of productive sounds with their origins, we might logically conclude that conscious arrangement creates their beneficial effects. Clearly, any music or sound made through composition and thought ultimately stems from consciousness. This means Baroque music, a bird's song, a barking dog, and even heavy metal music are rhythms of consciousness. The harmonics obtained by resonating notes vibrated at octaves with a variance of structured pitches and pacing at tempos consistent with nature's rhythms are undoubtedly preferred by living organisms. Such an orchestration could only logically exist through consciousness within a larger context.

As Newton's *Law of Motion* contends, *"an object in a state of uniform motion tends to remain in that state of motion unless an external force is applied to it."* Utilizing this principle, an external conscious source would be logical, given designated resonance within sound.

Conscious Sound

We can test the concept of conscious sound by observing ourselves as we hear a favorite song on a radio station intruded with static interference from poor reception. The common response is one of irritation at the lack of reception. Even with the interference, snippets of the song might still resonate with us. It might remind us of a time past or a person, for example. We might thus make an extra effort to improve the reception, putting up with the irritating static. Instead of finding a clear station, we may put up with the interference. We can assume from this that the irritation level is secondary to how the song harmonizes or resonates with us on a deeper conscious level. Obviously, a clear station would be less physically irritating without the static.

This situation is not a physical one. It is the consciousness memory of the sound that is the cause for our attraction—not the quality of the song. Possibly, for a new song we may seek quality first—until at least the song becomes part of our consciousness. Once the song has reached our conscious being—our memories, viewed by the inner self—it becomes more important than sensual music quality. This is also illustrated by the fact that heavy metal or heavy rock music can be soothing to certain people while irritating to others. The frequency, pitch, and tonal parameters of the heavy metal does not resonate well with nature, as we discovered in plant research. Rather, the resonance between the consciousness of the musician and the listener form a bond that then translates into a memory tied to consciousness.

Testing between human musicians and computer metronomes (music notation entered into a computer) has revealed that computerized sound facsimiles do not resonate with a listener in the same way a live musician playing the same song will. Converting a song to the precise computerized rhythm of a metronome simply does not translate the sense of rhythm, pace and consciousness human musical sound contains.

This reminds us of the techniques of memory building. Memory experts tell us that extending our memory can be accomplished by imaging funny and/or personal stories and connecting these stories to the items to be memorized. Furthermore, recent research on Alzheimer's indicates that association with a wide variety of people, family, and/or situations extends our ability to remember. These point to the understanding that memory is tied to consciousness and consciousness is intimately connected to relationships.

What we are talking about is a resonance between the consciousness of the sound source with the consciousness of the listener. Once the music player and the listener connect through the sound vibrations, the result upon the listener's rhythmic body is determined by the intersection of their conscious intentions. The intentions of the listener are just as important as the intentions of the musician. Among children, music may have more dramatic effects, because the consciousness of the child is more easily impressed. As we mature listening to various types of music, we begin to form biases based somewhat upon early exposures. This is the effect of consciousness. While classical music may encourage learning and cognition among younger people not biased against classical music, it may well be irritating to others who have been influenced by the consciousness effects of other music. This bias may be the result of the influences of others, or it may be an association between classical music and an unpleasant experience, for example. Either way, it is the still an effect of consciousness. Should we force a person averse to classical music to listen to it, we will likely find the opposite effect Dr. Lozanov discovered among Bulgarian children.

This effect can also be observed in popular music, as fans often look up to and respect—sometimes even worship— composers and musicians. This may continue many years after their childhood years—even quite fanatically. This worship is the result of the listeners further connecting with and harmonizing with the consciousness of the musicians. The music simply provides a medium for this conscious connection.

Occasionally we find this awe and respect is misplaced among popular musicians. We might discover the purveyors of the music we have come to appreciate over the years are not who we thought them to be. We might wonder what we ever saw in their music, as they succumb to their personal addictions or an uncontrolled desire for wealth and admiration. It is at these times it becomes obvious there is a deeper source of conscious harmonic being transmitted through their music. While the music may be physically coming from the musician, the effect that music had as it originally intersected with our consciousness begs a deeper source of consciousness.

This correlates well with music talent itself. Where does music and musical talent come from?

We emphatically propose—as did Pythagoras, Aristotle, Plato and Socrates—a deeper rhythm resonating throughout and within nature.

The fact that music has a deeper mechanism of consciousness becomes evident as we consider musical *savants*. *Savant Syndrome* usually refers to someone who has a form of autism, often affecting the left side of the brain. Savant Syndrome is symptomized by an extraordinary skill in some particular talent or science. A savant is a person with extraordinary talents, usually in science or literature. Musical savants are people who suddenly display extraordinary and untaught musical talents. These can emerge at any point in life, although often during childhood.

Musical savant syndrome illustrates that the resonating harmonics brought expressed through music come from a deeper dimension. Many musical savants are not only gifted with the knowledge of how to play particular instruments, but many have an innate ability to write advanced original compositions. One famous savant, Jay Greenberg, portrayed extensively in the media (*CBS 60 Minutes* 2006), illustrated the depth of the source of musical composition by writing five complex symphonies by age 13. By 15, his original and acclaimed compositions had been played by a number of acclaimed orchestras, including the London Symphony Orchestra. He writes all the notes for every instrument in the orchestra, requiring a working knowledge of the specifications and range of each, along with a profound working knowledge of how to write music and how to conduct it. Composer and Juliard professor Sam Zyman, who worked with Jay, stated that Jay's talent is at the level of Mozart and Mendelsohn. Neither Jay's mother nor father have any special musical talents, and neither does his 10-year old brother. At 2 years of age, Jay asked his parents if he could buy a small cello, never having seen one nor heard of one before.

The fact that Jay's songs are originals illustrates they are not arising from recollection. The fact that the talent came from a person with no prior training or experience illustrates that musical creativity transcends the physical body. Since musical savants cannot recall the source of their knowledge, this creativity is also beyond the physical mind. While the specific knowledge of instruments may be recalled from a previous lifetime, the source of the original composition and creativity surely must have a deeper source.

This indicates a deeper source of consciousness driving the fundamental rhythms of music. This reveals a deeper consciousness resonating throughout the physical dimension, as music appears to resonate with nature. This deeper consciousness is expressed through the chirping of birds; the soothing sounds of the ocean; the movement of wind through the trees; and through the music of humans.

When we watch one tuning fork resonate with another, we see the second tuning fork vibrate as though it was the original fork struck. While this second fork was never struck, it nonetheless vibrates precisely with the resonating harmonic of the note. Similarly, if we were to walk into a music store and play one key on one of the pianos in a room full of quality pianos and stringed instruments, the other instruments would begin vibrating that same note.

Conversation is a reflection of our consciousness. When we speak to others around us, our specific language, intonation and delivery all reflect how we feel at that moment. Even when we might desire to hide our true feelings as we speak, our true feelings still reflect in one way or another through our sound vibrations. While the naked ear may often be fooled, science has revealed several methods of breaking into the underlying consciousness behind our voices.

Our vocal cords move rapidly, generating fundamental frequencies of between 70 hertz and 300 hertz. An instrument called the *electroglottograph* or *EGG*, can filter the sounds made outside of the vocal cords either by digital calculation or by the positioning of electrodes onto the throat area. This allows a spectrographic imaging method to be used, focusing strictly on the vocal cord vibration converted to two-dimensional graph format. This allows for an analysis of the nuances of the vocal cord vibration. Vocal cord vibration can reveal issues such as stress, nervous disorders, and/or irritability. Often this process reveals tiny stress-related spikes in the voice waveforms, called voice-tremors, for example.

Another process called *Layered Voice Analysis,* patented by a company called Nemesysco®, claims the potential for lie detection. The process analyzes vocal activity together with brainwave activity. This allows an extraction process utilizing algorithms to narrow 120 emotional factors down to eight core markers. These markers include stress, arousal, attention, deception patterns, veracity, and so on. The algorithms allow for a high level of potential accuracy in terms of determining the intentions behind certain voice sounds. Intention is of course the key focal point of consciousness. The intention of the transcendental inner self is translated into the physical world through the rhythmic body, and this intention is in turn reflected by the waveforms arising from our vocal cords. While the expression *"actions speak louder than words"* may be true, certainly words speak loud as well.

It is important to recognize the effect our conversational sounds have not only upon others, but also upon our selves. Our intentions are reflected in one way or another by our vocal sounds, but these reflections also affect consciousness in a number of ways. First, when we speak, these words are heard by our ears, as well as are vibrated throughout the body via the vocal cords. These vibrations have the effect of cementing our consciousness in one way or another, through the various feedback systems of the rhythmic body. As we hear ourselves, the intent of our voices reflects upon the mind, and we mirror that intent. Since the inner self is already influenced by the illusion of identification with the temporary body, the reflection of certain words can further commit ones consciousness in this direction.

In times of little control over the voice, we may blurt out something we may not really be convinced about. This may come to invite the attention of others. Once heard, this blurting reflects upon the screen of our mind, committing our consciousness to what we say. Once this happens, we may well feel convinced of our particular point or opinion though it may not be true. We often hear the expression that someone becomes *"convinced of their own lies."* This is an overt example of this. It illustrates how the waveform-consciousness process can work.

We must also recognize the effect that our words and sounds have upon others who may hear these vibrations. Our voice-reflected intentions are mirrored onto the mind-screens of those who hear us. Once reflected, that inner self may become influenced in one way or another. We might think since so many people talk all day long that our words have little effect. This is far from the reality. While we may think we are not being heard, or perhaps the other person didn't react much to our statement, our statement will permanently be recorded into the history of sounds the inner self has perceived.

It is both this history and the immediate reflection of sound that affect the inner self. Often the affect will not be seen for years or even decades later. Often the person will even deny or try to erase the mindscreen's sound reflection. Nonetheless, the historical effect is there. Our brothers and sisters are permanently marked by each and every sound vibration emanating from our consciousness.

We may realize that angry or selfish words will affect anyone within listening range in a negative way. Those sound rhythms will vibrate our intentions through our mind and those around us. The impressions these rhythms create depend upon our consciousness at the time of our expression. Because the self, underneath all the false identification is a soft, impressionable entity, it is easy to affect the self with sound vibration. For this reason, radio and television provide strong influences among society. Hearing particular types of sound waveforms coming from violent or self-motivated consciousness for any length of time can have long-term effects, especially among those easily influenced. While we might assume the consciousness of children to be more impressionable due to less mental loading, every inner self is still impressionable.

Therefore, it is imperative for each of us to recognize the effect that our sound waves have upon those surrounding us. Loving, giving, and understanding sound vibrations will create an accumulated effect of encouraging a loving, giving, and understanding consciousness upon those around us. Even if that sound is one drop in an ocean of negative sound, it is important. While seemingly overwhelmed by the amplitude of negative sound, positive sounds can dramatically shift consciousness.

In the end, intent drives the effects of the sound rhythm. Actors understand this when they *"get into role."* If their script lines are accompanied by a true feeling emotion—even if based on an outside personal event— they will be more convincing in the role. Thus, actors work on internalizing the roles they play. Successful actors begin to relate the character to their personal experiences and thus commingle the script lines with personal consciousness. Otherwise, their lines will simply sound trite and shallow.

Likewise, public speakers receive a better response to speeches about which they have strong personal reasons. Belief ultimately arises from consciousness. The stronger the belief, the more we are consciously focused. Often we find politicians who speak powerfully but their actions reflect their dishonesty or lack of integrity. Misleading an audience with powerful speaking also comes from conscious intention. If the intention is focused and powerful enough, the speech will reflect the strength of those intentions. If there is a strong intention for gain at the cost of deceiving others, this strength may also be reflected in their speeches. This is what we might call *misdirection*. The audience is fooled into thinking the speaker has strong beliefs on the topic. Like the actor, he or she may have developed the ability to channel strong feelings about another topic into their speech.

Research is illustrating that our conversations have a tremendous effect upon our health. In a 35-year study completed at the Department of Psychology at the University of Arizona (Russek and Schwartz, 1996), Harvard University undergraduate men recorded their perception of the love and caring they received from their parents. Those using fewer positive words describing this had significantly more disease pathologies in mid-life than those who used more positive words. Various biases were removed to indicate the connection between the language used by the subjects.

Just as waves of wind move over and undulate the wheat in a wheat field without removing any wheat, it is the essence of a soundwave rather than its substance. It is the underlying conscious intent that travels through the sound waveform that matters.

As we translate this to wellness, surrounding ourselves with sounds coming from intentions of hatred, greed or anger will affect us negatively, while loving, giving and thoughtful sounds will affect us positively. Loving or giving intentions will uplift our consciousness and bring our rhythmic body into harmony with the better parts of our own consciousness.

In addition to focusing upon a particular sound itself, we can probe a bit deeper into the intention of the person creating that sound. Whether the intentions are related to money, status, or attention, we can often "hear" someone more accurately through a resonance with their consciousness. This can be especially fruitful when we find another's conscious intentions resonate with positive intentions we hold dear. Intentions that resonate with the goodness within us promote mental and physical wellness. Intentions resonating with our greedy side bring discord and illness.

Giving and caring for another, even if it is not returned, is the natural consciousness of the self. When we give and care for others without the need for return, we are doing what comes naturally. Words that express this natural inclination resonate with the self. Therefore, sounds from these intentions create harmony within and around us. This is why we all get a warm and fuzzy feeling when we hear of someone reaching out, caring, and helping someone else. For this reason, serotonin, dopamine, and oxytocin neurotransmitters are found more prevalent among those who are exchanging loving feelings (Esch and Stefano 2005). Feelings of love are harmonic with our true self. As a result, these feelings reduce stress and promote healthful waveforms—which in turn are expressed by particular neuro-chemicals.

As proposed by Plato, Socrates, and Pythagoras, musical harmony has universal acceptance. Among humans, birds, plants and animals—musical harmony resonates to a common and deeper cord occurring within. Where does this deeper cord come from?

As we observe and quantify the various resonating sounds within the universe, we must conclude the rhythmic world resonates with a deeper and pervasive consciousness. Just as instruments resonate to master harmonic cords, the various organisms and the living beings with them resonate to love and relationships. Every organism and living being seeks loving relationships. Whether it is centered around the core family unit as in many animals; spread among a hive or colony like bees and insects; or found in various forms among humans; living organisms live for loving relationships. For this reason, as we examine the various sounds vibrating through the universe, we repeatedly hear sound expressions relating to relationships. We hear birds calling their mates or protecting their family's territory with song. We hear dogs barking to please their masters. We hear lonely coyotes calling for mates. We hear humans seeking other humans through the electromagnetic waves of telephone, television, radio, and the internet.

While many of us hear the songs, barks, or howls of animals as indiscriminate and narrow, each sound from each organism is unique and individualistic. While the human language is certainly more complicated than many animals', to assume one song of a songbird species is identical with the song of another bird of the same species is short sighted. This is illustrated by the specific identification between penguins and their mates after an extended leave of absence. After the mother returns from a four-month march for food across the tundra—leaving the father behind to guard the egg—she identifies her mate by his call. While there are hundreds or even thousands of other identical-looking penguins making what sound to us to be identical sounds, the mothers can easily distinguish the sound of the father. Before the father leaves the newborn with the mother for his own extended hunt for food, he will have a verbal exchange with the newborn. This enables him to also be able to identify the newborn among the hundreds or thousands of possible offspring when he returns from his feeding trip. The voiceprint of the newborn is somehow cemented onto the consciousness of the parent.

Deep Sound

The most healing sounds originate from spiritual intention. This has been evidenced by a large body of research showing that prayer aids the process of healing, and many studies show prayer to be at least a viable adjunctive measure creating wellness. In one review of the medical literature, 130 studies showed that prayer not only had a remarkable beneficial effect in human studies, but also had significant effects upon animals, bacteria, and seed germination (Harding 2006; Krebs 2001).

At the same time, most Americans believe in the power of prayer. A Time/CNN poll of a few years ago discovered that 82% of people in the U.S. believe that prayer can cure illnesses of a serious nature. While this poll can hardly be presented as controlled research or evidence, it nevertheless illustrates there is an overwhelming confidence in a deeper form of sound: sound with spiritual intention.

Some might conclude that this overwhelming belief in the power of prayer would sway any research with bias. Legitimizing this concern, some research has shown conflicting results in the category of *intercessory prayer.* Intercessory prayer is when someone—sometimes even a stranger—prays for someone else—often without the patient being aware they are being prayed for.

However, there is little disagreement in medical care that supportive behavior and comforting sound waves from medical staff and family members are effective healing agents (Penson *et al.* 2006). The loving sounds of caring family members appear to have an even more profound healing effect. Comforting sounds of family members with strong relationships with patients have reversed or slowed the effects of comas, heart attacks, strokes, and other traumatic diseases (Johansen 2007). This is a primary reason doctors duly allow family members to visit with patients. The effects of loving conversations are significant and self-evident in hospital care.

While touching is an important aspect of these exchanges—as we discuss elsewhere—the sound waves coming from a consciousness of care and support dominate the process. Behind the loving words of a caring family member is a consciousness that reflects a loving relationship between the patient and the family member. Without that relationship, the sound waveforms would have no real content: If a stranger came into the room and said those same words without having the loving intentions or consciousness, the sounds themselves would have little or no effect. They might even have the opposite effect, as the patient may become upset at someone attempting to be sincere.

The latter has been experienced as well. A dying man surrounded by uncaring family who await their inheritance may well react negatively to the gathering of insincere family members.

As most of us have experienced in our lives, our most special feelings are connected with the exchange of loving feelings communicated through sound. Those special *"three words"* uttered from time to time between family members, spouses and significant others are remarkably common in all cultures. We have also seen in many circumstances those with a lack of loving sound exchange in their lives become despondent and/or even violent. To this end, the dissolution of a loving relationship also can have negative consequences (Langhinrichsen-Rohling *et al.* 2000). The exchange of loving sound expressions is quite simply the reflection of our natural consciousness requiring relationships.

Clinical research has confirmed that even loving relationships with pets can bring about positive health and mental benefits (Jorgenson 1997). We must investigate this further, however. Are the animals communicating special loving relationships to the patient? It is unlikely this is being transmitted via language, other than perhaps some purring here or there. Part of this effect is understood to be the physical effects of touching between the human and the animal. This we cannot deny. Outside of this, we notice that the human typically communicates loving words to the animal as they exchange touch. These words are certainly recognized by the animal, as evidenced by the animal's comfort level and show of physical affection. Otherwise, the use of loving words towards the animal by the human would also certainly benefit the human, as the human certainly hears those spoken words as well. It is undeniable that the communication of loving feelings affects both the sender and the receiver in a positive manner.

This is quite easy to test personally. We can simply speak affectionately towards our pet or a friend's pet and watch their reaction. Then we can speak with an angry tone and watch their reaction. By comparing the two reactions, we can conclude the pet is more comfortable with affectionate sounds. This is a nice experiment because they likely have no idea what we are saying. We could say "dirt is brown" in an affectionate manner and they will likely respond the same. This makes for an interesting personal experiment, with the subject adequately "blinded."

The central issue surrounding deep sound is that both prayer and loving sounds come from a deeper part of our identity. While we might cry for food or moan if our sexual needs are not satisfied, these are reflections of physical identity and have little to do with our deeper need for relationships and love. As evidenced by our relationships with others not within our sexual range or even animals, our need for loving relationships quite frankly transcends the physical issues of the survival of the body.

Deep sound is by no means a new science, however. Rather, it is one of the oldest sciences. The ancient texts of Tibet; the Sanskrit texts of the Vedas; the Egyptian hieroglyphs; the writings of the Mohammedans; the Jewish Psalms; the teachings of the Greek philosophers; the ceremonies of the North American Indians and the Polynesians; and many other ancient traditions all report the use of soulful and prayerful chanting. The use of special hymns, chants, or mantras, intonated precisely with focused intention or in special ceremonies of celebration and worship, has been clearly documented as one of the prime facilities of the ancient ceremony.

The use of deep chants and mantras were considered extremely powerful tools during the ancient times. The science of sound was highly respected in most cultures. Those who developed a command and expertise of the meaning, tempo, frequency, pitch, and tone of these special hymns, chants and mantras were exalted and respected. This was because it was recognized that sound—if utilized properly—had tremendous power.

We find the utterance of deep rhythmic chants and mantras one of the most important rites among various religious ceremonies even today. We find deep rhythmic chants among Taoists, Buddhists, Vedantists, Judaists, Kabbalists, Christians, and Muslims alike. Special worshipful chants are used today by billions of people. While some chants are performed individually in solitude, many are done congregationally. Some deep sounds are done with song, while others are chanted over public address systems. Most of these chants are highly reverential and devotional in context, as they praise or call upon the Almighty. Examples of powerful rhythmic devotional hymns invoking the Almighty still in use today include the _Psalms_, Gregorian chants, daily Muslim prayer to Allah, the devotional mantras of the _Hindu_ faith, and of course _The Lord's Prayer_ of Jesus.

Many of these deep rhythmic chants and mantras are counted for remembrance, rhythm, and consistency. Among the Baha'i faith, we find the chanting of God's name _Allah-u-Abha_ counted on 95 beads. In Buddhism, we find the _Japa Mala_ beads of 27 for counting prayer. In Christianity, we find rosary of _Mother Mary_ or Anglican prayer beads, ranging from 15 to 33 beads. In the Vedic faith, we find again the _Japa Mala_ prayers to the Almighty recited on 108 beads. In the Islam faith we find 33 chanting beads—called _Tasbih_—used to submit _"Glory be to God."_ (Saint Muhammad used his fingers to count these prayers). Sikhism promotes the use of 99 tied knots on a wool string to count prayers. In all these cases, the counting of prayers or chants has been considered the ultimate in sound vibration—its waveforms resonating with the highest levels of spiritual transcendence.

As we compare these spiritually focused sound rhythms with the loving vibrations occurring between family members and loving friends, we can empirically establish that particular sounds—those related to a deeper love and devotion—are held most dear to the consciousness of the living being. Those sounds expressing our intention to love and serve in devotion reflect the most intimate part of our natures: We all seek devotion and love from deep within our consciousness.

Harmonization: Surround Sound

❖ We can easily surround ourselves with the music and sounds with which we best resonate. We can change the beat and tempo of the music as our moods or activities change. When we are energetic and about to participate in a physically challenging situation, a fast beat—faster than our heartbeat—with higher frequency beats and tones—can provide increased energy and enthusiasm. When we wish to lower our stress and slow down, music with a slower tempo with a lower pitch and longer wavelengths (lower frequencies) may be selected.

❖ We can also consider the conscious intentions of the music and its musician(s). By selecting musicians and lyricists we trust and resonate with, we can be hearing sounds that cultivate and support the consciousness we seek to maintain or reach. In most cases, we have the control over the sounds that may raise our consciousness or lower our consciousness. A song praising the beauty of nature or a loving side of a person can be more therapeutic and invigorating than a song about hatred or condemnation, for example. Acoustic music without lyrics can be wonderfully soothing because even with increases in tempo, the instrument's grounding with raw elements such as wood renders a sound more harmonic with nature. By no coincidence, acoustic lyrics are also usually grounded with a soberness and rationality that bears thoughtfulness. Fortunately, many popular acoustic lyrics are vague enough to apply our own interpretations. This can allow us to wander within our deeper consciousness as we listen to the music.

❖ Sounds of nature can easily be applied to our lives. The means is simply listening while in a natural environment. While others may wear their MP3 players as they walk or run through nature, we can listen to the rhythmic waveforms of the wind, the birds, the crickets, the water, and other natural sounds with wonder and awe. We can also open our windows at night to listen to nature's night sounds. In the morning, we will have a symphony of birds playing to our waking ears. If we are not in a location where we can hear birds during the morning and crickets during the night from our window, we are most likely living in the wrong place. As we have discussed, nature's rhythmic sounds play with a beat and rhythm that resonate with the rhythms of our cells, nerves, organs, and brain waves. They can also remind us of our humble position within the majestic beauty of the rhythmic world of waveforms. They can remind us that the natural world is a well-tuned orchestration needing little improvement by us.

❖ As we listen to the sounds of nature from a deeper context, we can ponder how every one of us is striving to connect with one another through sound. The mechanisms of sound provide the conduit for birds, crickets, dogs, cats, humans, and so many other organisms who reach out for others through sound vibration. Through connections made through sound, we find organisms undergo various types of experiences and lessons related to the consequences of their actions and sounds. As we listen to their sounds, we can ponder the depth of the mechanism providing the connection of consciousness through the transmission of sound. Like a broadcasting network joining members of common species together, sound enables those beings a facility to share and evolve.

❖ Recognizing the power of sound, we can carefully select our environment as it pertains to hearing the conversations of others. Choosing acquaintances who intend to share conversation and extend relationships with their sound rhythm can be therapeutic. Sharing conversation with faultfinders and gossipers can bring about increased levels of fatigue and stress. These latter sound rhythms burden our psyche and interfere with our own rhythms. Being a faultfinder or gossiper can both burden our selves and others.

❖ Public events with a preponderance of hostile or conflicting sounds can also wear upon our psyche. Going to a boxing match or wrestling match can immerse one into a sound frenzy of hostility and violent consciousness, as the crowd wishes to see one boxer hurt another. On the other hand, a thoughtful public performance of music or a lecture of higher learning provides obvious therapeutic rhythms and gains in consciousness.

❖ We can also create a therapeutic atmosphere by speaking with honesty, positive intention, and care for others. We can provide healing and gentle sounds that promote growth and reflect kindness. Sounds that reflect feelings of hostility, anger, frustration, resentment, or greed will not only be heard by our own selves. They will be heard by others, affecting everyone within hearing range. These sounds create stress amongst others. They create conflict, interfering with natural rhythms stemming from our deeper selves, trending towards love and devotion. Should our use of sound reflect genuine feelings of humility and compassion, these rhythms will also resonate with our deeper identity and those of others in our midst.

❖ The inner self—residing within the physical body—can mitigate almost all the negative effects of destructive sound interference simply with the application of a deeper intention to learn and give during this physical lifetime. This intention will in most cases, neutralize the potential negative influence of certain sounds. Our conscious intentions are quite powerful. They act somewhat like a strong immune system—ready to counteract interference.

❖ As we reach deeper, we can consider the spiritual chants, hymns, and devotional prayers that most resonate with our particular philosophy or faith. Praising the Almighty by invoking His Name and characteristics has provided a powerful and meditative channel for growth and fulfillment in many traditions and cultures for thousands of years.

Chapter Eleven
Waves of Consciousness

Waves originate from consciousness. Without consciousness, there would be no inertia to cause the orchestration required for creating an organized information-containing pulse. This requirement is illustrated through the sounds of nature. Nature's waveforms all contain coherency, resonance, organization, and information. Without such, the sounds of nature would have no consistency. They would convey no means for communication. Communication relies upon consistency. If we tapped three times one day to communicate something, we would have to tap three times again the next day to communicate the same thing. This need for consistency negates chaos. In other words, communication negates chaos.

The communications of birds, crickets, bees, dogs; and even our mechanical noises of traffic, airplanes, radios and so on are all stimulated by living consciousness. There are conscious beings behind each of these sounds. When we consider airplane noise, although the airplanes are not living beings, they are designed, built, and flown by conscious living beings.

While other sounds of nature like the babbling of a stream or the crack of lightening might be hard to relate to a living source immediately, we simply do not have the scope and perception to recognize their living source with our tiny eyes and minds. Consider an ant crawling across the kitchen floor. The ant might see lights in the sky and might feel thunderous (seismic) vibrations on the floor. Because the ant does not have the scope of perception to recognize the vastness of the human house, he cannot perceive the consciousness behind these sounds. The ant cannot perceive humans turning on and off the lights. The ant cannot perceive a human's walking across the floor as the cause of its perception of what might feel rather like earthquakes.

Consider humans in the place of the ants. We cannot see living consciousness on the grander scale, even with our big radioscopes and telescopes. This is because our scope of perception, and the images from these 'scopes' are converted into images our eyes and minds can comprehend and recognize. Anything outside that normal scope cannot be recognized.

We can see symptoms of living consciousness with intelligence and logic. We can scientifically establish, as Dr. Lovelock has with his *Gaia hypothesis*, the body of the earth, with its mysteriously hot core, volcanoes, tectonic movement, and planetary metabolism as a living organism.

While modern astrophysicists have proposed many accidental creation theories ranging from big bangs and singularity principles to string theories, none of these explains the existence of consciousness. Consciousness is a force separate from the atoms and molecules scientists so intently focus on. When the force of consciousness leaves the living body so does life. Yet the living body retains all the atoms and molecules—none are missing. This intelligently points to the reality that the source of consciousness does not lie within the physical atoms and molecules. This force of consciousness is separate from the physical body and its bevy of biochemicals.

It is illogical to assume the organization and systematic conscious mechanisms displayed by nature and the living organisms within it originate from chaos accidentally: Can organization arise from chaos? Can something arise from nothing? The precision we have discovered among nature though the various laws of physics, chemistry, and electromagnetic radiation all point to an organized and independent source at the root of existence.

When we expand the horizon to consciousness, it is not logical to assume that consciousness accidentally arises from nothing. The very fact that each of us—each conscious entity—has an inherent need for purpose. A need for purpose indicates we have a destination. Any destination requires an origin. Therefore, it is logical that each conscious being has an origin.

A larger consciousness is illustrated by the dependent interrelationship between the earth's activities and the other organisms on the earth. The thermal heat of the earth provides protection to most life on earth and seas. The circulation of the earth's bodily fluids provides hydration to the earth's living organisms. The circulation of air and temperature keeps the earth's living organisms from burning up or freezing, or being bombarded by radiation. The various circulatory ecosystems movements of the earth are needed for the planet's sustenance. All of these movements of the earth are rhythmic. They are organized. They are informational. They are indicative of a consciousness larger than ourselves replete with a purpose of design.

We can extend this same reasoning to the rest of the universe. While we could (and physicists have) come up with hundreds of creative theories for how the universe accidentally came into existence and will accidentally cease existing one day, we cannot avoid the organized harmonic and holographic tendencies of the universe from top to bottom. This is not indicative of an accident.

Harmonic Perception

As we review the scientific discoveries regarding electromagnetic radiation, we must not miss the obvious. As Edgar Allan Poe's *The Purloined Letter* illustrated over a century ago, what we are looking for may actually be in plain sight.

The nature of the universe and all that we sense around us must be looked at with objectivity and reason. Modern physics has shown us scientifically that what we see around us with our eyes, neurons and mind map is not as it appears. The world may appear to be solid, but it is not. The world and all of these shapes and forms may appear to be permanent, but they are not. While we recognize through all this research that matter is cyclic and wave-like (with constant interchanging atomic arrangement), our scientific theories still hold on to solid ball-and-stick atomic models to explain and visualize matter. Although the evidence is clear that the universe is built of waveforms, we prefer to see it as solid and permanent. Likewise, while we know the human body is degrading and being rebuilt (or shall we say replicated) constantly, we prefer to see the body as a permanent structure.

The waveform nature of matter illustrates our senses' lack of accuracy. The eyes only see a wavelength range of 380-760 nanometers. Our ears only hear a range of 20 to 20,000 hertz, though sounds between 1000 and 4000 are typically heard. Our skin only senses a thermal range of 29-49 degrees Celsius. Our tongue only tastes a narrow range of chemical organoleptic signaling while our olfactory senses only pick up certain odorous waveforms.

We are also limited by the scope of our perception. The components of frequency range do not incorporate amplitude. We may be able to observe a wide range of frequencies with either gross senses or technical instrumentation. However, waveforms with amplitudes outside our perception range will not be perceivable. We will not have the scope of perception. As an analogy, we may easily observe a small wave breaking on the beach, and even a ten-foot wave breaking on the beach. Could we recognize a wave if it were one billion feet high? Could we even see the top and bottom of the wave well enough to tell it was actually a wave? It would be unlikely. We might perceive a massive inundation of water. Because we did not have the scope of perception to see the top and bottom, or beginning and end of the wave, we would have no way of seeing its form or shape in its completeness.

Consciousness appears in our universe at all frequency ranges and amplitudes. Conscious organisms can exist at subtle frequencies with amplitudes so small we can hardly observe their existence. The most obvious of these are paramecia, which we can only observe with the help of microscopes. While these single-celled creatures appear visible in the same frequencies our eyes can observe, their tiny sizes are outside the scope of our perception. They require the heavy magnification of a microscope—which in effect increases the scope of the amplitude for these waveforms. If they were larger, we would be able to see them with our eyes—which is why we can see them with simple magnification. Certainly the existence of other organisms—both smaller and larger—may certainly exist out of our normal scope of amplitude.

This of course assumes the organisms have the same frequencies as well. While we have observed with spectroscopy that different molecular combinations exhibit different frequencies, we assume that an atmosphere made of different molecular matter will exhibit organisms within our visible range of frequencies. This is obviously a misguided assumption. If different molecular combinations emit different frequencies, then those organisms living within a different atmosphere will certainly not be readily observed in the range of frequencies available to eyes made of our molecular atmosphere. For this reason, we see on other planets obvious indentations indicating the flow of liquids like oceans and rivers, yet we cannot observe those liquids: They are outside of our range of vision or our instruments—also made of the molecular basis of this atmosphere. While we can assume that certain molecular combinations and densities arranged for this atmosphere cannot exist at the temperatures and atmospheric pressure of another planet, this does not mean that planet cannot contain molecular forms designed for that atmosphere. Furthermore, we would have no visible entry into that atmosphere, as it has its own range of molecular combinations, densities, frequencies, and amplitudes. For this same reason, many aquatic lifeforms cannot observe our physical presence.

The researchers further analyze the world around us using various instruments, the more it becomes obvious that matter is not what it appears to be from our senses. We must begin to abandon our assumption of solid matter and assume a more harmonic perspective: A perspective of consciousness.

Information Waves

When we watch a wave breaking on the beach we do not think the wave is carrying much information. Most of us see this a wave as a lifeless pulse of water. Yet a wave with almost the same wave shape with a different amplitude and frequency will oscillate from the larynx of another person. This wave will have a meaning to us as its waveform is interpreted. We know from this meaning that this wave communicates information. It carries information sent from another person with a conscious purpose. The waveform might be part of a larger communication about a particular subject matter, or it might be a final "yes" or "no," reflecting a conscious final decision. Nonetheless, we naturally consider these particular sound waveforms to be *information waves*.

Information is not limited to moving through sound waves. Other information waves include spectral waves that transmit particular colors; reflective waves that transmit particular shapes; radiowaves that communicate sounds and pictures over long distances; and many other waveform types. These waves communicate the activities of the world around us. Tactile vibrations would also be considered information waves. Tactile (or seismic) waves can also transmit deeper information such as emotions, for example. This is quite easily evidenced by a loving embrace or a special handshake. Communication through touch is exercised quite effectively by blind people, for example.

These informational waveforms are not unlike ocean waves in many waves. In face, an ocean wave breaking on the beach certainly communicates an event that took place in the middle of the ocean. Some type of storm or wind patter is being reflected in the amplitude and frequency of the wave. Larger waves will communicate a storm out at sea. A series of short wavelength waves might indicate a boat having passed nearby.

A surfer might see reef- or beach-break waves with yet another perspective. Larger wave sets might communicate a coming swell and an opportunity to surf fun waves. Is this all ocean waves communicate however? How do we know there is not deeper, conscious information contained within this wave? We do not. Consider the ant walking over the cover of speaker. There will be large seismic waves (in the ant's perspective) that would significantly undulate the ant body, not unlike a surfer surfing a large ocean wave. Surely the ant could not perceive the information contained in those sound waves. While we know ants can communicate quite effectively using their antennae, mandibles, and various scents, they would not be able to understand the information being transmitted through our music speaker as he walks across it. This is because the forms of these waves are outside of the ant's sense perception range.

It is illogical to assume that all these organisms have limited sense perception and humans have the gold standard. We too have limitations. Assuming this, it is logical to also assume that ocean waves and other waveforms are also informational in context. It is not only logical because of the limitations we see among other organisms, but also because the shape and structure of these larger, natural waveforms have definite similarities with the waveforms we know are informational in context. While we can see these waveforms visually, we do not necessarily have the ability to perceive the information they transmit.

While our scientists assume organized natural waveforms are occurring without cause or meaning, we can logically determine otherwise. As we consider waveforms such as nature's ecological cycles of decomposition and formation of natural commodities, we can easily see these larger waveforms are not only well designed, but they are repetitive yet accommodating. As nature's elements become polluted, the ecosystem begins accommodating those changes in an effort to clear up the problem. Nature is doing this now with changing weather cycles to accommodate and repair the dramatic increase of carbon. Part of this is a warming effect. Our bodies do the same thing with the advent of a fever to increase metabolism to clear out an infective agent. In the case of the planet, unfortunately humans are the infective agent, and our misuse of nature's resources is the infection.

Various spectrometry tests reveal electromagnetic properties at the atomic levels of matter with waveform properties similar to light, sound and radiowaves. The latter two waveforms are known to be informational, as we watch our televisions and listen to our family members. Is it logical to assume the others are not informational?

Directional Waves

One of the most basic yet profound qualities regarding the movement of waves through our dimension is the quality of direction. As we look around us at the basic elements of our natural world, we see elements in not only motion, but also having directional capacity. We see electromagnetic radiation moving from the sun outward; pulsing with direction and speed to the surrounding solar system. We see water moving in currents, tides, and waves, in one direction or another. We see wind moving directionally. Sometimes the wind comes from the east and sometimes it comes from the west, north, south, or a combination thereof. Every location on the map has a typical prevailing wind direction. We see air currents moving with direction as well, often moving with temperature, wind, pressure and other effects. We see migratory animals moving with invisible magnetic compasses and viewfinders, always in one direction or another. In the northern hemisphere they usually move north in the summer and south in the winter, and the opposite direction in the southern hemisphere, and typically back to their birthing grounds for reproduction.

When we look deeper into the smaller units of matter, we also see direction. We see direction in the movement of molecules. We see movement of ions and we see movement of electrons, nuclei and other basic elements of the atomic world. These tiny movements of direction harmonize with their prevailing electromagnetic qualities. We see movement of these forces not only within an axis, but replete with spin and other related trajectories.

We have applied a number of quantum characteristics to this organized movement and trajectory. Should this translate to a simple, chaotic source? Rather, it would be logical to conclude that organization with complexity and direction is more likely coming from a complex source with purpose. According to current quantum theory, quarks and leptons are the primary types of matter, while antimatter originates from antiquarks. Quarks have been attributed properties such as *charm, strange, up, down, top* and *bottom*. Quantum theorists contend this is due to simple confinement within subatomic particles. The question this bears is how this confinement regimen came to be established and how is it being maintained? While the assumption of complex, organized waveforms somehow originating from a simpler chaotic environment is prevailing among scientists, we know that the laws of nature do not support such a organization-from-chaos assumption. In nature, we can always trace organization to a source of design and purpose.

The past forty years of intense research on the properties of waves and electromagnetism have revealed that waves and their various interference patterns contain vast amounts of data on the smallest levels. Waves of different forms only contain directional motion, but they organize and store data through their various interactions. This is increasingly being illustrated as we discover previously unknown reasons for patterns of natural waves occurring within our atmosphere.

We also know that within the chemical periodic table we find complex organization in a series containing directional data. We find as atomic weight and electron count proceeds in the positive direction, particular combinations occur with a harmonic basis equivalent to the valence shells. We find that harmonic rhythm exists in molecular combinations, as their electron valence numbers become combined. We see these complex molecular combinations geometrically and proportionately directional. They form complex and structured bonding shapes, which form complex lattices. These lattice structures formulate the shape and directional matter we see around us.

This organized complexity would be logical only if subatomic particles are reflecting a conscious source of purpose. The only reasonable explanation for why matter at the tiniest levels is organized at these levels rather than chaotic is because their tiny rhythms reflect a grander conscious purpose. We can observe this purpose as we find these atoms combining in special ways to create larger structures, which move with direction, precision, and design. The only logical explanation is that they are coming from a state of organized intention, and their movements reflect a controlled purpose.

We might compare this to some ripples created in a pond by the movement of a boat. A series of waves will move outward from the boat's motion: If the boat had an organized speed and direction—reflecting a destination and purpose of the driver—the boat's waves would be consistent and organized. If the boat were to zip around with a random direction and speed, however, it would be a different situation. The waves would be unstructured and disorganized. They would indicate a lack of direction. They would have erratic waveforms.

What we see in nature is the former rather than the latter case. The typical sine wave—which we see in so many forms and structures throughout our environment—is repetitive and organized. It is structured. We see so many other geometrically aligned structures in nature, from the golden section and spiral to the Fibonacci sequence; we see repeating structure and organization.

Conscious Waves

We might compare nature's waveforms to the stroking of a canoe paddler. Because the paddler consistently strokes the paddle through the water, the waves on the lake have both direction and organization. They also communicate something about the paddler. They communicate where he is going and how fast he is going, for example. These two conditions will also tell us more. They will indicate the paddler wants to get somewhere—such as to the other side of the lake. The waves will also indicate whether the paddler is an efficient paddler or a sloppy paddler. They will indicate whether the paddler is careful or random in his direction. They will also indicate whether there is navigation behind the stroking of the paddler. If the paddle strokes are in a straight line, for example, pointing toward a dock on the other side of the lake, the paddler obviously is navigating his course across the lake.

A navigated course requires several components. Firstly it requires purpose: A destination. Secondly, it requires a rationale for that destination: Consciousness.

Conscious intention has been the subject of scientific research over the last forty years. There have been a number of tightly controlled experiments studying intentional thoughts. In one study done at the McGill University in Montreal in 1959, Dr. Bernard Grad studied the effects of depressed patients on the growth rates of barley plants versus non-depressed subjects. The subjects would hold a sterile glass of water for a specific amount of time. The water was then used to feed young barley plants. Along with the treated barley plant groups there was a control group of barley plants watered with sterile water not subject to any holding. Dr. Grad found that the barley plants treated with water held by the depressed patients grew the least. The barley plants held by the normal subjects grew the most—greater than both the control group (not touched) and the depressed group. This research substantiated the effects of consciousness (Brad 1964).

Benor (1992) described research linking distant healing intentions to significant improvements among forty AIDS patients. While the treatment group had significantly better survival rates (ten of the control group died and none of the healing group died), the treatment group had better T-cell counts and fared better with AIDS-defining illnesses. In a follow-up study attempting to replicate these effects with more patients, a cross-sectional group of healers from various philosophies treated a group of patients remotely. When the treated group was compared with the control group, the treated group had significantly less severe illnesses, fewer doctor visits, fewer hospitalization days and greater survival rates than the control group.

The rhythmic effects of intention have been observed in various other ways. Multiple studies (such as Routasalo and Isola 1996; and Bottorff 1993) have confirmed that touching by hospital staff and families is important to healing. This type of touching is a reflection of the intention of care. As we can easily observe when we stroke a pet dog or cat, touching is needed by all living organisms because it is the expression of a conscious attempt to love.

If we think about the word *particle*, the root of the word is *part*. It signifies that it is a tiny *piece of something larger.* What is the "something" that is larger?

When we consider the many diverse rhythms of different densities moving through the universe—such as the thermals, liquids, solids, gases and electromagnetics, they come together to form the ground we walk on, the foods we eat, the air we breath and the water we drink. These elements move with rhythms resonating together harmonically. We can experience this personally and daily: When we take a walk in the forest, we usually feel calm and relaxed. This is because the harmonics existing within the natural elements are harmonically resonating with our physical bodies.

Coherent Consciousness

We know from observation that a waveform sent out from a particular source into a field of other waveforms will interfere with the other waves. This may result in the waves constructively or destructively interfering. As long as the waves are coherent, they will result in either a harmonic resonance or destructive disharmonic interference pattern. Either way, their coherence creates a natural result. A large non-coherent waveform interference will theoretically conflict and create various incoherent interference waveforms as a result, creating disruptive physical effects. We note that it is unproven that incoherent waves actually exist. While we may not see a coherence between waves within our perception, a particular wave seeming incoherent could well be coherent with waves we cannot currently perceive.

The proposal here is that *non-coherent waves do not exist*. Every wave, even if created with disruptive motives, has a coherency on some level. We can see this within our ecosystem. A plastic created by a chemical manufacturer may not degrade within our lifetimes. But it will degrade at some point. Some analyses have illustrated some five hundred years. So while plastic is not coherent within the context of our natural dimension during our lifetimes, it certainly will resonate with a longer scope. This is not to condone plastic use. This is to say that the components of plastic did come from natural elements of the planet. The fact that they were synthesized under extraordinary circumstances only means that nature will also have to exert extraordinary ecological strategies to decompose it—another word for resonation. The problem is that part of that extraordinary strategy may include the extinction of humans on the planet.

In the 1980s, German researcher Dr. Walter Schempp determined that radiowaves received from distant stars could be converted mathematically into three-dimensional complex holograms using calculations that convert waves into information. His applications soon became focused on magnetic resonance imaging, which had been developed to create three-dimensional images of human anatomy. Through the application of MRI to distant radiowaves, Dr. Schempp realized that the electronic rhythmic bonds of atomic and molecular structures contained unique information: Holographic information reflecting larger structures. Because DNA holographically reflects the information of the entire body, the rhythmic information of the electron waves of atoms should also reflect the larger information of structures they composed.

Remembering our canoe analogy, this would make perfect sense: Every tiny ripple in the water somehow reflects the stroking of the paddle and the boat moving through the otherwise still lake.

The rhythmic world around us is continuously supplying each of us with learning and decision-making situations. The world's rhythmic effects presents us with choices between good or bad, love or hate, greed or giving, persistence or quitting, kindness or violence, and so on. Accompanying each and every choice is a corresponding result, programmed with a number of variables. Should we smile at another person in the right situation, that positive greeting will usually be responded to with a smile in return. Should we thank another person we will usually receive positive feedback like *"you're welcome."* Should we help others around us, we will usually find that others will not only be grateful, but also come around to help us when we are in need. In other words, activities receive responses *in kind.*

On the other hand, should we approach the world greedily—perpetually thinking in terms of what we will get out of it—the world's rhythms will constantly challenge us as they accommodate our approach. Matter will arrange constant frustrations for us. Should we be reduced to yell and scream at the others around us, we will find them yelling and screaming right back at us—or deeming us neurotic and locking us up. Should we unfairly judge others—communicating unneeded disappointment in their efforts; we will likely receive a measured negativity in return. Others will watch our every action, applying the same stringent judgmental negativity we applied to others.

We can see these reflective effects everyday as we watch the people around us. We can watch this taking place at our places of work, as our bosses and co-workers each deal with their varying approaches to others. We can see this effect on the news, as politicians deal with the public's reactions to their unabashed confidence. We can see this effect with celebrities as they deal with the reflective responses of their once-intense desire for attention. Once they achieve the stardom they once desired, they are now hounded by critical media and infringing *paparazzi.* Now they wish they were out of the public eye.

The ironies and consequences of nature's responses are orchestrated through interference patterns evolving from consciousness. When we convert all of these activities to waveform interference, we must see that in order for our activities to produce an organized and coherent lesson in response, they must be met with a waveform response system coming from a coherent source: In other words, the natural waveforms must be coherent with the waveforms of conscious activity. They must also be coming from consciousness.

We might ask ourselves why nature presents us with lessons reflecting our prior choices. Why are we met with challenges that teach us particular lessons related to our decisions? Interestingly, we can often ponder life's lessons on both a real-time basis and a symbolic basis. Could these multi-layered lessons of life merely be accidental hallucinations of random bio-chemical brain cells?

Surely, we could suppose this if we choose not to learn the lessons life is trying to teach us: This conclusion, however, would be nonsensical. The only rational reason for nature's intricate reflective response system lies in the fact that the source of nature is conscious and intentional, and the rhythms of cause and effect have the ultimate purpose of teaching us to grow: An intentional school of sorts.

We have discussed atomic and molecular *bonding*—the force that exists between theoretical particles. Let us for a moment consider another type of bonding. This is the bonds existing between conscious living beings. We can see these sorts of bonds throughout the animal, marine, and in-sect kingdom. Organisms may bond together to procreate and defend their offspring. They may bond together to stroke each other and teach each other. They may bond as they play together, as sibling cubs might. The bond that generally keeps two people together is a bond of mutual care and concern. We often refer to this type of bond as love. In fact, as time proceeds, this type of bond often results in various related bonds: Those between mother and son, mother and daughter, father and son, father and daughter, and otherwise between people related or mutual friends. These types of bonds tend to keep groups bound together, within the same household or community for a while, even later through phone calls and holiday visits.

Other types of bonds may keep people together. These may be bonds of mutual goals, fellow employees, community groups, churches, and even bonds of self-perpetuation such as music bands, movie productions, fash-ion shows, and parties.

The glue of these bonds is undoubtedly consciousness. People con-sciously bind together for an intended purpose. They may or may not have complete control over those bonds. Yet these bonds are nonetheless consciously oriented. We can also accurately describe these bonds as being coherent, because there is a necessary level of communication required to maintain the bond. Communication in this case is a form of waveform interference: The expressions of one person are received by another. This reception creates an interference pattern, which creates an understanding of some sort between the two. Communication interference may occur through speaking, touching, emails, phone calls, and so on.

This does not mean the bond is always a positive one. In other words, the interference pattern between the two parties may be constructive—

leading to goodness or health—or it may be destructive—leading to negative habits or violence. The bonds created by this later form of
interference is still coherent. Coherency is independent of its result. We
would propose, in fact, that the ultimate effect of coherency is still positive. Whether there was violence or goodness, there is still learning. We
might propose that coherent bonds are simply tools for learning.

While we do not propose atoms are living organisms in themselves,
we can compare these human bonds to the bonding between subatomic
particles and atoms within molecules. We see the same types of interference patterns between these bonding of combinations. We see a
synchronicity between the bonding structures of nature and the bonding
structures of living organisms. We also see living beings existing within
molecular wave structures, utilizing their properties for expression and
communication. This points to the real possibility that atomic matter is
simply an extended expression of consciousness: A subset of conscious
intention.

Nature's Arrangement

As we evaluate any waveform, we naturally consider its source. A
pebble dropped into a pond will result in ripples concentrically moving
outward. The ripples are obviously connected to the pebble being
dropped. No pebble—no ripple. Similarly, if we hear a loud noise in the
distance we may turn immediately toward the source of the noise to see
what caused it. Should we hear a honk while driving we might look
around us to see who honked and why.

We assume these waveforms have sources because this is consistent
with physical existence: Every unique rhythm has a unique source. Once
created, the waveform moves outward, affecting waves and mediums in its
path. Is this an accidental journey?

The reason mathematicians and physicists can create formulas that
can be consistently applied to natural waveform motion such as gravity,
momentum, energy, light, velocity and other elements is quite simple:
Nature contains an organized structure of specificity and designation.
These structures are expressed in the various sinusoidal motions of waveforms from the largest to the smallest of natural elements. They are
illustrated by the precision and measurability of spiraling Fibonacci-
sequenced tree and plant branching; the rotational orbits of planets, stars
and galaxies; the frequency and power of solar storms; the spiraling of
hurricanes, tornadoes, and the cross sections of ocean waves; along with
so many other mathematically precise and congruent oscillations occurring in nature.

Because rhythmic matter has arrangement and specificity, it must have design and programming. Since we measure matter as relationally functional, this precision must be programmed. For example, if we walked into a room full of chairs arranged in a circle, we would automatically assume the chair positions were arranged by someone. We would assume that someone arranged the chairs for a particular purpose. Possibly the chairs were arranged in a circle so a group could hold a meeting. Otherwise, why would these chairs be so arranged? While we could claim this arrangement of chairs might have come about accidentally—this assumption would also require that the pieces of wood also just happened to fall together into chair forms as well.

Let us consider a room where the furniture not only was arranged when we walked in, but continually rearranged itself to accommodate the group and its members' various decisions and desires. This would present us with a situation of initial ordered arrangement as well as programmed responsiveness. If something responds one way to one thing and another way to another thing, then this is programming: A responsive arrangement of cause and effect.

Now if the furniture constantly rearranged itself to accommodate our movements throughout our stay in the room, we would soon realize the programming was intentionally set up for a particular purpose. This constant pattern of programming indicates functionality from conscious intention.

We know from electricity that continuous current must come from a power source. A continuous electrical oscillation indicates an ongoing source of power. If a waveform is not only continuous, but it is precisely rhythmic—producing consistent amplitude and frequency—then the source of the wave must be a consistent source of power. Otherwise the waveform could not be consistent or precise. Considering the organized synchronization of these rhythmic forces throughout nature, we can understand that the power source is not only organized, but must exist outside the system.

We know the activities of SCN cells do not provide the clock, but rather reflect a greater biological clockwork system. Research has indicated that SCN cells are stimulated by light. This means the biological activities of these SCN cells reflect the rhythms of light waveforms generated by the sun and the moon, among other natural forces. As the timing of nature synchronizes these body clock cells, neurochemical release is stimulated to catalyze bodily functions. Therefore, it is clear that these suprachiasmatic nuclei cells are merely tuning or harmonizing the body to the larger clockworks of the universe. They are not the clock itself. Like musical instruments, our bodies are simply being tuned to the rhythms of a greater harmonic.

In a symphony orchestra the violins, flutes, drums, oboes, trombones and many other instruments all play in synchronized harmony, interweaving their various melodies and beats to form a composition or musical performance. In order to play a specific part of the composition, each musician reads from a sheet of music written specifically for that instrument, enabling a coordinated harmony between each instrument and the remaining symphony orchestra.

The healthy human body vibrates with a similar type of orchestration: The heart pumps blood at one rhythm while the lungs take in oxygen and dispel carbon dioxide at another rhythm. The brainwaves ebb and flow to another rhythm while the four ventricles of the brain pulse cerebral spinal fluid to yet another rhythm. The body's endocrine system releases cortisol, growth hormone, follicle stimulating hormone, insulin, vasopressin, rennin, oxytocin and many other activating hormones—all in rhythmic cycles, synchronized to the time of day and the events occurring around us. All of these various cycles ebb and flow together in synchronicity, like instruments playing together within a tuned orchestra:

The instruments in an orchestra must be tuned to a particular pitch of notes and octaves before they can play harmonically. Once tuned together, the orchestra is prepared to blend their sounds together within a resonance immediately palpable by the audience. Should even one instrument be poorly tuned, the audience will likely immediately sense a deficit among the orchestra.

In any harmonic orchestra performance, there must be a song written up and a conductor to guide the performance.

Intentional Duality

The various mathematical formulas derived from natural science provide an indication that harmonic energy flows originate from consciousness. This is because nature exhibits intentional response. In other words, a specific response is created by a specific action. The primary intention expressed in this systematic cause-and-effect cycle is education. As we make choices and take action, the results of those decisions and actions cause us to learn. Any good schoolteacher will tell us that a well-organized lesson plan is critical to providing a good platform for teaching any subject. A chaotic classroom full of wild kids playing games hardly provides a sound platform for learning to read and write, let alone learn math, for example. Learning requires an organized strategy and sometimes a disciplinarian to keep the instruction moving forward.

The duality of the universe has slowly become evident. Over the past couple of centuries we have confirmed that physical matter is composed of a combination of two waveforms: Electronic and magnetic. These two waveforms are termed 'electromagnetic' because these complementary vectors appear to be moving in perpendicular motion yet revolve around each other with a duality of direction and motion.

The two forces of the electromagnetic, oscillating in synchronicity through the physical world, provide the basic harmonic that all other waveforms and their reflective states are built upon. Though they are dramatically different energies and waveforms, they are compatible and co-functional. They are dualistic.

We have become familiar with the peculiar effect of magnetism as a polarizing force. This renders another dual force; one of positive and negative, north and south, up and down, east and west.

Yin and Yang

The principles of *Yin* and *Yang* have played a significant role of some of the oldest natural sciences for thousands of years. They are also commonsense and practical. We see duality evident on many levels of everyday observation. Within nature exists male and female qualities, each with distinct traits and effects; each balancing the other. We see this balance of duality resonating within the structures of society including marriage, friendships, businesses, politics, and social groups.

The traditional concept of *yin* is dark, mysterious, passive, yielding, cool, soft, and feminine. *Yang* on the other hand is illuminated, aggressive, hot, hard, and of course masculine.

We observe the dual nature of *yin* and *yang* throughout our scientific explorations. We see the logical left-brain activity balancing the more creative right brain activity. We observe dualities among the elements, expressed as hot and cold; black and white; left and right; light and dark; north and south; in and out; push and pull; and even greed and love. In mathematics we observe positive and negative numbers; even and odd numbers; and rational and irrational numbers. We see two sides of every mathematical equation, separated by an equal sign—balancing the *yin* of that equation with its *yang*.

Each electromagnetic waveform moves with perpendicular forces of *yin* and *yang*, revolving in a four-dimensional (including time) manner. We see every waveform reaching downward into a trough, only to climb upward into a crest, each alternating with *yin* with the *yang*. As the structure of the nucleus of the living cell unfolded to us, the swirling DNA molecule has been unveiled. Within DNA we observe a complementary double-helix of duality. We observe two complementary *yin-yang* strands of phosphate-ribose, bound together by dual pairs of bases—each pair expressing a specific amino acid within its *yin-yang* balance. As we have explored the activities of DNA, we have discovered the *yin* existence of RNA. RNA complements the existence of the *yang* DNA by wrapping itself with DNA in transcription, producing duplicates to spread the coding instructions.

Within the *yin* and *yang* symbol we see the inter-relationship between the duality of nature. While the *yin* tear contains a hole with the illuminated color of the *yang*, the *yang* tear contains a hole with the darkened color of the *yin*. This exchange between the tears represents the intertwining harmonic of nature. Within each male, we find some essence of female; and within each female, some masculinity. Within each hurricane there is an eye of stillness. Within each crashing wave there is an empty tube or barrel of calm.

The *yin* and *yang* model also discloses a rhythmic relationship. Just as the *yin* edge closes in to a point, the *yang* is rounding into full focus. As the *yang* aspect is waning into a point; the *yin* quality reins fully. The rounded concave edge of each aspect produces a symmetrical wave-like structure—its point not unlike an ocean wave preparing to crash onto the beach. As they each move together, each side gathers into a wave-like shape to create an interactive spiral effect—with each side moving perpendicular to the other side yet rotating together into the middle. Each point displays the spiraling arms familiar to the natural spirals of the cosmos and weather systems. Should we extend the points of each of the *yin* and *yang* aspects outward, we would see a rhythmic spiraling effect rotate each of the spiral arms around the inner core. What we will see is *Fermat's spiral*: a spiral representing the spiraling of flowers, plants and other natural movements.

Obviously the *yin* and *yang* representation of nature stands as further evidence of a harmonic duality within nature. When we see balance and congruity between two so many separate events or processes, we consider this no different from a complex equation proof. Numbers are simply symbolic representations of nature's balanced forces anyway. The reason complex formulations and proofs balance with an equal sign is because our scientists and mathematicians have inherently recognized nature is balanced in duality.

As we take notice of this congruent duality, we must incorporate the conscious components. The difference between the physical body and the inner self or personality is that one is driven by consciousness and the other *is* consciousness. This in itself creates a duality between life and chemistry. This duality is reflected in the observations of life and death;

movement and stillness; happy and sad; empty and full; complete and incomplete; spiritual and material; and of course heaven and hell. The 'driving' of the chemical body takes place through conscious intention. A dead body displays no consciousness because the driving component of the inner self is absent. Without this component of consciousness, there is no balance within the body.

Purusa and Prakriti

The ancient Sanskrit texts of Ayurveda reference a duality between even the conscious forces of nature. Ayurveda points to consciousness stemming from an original Conscious Being or *Purusa*. From the *Purusa* comes the creative force, producing the *created*. This created force is referred to as *prakriti*. These two forces, the *Creative Purusa* and the *created prakriti*, are said to intermix within nature to provide the forces of action and consciousness.

These same principles are mirrored in many ancient texts from different cultures, eras, and languages.

Prakriti in the broader sense also includes us in the concept of Ayurveda, because we are also *created* elements of the *Purusa* according to Ayurvedic science. In this sense, the *Purusa* is the male, dominating consciousness of the universe, and we individuals—the *prakriti*—are the female, created conscious forms. These two conscious forces are thus considered the basis for relationships and the existence of love. Within the concept of love, we observe the ultimate intertwining of duality. We observe giving and receiving. We observe togetherness and separation. When the *Purusa* and the *prakriti* are in harmony, there is love, and the ultimate activity of healthy consciousness according to Ayurveda revolves around loving service.

The Conscious Conclusion

Each conscious being reflected onto this holographic physical world of waveforms is an individual. This is confirmed by our unique individuality of expression and bodily features. This is why each of us strives to differentiate ourselves from others. We seek individual success, individual attention, and individual love. Each of us has a unique personality, and this individual personality is reflected through the body by unique fingerprints, unique DNA, unique irises, and unique activities.

Meanwhile our permanent existence is hidden from our physical view, just as the movie theater environment prevents the moviegoer from seeing the real world outside the theater. We are left to a world where we can temporarily have separate identities and separate quarters from our permanent homes and identity.

Here we have a physical dimension of holographic duality. It is this separateness from our permanence that expresses the duality of the holographic physical universe—with its dual forces of good and bad; black and white; and cold and hot. Formed by the Creative, the world of duality renders a separate and temporary "double" identity for each of us, much as an actor might assume a separate identity for a short time during a stage or screen performance.

The ancient Ayurvedic science explains that the conscious expression of duality translates through nature into the three *gunas* or modes: Goodness, ignorance and passion. The mind is said to form around these three modes, and the five gross elemental layers (solid, gas, liquid, heat, and electromagnetic space) take shape around the mind. These elements are each reflected by the five types of sensory waveform perceptions (touch, hearing, taste, vision, and smell). These senses and sense perceptions are further expressed into the five gross chakras and the two subtle chakras, circulating through the physical body and its layered physiology, which conducts sense perception, movement, and so on into a brainwave mapping system.

While the three *gunas* reflect these natural elements and senses, according to Ayurveda they also express themselves into the three *doshas* of *pitta, kapha* and *vata*. While there is an amalgam of influences here, the *pitta dosha* reflects active creativity, the *kapha dosha* reflects stability and groundedness, and the *vata dosha* reflects learning and maturity. Each of these *doshas* is also expressed as particular body features such as our bone size, face shape, eye color, metabolism and so on. According to Ayurveda, while each of our bodies express a mixture of these three *doshas*, each of us has a unique tendency towards a particular combination thereof. For this reason, Ayurvedic physicians play close attention to a person's particular *dosha* in determining what might be out of balance. Their treatments subsequently tend to focus upon bringing that person's *dosha* balance back to their normal *prakriti*.

While these concepts of *yin* and *yang* or *Purusa* and *prakriti* might appear abstract as we consider our personal search for harmony, they do indicate the need to include consciousness within our view of life. While scientists tend to look purely at the natural laws of physics and focus on pragmatic research, the conscious elements that drive nature are missed. These conscious elements explain personality, choice, and desire. Without consciousness, these elements cannot be adequately explained through purely physical science.

Finding our own harmonic within the universe must include a sense of our own consciousness. We might try to solve our quest for fulfillment through the mixing and matching of physical possessions, diet, exercise, and various sensual explorations. As we have discovered from our experiences and the experiences of those with excessive amounts of these assets, these physical elements do not bring fulfillment. A deeper level of fulfillment is needed, because consciousness exists on a deeper level. For the conscious being, a solution must contain consciousness. For this reason, it is commonly accepted in our society that love is the ultimate fulfillment of humanity. Love and humility, it seems, is the purest expression of consciousness, especially when it is practiced without condition.

Using the evidence presented here—ranging from the most ancient of sciences to the latest experimental data—we can logically conclude that consciousness completes the activity of the universe. Love is the purest fulfillment of consciousness. Love, however, requires a relationship between beloved and lover, the dual nature of consciousness. Herein lies the mystery of existence: Achieving the ultimate loving relationship within the pure consciousness of knowing our intended beloved. Humbly respecting and harmonizing with the laws of nature are integral and necessary parts of the process of ultimately achieving personal perfection for each of us.

References and Bibliography

Abdou AM, Higashiguchi S, Horie K, Kim M, Hatta H, Yokogoshi H. Relaxation and immunity enhancement effects of gamma-aminobutyric acid GABA. *Biofactors.* 2006;26(3):201-8.

Ackerman D. *A Natural History of the Senses.* New York: Vintage, 1991.

Airola P. *How to Get Well.* Phoenix, AZ: Health Plus, 1974.

Aissa J, Harran H, Rabeau M, Boucherie S, Brouilhet H, Benveniste J. Tissue levels of histamine, PAF-acether and lysopaf-acether in carrageenan-induced granuloma in rats. *Int Arch Allergy Immunol.* 1996 Jun;110(2):182-6.

Aissa J, Jurgens P, Litime M, Béhar I, Benveniste J. Electronic transmission of the cholinergic signal. *FASEB Jnl.* 1995;9:A683.

Aissa J, Litime M, Attias E, Allal A, Benveniste J. Transfer of molecular signals via electronic circuitry. *FASEB Jnl.* 1993;7:A602.

Aissa J, Litime MH, Attis E., Benveniste J. Molecular signalling at high dilution or by means of electronic circuitry. *J Immunol.* 1993;150:A146.

Aissa J, Nathan N, Arnoux B, Benveniste J. Biochemical and cellular effects of heparin-protamine injection in rabbits are partially inhibited by a PAF-acether receptor antagonist. *Eur J Pharmacol.* 1996 Apr 29;302(1-3):123-8.

Albrechtsen O. The influence of small atmospheric ions on human xell-being and mental performance. *Intern. J. of Biometeorology.* 1978;22(4):249-262.

American Chemical Society (2007, December 30). Culinary Shocker: Cooking Can Preserve, Boost Nutrient Content Of Vegetables. *ScienceDaily.* http://www.sciencedaily.com /releases/2007/12/ 071224125524.htm. Accessed 2007 Dec.

American Conference of Governmental Industrial Hygienists. *Threshold limit values for chemical substances and physical agents in the work environment.* Cincinnati, OH: ACGIH, 1986.

Anderson M., Grissom C. Increasing the Heavy Atom Effect of Xenon by Adsorption to Zeolites: Photolysis of 2,3-Diazabicyclo[2.2.2]oct-2-ene. *J. Am. Chem. Soc.* 1996;118:9552-9556.

Anderson RC, Anderson JH. Acute toxic effects of fragrance products. *Arch Environ Health.* 1998 Mar-Apr;53(2):138-46.

Anderson RC, Anderson JH. Respiratory toxicity of fabric softener emissions. *J Toxicol Environ Health.* 2000 May 26;60(2):121-36.

Anderson RC, Anderson JH. Respiratory toxicity of mattress emissions in mice. *Arch Environ Health.* 2000 Jan-Feb;55(1):38-43.

Anderson RC, Anderson JH. Toxic effects of air freshener emissions. *Arch Environ Health.* 1997 Nov-Dec;52(6):433-41.

Armas LA, Hollis BW, Heaney RP. Vitamin D2 is much less effective than vitamin D3 in humans. *J Clin Endocrinol Metab.* 2004 Nov;89(11):5387-91.

Armstrong BK, Kricker A. Sun exposure and non-Hodgkin lymphoma. *Cancer Epidemiol Biomarkers Prev.* 2007 Mar;16(3):396-400.

Asimov I. *The Chemicals of Life.* New York: Signet, 1954.

Askeland D. *The Science and Engineering of Materials.* Boston: PWS, 1994.

Aspect A, Grangier P, Roger G. Experimental Realization of Einstein-Podolsky-Rosen-Bohm Gedankenexperiment: A New Violation of Bell's Inequalities. *Physical Review Letters.* 1982;49(2):91-94.

Aton SJ, Colwell CS, Harmar AJ, Waschek J, Herzog ED. Vasoactive intestinal polypeptide mediates circadian rhythmicity and synchrony in mammalian clock neurons. *Nat Neurosci.* 2005 Apr;8(4):476-83.

Atsumi T, Tonosaki K. Smelling lavender and rosemary increases free radical scavenging activity and decreases cortisol level in saliva. *Psychiatry Res.* 2007 Feb 28;150(1):89-96.

Avanzini G, Lopez L, Koelsch S, Majno M. The Neurosciences and Music II: From Perception to Performance. *Annals of the New York Academy of Sciences.* 2006 Mar;1060.

Azar JA, Conroy T. Measuring the effectiveness of horticultural therapy at a veterans administration medical center: experimental design issues. In Relf, D. (ed) *The Role of Horticulture in Human Well-Being and Social Development: A National Symposium.* Portland: Timber Press. 1992:169-171.

Bach E. *Bach Flower Remedies.* New Canaan, CN: Keats, 1997.

Bach E. *Heal Thyself.* Walden: Saffron CW Daniel, 1931-2003.

Backster C. *Primary Perception.* Anza, CA: White Rose Millennium Press, 2003.

Baker SM. *Detoxification and Healing.* Chicago: Contemporary Books, 2004.

Balch P, Balch J. *Prescription for Nutritional Healing.* New York: Avery, 2000.

Ballentine R. *Diet & Nutrition: A holistic approach.* Honesdale, PA: Himalayan Int., 1978.

Ballentine RM. *Radical Healing.* New York: Harmony Books, 1999.

Barber CF. The use of music and colour theory as a behaviour modifier. *Br J Nurs.* 1999 Apr 8-21;8(7):443-8.

Barker A. *Scientific Method in Ptolemy's Harmonics.* Cambridge: Cambridge University Press, 2000.

Basnyat B, Sleggs J, Spinger M. Seizures and delirium in a trekker: the consequences of excessive water drinking? *Wilderness Environ. Med.* 2000;11:69-70.

Bastide M, Daurat V, Doucet-Jaboeuf M, Pélegrin A, Dorfman P. Immunomodulator activity of very low doses of thymulin in mice, *Int J Immunotherapy*. 1987;3:191-200.

Bastide M, Doucet-Jaboeuf M, Daurat V. Activity and chronopharmacology of very low doses of physiological immune inducers. *Immun Today*. 1985;6:234-235.

Bastide M. Immunological examples on ultra high dilution research. In: Endler P, Schulte J (eds.): *Ultra High Dilution. Physiology and Physics*. Dordrech: Kluwer Academic Publishers, 1994:27-34.

Bates M. *The Forest and the Sea*. Alexandria, VA: Time-Life, 1980.

Batmanghelidj F. *Your Body's Many Cries for Water*. 2nd Ed. Vienna, VA: Global Health, 1997.

Beaulieu A, Fessele K. Agent Orange: management of patients exposed in Vietnam. *Clin J Oncol Nurs*. 2003 May-Jun;7(3):320-3.

Beauvais F, Bidet B, Descours B, Hieblot C, Burtin C, Benveniste J. Regulation of human basophil activation. I. Dissociation of cationic dye binding from histamine release in activated human basophils. *J Allergy Clin Immunol*. 1991 May;87(5):1020-8.

Beauvais F, Burtin C, Benveniste J. Voltage-dependent ion channels on human basophils: do they exist? *Immunol Lett*. 1995 May;46(1-2):81-3.

Beauvais F, Echasserieau K, Burtin C, Benveniste J. Regulation of human basophil activation; the role of Na+ and Ca2+ in IL-3-induced potentiation of IgE-mediated histamine release from human basophils. *Clin Exp Immunol*. 1994 Jan;95(1):191-4.

Beauvais F, Shimahara T, Inoue I, Hieblot C, Burtin C, Benveniste J. Regulation of human basophil activation. II. Histamine release is potentiated by K+ efflux and inhibited by Na+ influx. . *J Immunol*. 1992 Jan 1;148(1):149-54.

Becker R. *Cross Currents*. Los Angeles: Tarcher, 1990.

Becker R. *The Body Electric*. New York: Morrow, Inc., 1985.

Beecher GR. Phytonutrients' role in metabolism: effects on resistance to degenerative processes. *Nutr Rev*. 1999 Sep;57(9 Pt 2):S3-6.

Beer FP, Johnston ER. *Vector Mechanics for Engineers*. New York: McGraw-Hill, 1988.

Benedetti F, Radaelli D, Bernasconi A, Dallaspezia S, Falini A, Scotti G, Lorenzi C, Colombo C, Smeraldi E. Clock genes beyond the clock: CLOCK genotype biases neural correlates of moral valence decision in depressed patients. *Genes Brain Behav*. 2007 Mar 26.

Bennet LW, Cardone S, Jarczyk J. Effects of therapeutic camping program on addiction recovery. *Journal of Substance Abuse Treatment*. 1998;15(5):469-474.

Bensky D, Gable A, Kaptchuk T (transl.). *Chinese Herbal Medicine Materia Medica*. Seattle: Eastland Press, 1986.

Bentley E. *Awareness: Biorhythms, Sleep and Dreaming*. London: Routledge, 2000.

Benveniste J, Aïssa J, Guillonnet D. A simple and fast method for in vivo demonstration of electromagnetic molecular signaling (EMS) via high dilution or computer recording._*FASEB Jnl*. 1999;13:A163.

Benveniste J, Aïssa J, Guillonnet D. Digital biology : Specificity of the digitized molecular signal. *FASEB Jnl*. 1998;12:A412.

Benveniste J, Aïssa J, Guillonnet D. The molecular signal is not functional in the absence of "informed" water. *FASEB Jnl*. 1999;13:A163.

Benveniste J, Aissa J, Litime MH, Tsaegaca GT, Thomas Y. Transfer of the molecular signal by electronic amplification. *FASEB J*. 1994;8:A398.

Benveniste J, Arnoux B, Hadji L. Highly dilute antigen increases coronary flow of isolated heart from immunized guinea-pigs. *FASEB J*. 1992;6:A1610.

Benveniste J, Davenas E, Ducot B, Spira A. Basophil achromasia by dilute ligand: a reappraisal. *FASEB Jnl*. 1991;5:A1008.

Benveniste J, Ducot B, Spira A. Memory of water revisited. *Nature*. 1994 Aug 4;370(6488):322.

Benveniste J, Guillonnet D. QED and digital biology. *Riv Biol*. 2004 Jan-Apr;97(1):169-72.

Benveniste J, Jurgens P, Aïssa J. Digital recording/transmission of the cholinergic signal. *FASEB Jnl*. 1996;10:A1479.

Benveniste J, Jurgens P, Hsueh W, Aïssa J. Transatlantic transfer of digitized antigen signal by telephone link. *Jnl Aller Clin Immun*. 1997;99:S175.

Benveniste J, Kahhak L, Guillonnet D. Specific remote detection of bacteria using an electromagnetic / digital procedure. *FASEB Jnl*. 1999;13:A852.

Benveniste J. Benveniste on Nature investigation. *Science*. 1988 Aug 26;241(4869):1028.

Benveniste J. Benveniste on the Benveniste affair. *Nature*. 1988 Oct 27;335(6193):759.

Benveniste J. Diagnosis of allergic diseases by basophil count and in vitro degranulation using manual and automated tests. *Nouv Presse Med*. 1981 Jan 24;10(3):165-9.

Benveniste J. Meta-analysis of homoeopathy trials. *Lancet*. 1998 Jan 31;351(9099):367.

Bergner P. *The Healing Power of Garlic*. Prima Publishing, Rocklin CA 1996.

Berman S, Fein G, Jewett D, Ashford F. Luminance-controlled pupil size affects Landolt C task performance. *J Illumin Engng Soc*. 1993;22:150-165.

Berman S, Jewett D, Fein G, Saika G, Ashford F. Photopic luminance does not always predict perceived room brightness. *Light Resch and Techn.* 1990;22:37-41.

Bernardi D, Dini FL, Azzarelli A, Giaconi A, Volterrani C, Lunardi M. Sudden cardiac death rate in an area characterized by high incidence of coronary artery disease and low hardness of drinking water. *Angiology.* 1995;46:145-149.

Berry J. Work efficiency and mood states of electronic assembly workers exposed to full-spectrum and conventional fluorescent illumination. *Diss Abstr Internl.* 1983;44:635B.

Berteau O and Mulloy B. 2003. Sulfated fucans, fresh perspectives: structures, functions, and biological properties of sulfated fucans and an overview of enzymes active toward this class of polysaccharide. *Glycobiology.* Jun;13(6):29R-40R.

Bertin G. *Spiral Structure in Galaxies: A Density Wave Theory.* Cambridge: MIT Press, 1996.

Besset A, Espa F, Dauvilliers Y, Billiard M, de Seze R. No effect on cognitive function from daily mobile phone use. *Bioelectromagnetics.* 2005 Feb;26(2):102-8.

Bierman DJ. Does Consciousness Collapse the Wave-Packet? *Mind and Matter.* 1993;1(1):45-57.

Bishop ID, Rohrmann B. Subjective responses to simulated and real environments: a comparison. *Landscape and Urban Planning.* 2003;65(4):261-277.

Bittman BB, Berk LS, Felten DL, Westengard J, Simonton OC, Pappas J, Ninehouser M. Composite effects of group drumming music therapy on modulation of neuroendocrine-immune parameters in normal subjects. *Altern Ther Health Med.* 2001 Jan;7(1):38-47.

Bjerregaard C. Plato and the Greeks on Music as an Element in Education. *The Word.* 1913 Feb.

Bodnar L, Simhan H. The prevalence of preterm birth varies by season of last menstrual period. *Am J Obst and Gyn.* 2003;195(6):S211-S211.

Boivin DB, Czeisler CA. Resetting of circadian melatonin and cortisol rhythms in humans by ordinary room light. *Neuroreport.* 1998 Mar 30;9(5):779-82.

Boivin DB, Duffy JF, Kronauer RE, Czeisler CA. Dose-response relationships for resetting of human circadian clock by light. *Nature.* 1996 Feb 8;379(6565):540-2.

Boray P, Gifford R, Rosenblood L. Effects of warm white, cool white and full-spectrum fluorescent lighting on simple cognitive performance, mood and ratings of others. J Environl Psychol. 1989;9:297-308.

Borchers AT, Hackman RM, Keen CL, Stern JS, Gershwin ME. Complementary medicine: a review of immunomodulatory effects of Chinese herbal medicines. Am J Clin Nutr. 1997 Dec;66(6):1303-12.

Bose J. Response in the Living and Non-Living. New York: Longmans, Green & Co., 1902.

Boston University. Effects Of Vitamin D And Skin's Physiology Examined. ScienceDaily. 2008 February 24. Retrieved February 24, 2008, from http://www.sciencedaily.com/releases/ 2008/02/ 080220161707.htm

Bottorff JL. The use and meaning of touch in caring for patients with cancer. *Oncol Nurs Forum.* 1993 Nov-Dec;20(10):1531-8.

Boucher RC. Evidence for airway surface dehydration as the initiating event in CF airway disease. *J Intern Med.* 2007 Jan;261(1):5-16.

Bourre JM. Effects of nutrients (in food) on the structure and function of the nervous system: update on dietary requirements for brain. Part 1: micronutrients. *J Nutr Health Aging.* 2006 Sep-Oct;10(5):377-85.

Boyce P, Rea M. A field evaluation of full-spectrum, polarized lighting. Paper presented at the 1993 Annual Convention of the Illuminating Engineering Society of North America, Houston, TX. 1993 Aug.

Boyce P. Investigations of the subjective balance between illuminance and lamp colour properties. *Light Resch and Technol.* 1977;9:11-24.

Boyd B, Zungoli P, Benson E. 2006. *Dust Mites.* HGIC 2551; Clemson University. http://hgic.clemson.edu.

Braunstein G, Labat C, Brunelleschi S, Benveniste J, Marsac J, Brink C. Evidence that the histamine sensitivity and responsiveness of guinea-pig isolated trachea are modulated by epithelial prostaglandin E2 production. *Br J Pharmacol.* 1988 Sep;95(1):300-8.

Brenner D, Hall E. Computed Tomography — An Increasing Source of Radiation Exposure. *NE J Med.* 2007;357(22):2277-2284.

Brodeur P. *Currents of Death.* New York: Simon and Schuster, 1989.

Brody J. *Jane Brody's Nutrition Book.* New York: WW Norton, 1981.

Brown V. *The Amateur Naturalists Handbook.* Englewood Cliffs, NJ: Prentice-Hall, 1980.

Brownlee C, Perkins S, Goho A. Nobel prizes: The sweet smell of success. *Science News.* 2004;166(Oct. 9):229.

Brownstein D. *Salt: Your Way to Health.* West Bloomfield, MI: Medical Alternatives, 2006.

Bruck SD, Mueller EP. Materials aspects of implantable cardiac pacemaker leads. *Med Prog Technol.* 1988;13(3):149-60.

Bruseth S, Tveiten D. Homeopathy—the past or a part of future medicine? *Tidsskr Nor Laegeforen.* 1991 Dec 10;111(30):3692-4.

Buck L, Axel R. A novel multigene family may encode odorant receptors: A molecule basis for odor recognition. *Cell.* 1991;65(April 5):175-187.

Buijs RM, Scheer FA, Kreier F, Yi C, Bos N, Goncharuk VD, Kalsbeek A. Organization of circadian functions: interaction with the body. *Prog Brain Res.* 2006;153:341-60.

Bulsing PJ, Smeets MA, van den Hout MA. Positive Implicit Attitudes toward Odor Words. *Chem Senses.* 2007 May 7.

Burton GW, *et al.* Human plasma and tissue alpha-tocopherol concentrations in response to supplementation with deuterated natural and synthetic vitamin E. *Am J Clin Nutr.* 1998;67:669-84.

Buscemi N, Vandermeer B, Pandya R, Hooton N, Tjosvold L, Hartling L, Baker G, Vohra S, Klassen T. Melatonin for treatment of sleep disorders. *Evid Rep Technol Assess.* 2004 Nov;(108):1-7.

Busch A. Hydrotherapy improves pain, knee strength, and quality of life in women with fibromyalgia. *Aust J Physiother.* 2007;53(1):64.

Cahill RT. A New Light-Speed Anisotropy Experiment: Absolute Motion and Gravitational Waves Detected. *Progress in Physics.* 2006; (4).

Cai L, Mu LN, Lu H, Lu QY, You NC, Yu SZ, Le AD, Zhao J, Zhou XF, Marshall J, Heber D, Zhang ZF. Dietary selenium intake and genetic polymorphisms of the GSTP1 and p53 genes on the risk of esophageal squamous cell carcinoma. *Cancer Epidemiol Biomarkers Prev.* 2006 Feb;15(2):294-300.

Cajochen C, Jewett ME, Dijk DJ. Human circadian melatonin rhythm phase delay during a fixed sleep-wake schedule interspersed with nights of sleep deprivation. *J Pineal Res.* 2003 Oct;35(3):149-57.

Caldwell MM, Bornman JF, Ballare CL, Flint SD, Kulandaivelu G. Terrestrial ecosystems, increased solar ultraviolet radiation, and interactions with other climate change factors. *Photochem Photobiol Sci.* 2007 Mar;6(3):252-66.

Capitani D, Yethiraj A, Burnell EE. Memory effects across surfactant mesophases. *Langmuir.* 2007 Mar 13;23(6):3036-48.

Caple G, Sands DC, Layton G, Zucker V, Snider R. Biogenic ice nucleation—could it be metabolically initiated? *Journal of Theoretical Biology.* 1986;119:37-45.

Carroll D. *The Complete Book of Natural Medicines.* New York: Summit, 1980.

Carson R. *Silent Spring.* Houghton Mifflin: Mariner Books, 1962.

CDC. Hyponatremic seizures among infants fed with commercial bottled drinking water - Wisconsin, 1993. *MMWR.* 1994;43:641-643.

Cengel YA, *Heat Transfer: A Practical Approach.* Boston: McGraw-Hill, 1998.

Chaitow L, Trenev N. *Pro Biotics.* New York: Thorsons, 1990.

Chaitow L. *Conquer Pain the Natural Way.* San Francisco: Chronicle Books, 2002.

Cham, B. Solasodine glycosides as anti-cancer agents: Pre-clinical and Clinical studies. *Asia Pac J Pharmac.* 1994;9:113-118.

Chaney M, Ross M. *Nutrition.* New York: Houghton Mifflin, 1971.

Characterization and quantitation of Antioxidant Constituents of Sweet Pepper (Capsicum annuum - Cayenne). *J Agric Food Chem.* 2004 Jun 16;52(12):3861-9.

Chast F. Homeopathy confronted with clinical research. *Ann Pharm Fr.* 2005 Jun;63(3):217-27.

Chen-Goodspeed M, Cheng Chi Lee. Tumor suppression and circadian function. *J Biol Rhythms.* 2007 Aug;22(4):291-8.

Chilton F, Tucker L. *Win the War Within.* New York: Rodale, 2006.

Chong NW, Codd V, Chan D, Samani NJ. Circadian clock genes cause activation of the human PAI-1 gene promoter with 4G/5G allelic preference. *FEBS Lett.* 2006 Aug 7;580(18):4469-72.

Christopher J. *School of Natural Healing.* Springville UT: Christopher Publ, 1976.

Cimetidine inhibits the hepatic hydroxylation of vitamin D. Nutr Rev. 1985;43:184-5.

Citro M, Endler PC, Pongratz W, Vinattieri C, Smith CW, Schulte J. Hormone effects by electronic transmission. *FASEB J.* 1995:Abstract 12161.

Citro M, Smith CW, Scott-Morley A, Pongratz W, Endler PC. Transfer of information from molecules by means of electronic amplification. In P.C. Endler, J. Schulte (eds.): *Ultra High Dilution. Physiology and Physics.* Dordrecht: Kluwer Academic Publishers. 1994:209-214.

Cochran ES, Vidale JE, Tanaka S. Earth Tides Can Trigger Shallow Thrust Fault Earthquakes. *Science* 2004 Nov 12: 306(5699);1164-1166.

Cocilovo A. Colored light therapy: overview of its history, theory, recent developments and clinical applications combined with acupuncture. *Am J Acupunct.* 1999;27(1-2):71-83.

Coles JA, Yamane S. Effects of adapting lights on the time course of the receptor potential of the anuran retinal rod. *J Physiol.* 1975 May;247(1):189-207.

Conely J. Music and the Military. *Air University Review.* 1972 Mar-Ap.

Contreras D, Steriade M. Cellular basis of EEG slow rhythms: a study of dynamic corticothalamic relationships. *J Neurosci.* 1995 Jan;15(1 Pt 2):604-22.

Cook J, The Therapeutic Use of Music. *Nursing Forum.* 1981;20:3:253-66.

Cook N, Freeman S. Report of 19 cases of photoallergic contact dermatitis to sunscreens seen at the Skin and Cancer Foundation. *Australas J Dermatol*. 2001 Nov;42(4):257-9.

Cooper K. *The Aerobics Program for Total Well-Being*. New York: Evans, 1980.

Couzy F, Kastenmayer P, Vigo M, Clough J, Munoz-Box R, Barclay DV. Calcium bioavailability from a calcium- and sulfate-rich mineral water, compared with milk, in young adult women. *Am J Clin Nutr*. 1995 Dec;62(6):1239-44.

Craciunescu CN, Wu R, Zeisel SH. Diethanolamine alters neurogenesis and induces apoptosis in fetal mouse hippocampus. *FASEB J*. 2006 Aug;20(10):1635-40.

Crawley J. *The Biorhythm Book*. Boston: Journey Editions, 1996.

Crick F. *Life Itself: Its Origin and Nature*. New York: Simon and Schuster, 1981.

Cummings M. *Human Heredity: Principles and Issues*. St. Paul, MN: West, 1988.

Cummings S. Ullman D. *Everybody's Guide to Homeopathic Medicines*. New York: Tarcher/Putnam, 1997.

Curtis *et al.* and Steel, K. Too Many Elderly Are Taking Dangerous Drugs. *Arch Internal Med*. 2004;164:1621-1625, 1603-1604.

Cuthbert SC, Goodheart GJ Jr. On the reliability and validity of manual muscle testing: a literature review. *Chiropr Osteopat*. 2007 Mar 6;15:4.

Dantas F, Rampes H. Do homeopathic medicines provoke adverse effects? A systematic review. *Br Homeopath J*. 2000 Jul;89 Suppl 1:S35-8.

Darby S, Hill D, Auvinen A, Barros-Dios JM, Baysson H, Bochicchio F, *et al.* Radon in homes and risk of lung cancer: collaborative analysis of individual data from 13 European case-control studies. *BMJ*. 2005 Jan 29;330(7485):223.

Darrow K. *The Renaissance of Physics*. New York: Macmillan, 1936.

Das UN. A defect in the activity of Delta6 and Delta5 desaturases may be a factor predisposing to the development of insulin resistance syndrome. *Prostagl Leukot Essent Fatty Acids*. 2005; May;72(5):343-50.

Davenas E, Beauvais J, Oberbaum M, Robinzon B, Miadonna B, Tedeschi A, Pomeranz B, Fortner P, Belon P, Sainte-Laudy J, Poitevin B, Benveniste J. Human basophil degranulation triggered by very dilute antiserum against IgE. *Nature*. 1988;333:816-818.

Davenas E, Poitevin B, Benveniste J. Effect on mouse peritoneal macrophages of orally administered very high dilutions of silica. *European Journal of Pharmacology*. 1987;135:313-319.

Davidson T. *Rhinology: The Collected Writings of Maurice H. Cottle, M.D.* San Diego, CA: American Rhinologic Society, 1987.

DaVinci L. (Dickens E. ed.) *The Da Vinci Notebooks*. London: Profile, 2005.

Davis S, Kaune WT, Mirick DK, Chen C, Stevens RG. Residential magnetic fields, light-at-night, and nocturnal urinary 6-sulfatoxymelatonin concentration in women. *Am J Epidemiol*. 2001 Oct 1;154(7):591-600.

Davis-Berman J, Berman DS. The widlerness therapy program: an empirical study of its effects with adolescents in an outpatient setting. *Journal of Contemporary Psychotherapy*. 1989;19 (4):271-281.

de Vries E, Coebergh JW, van der Rhee H. Trends, causes, approach and consequences related to the skin-cancer epidemic in the Netherlands and Europe. *Ned Tijdschr Geneeskd*. 2006 May 20;150(20):1108-15.

Dean E, Mihalasky J, Ostrander S, Schroeder L. *Executive ESP*. Englewood Cliffs, NJ: Prentice-Hall, 1974.

Dean E. Infrared measurements of healer-treated water. In: Roll W, Beloff J, White R (Eds.): *Research in parapsychology 1982*. Metuchen, NJ: Scarecrow Press, 1983:100-101.

Del Giudice E, Preparata G, Vitiello G. Water as a free electric dipole laser. *Phys Rev Lett*. 1988;61:1085-1088.

Del Giudice E. Is the 'memory of water' a physical impossibility?. In P.C. Endler, J. Schulte (eds.): *Ultra High Dilution. Physiology and Physics*. Dordrecht: Kluwer Academic Publishers, 1994:117-120.

Dement W, Vaughan C. *The Promise of Sleep*. New York: Dell, 1999.

Der Marderosian A. Understanding homeopathy. *J Am Pharm Assoc*. 1996 May;NS36(5):317-21.

Deutsche Gesellschaft für Ernährung. Drink distilled water? *Med. Mo. Pharm*. 1993;16:146.

Devaraj TL. *Speaking of Ayurvedic Remedies for Common Diseases*. New Delhi: Sterling, 1985.

Donato F, Monarca S, Premi S., and Gelatti, U. Drinking water hardness and chronic degenerative diseases. Part III. Tumors, urolithiasis, fetal malformations, deterioration of the cognitive function in the aged and atopic eczema. *Ann. Ig*. 2003;15:57-70.

Dotolo Institute. *The Study of Colon Hydrotherapy*. Pinellas Park, FL: Dotolo, 2003.

Drake AJ, Howells RJ, Shield JP, Prendiville A, Ward PS, Crowne EC. Symptomatic adrenal insufficiency presenting with hypoglycaemia in children with asthma receiving high dose inhaled fluticasone propionate. *BMJ*. 2002 May 4;324(7345):1081-2.

Dudley M. *Microwaved water and plants*. 2006; http://www.execonn.com/sf/. Accessed: 2007 Dec.

Duke J. *The Green Pharmacy*. New York: St. Martins, 1997.

Duke M. *Acupuncture*. New York: Pyramid, 1973.

Dunlop KA, Carson DJ, Shields MD. Hypoglycemia due to adrenal suppression secondary to high-dose nebulized corticosteroid. *Pediatr Pulmonol*. 2002 Jul;34(1):85-6.

Dunne B, Jahn R, Nelson R. Precognitive Remote Perception. Princeton Engineering Anomalies Research Laboratory Report. Princeton. 1983 Aug.

Dunne BJ, Jahn RG. Consciousness, information, and living systems. Cell Mol Biol 2005 Dec 14;51(7):703-14.

Durlach J, Bara M, Guiet-Bara A. Magnesium level in drinking water: its importance in cardiovascular risk. In: Itokawa Y, Durlach J: *Magnesium in Health and Disease*. London: J.Libbey, 1989:173-182.

Ebbesen F, Agati G, Pratesi R. Phototherapy with turquoise versus blue light. *Arch Dis Child Fetal Neonatal Ed.* 2003 Sep;88(5):F430-1.

Eden D, Feinstein D. *Energy Medicine*. New York: Penguin Putnam, 1998.

Edris AE. Pharmaceutical and therapeutic potentials of essential oils and their individual volatile constituents: a review. *Phytother Res.* 2007 Apr;21(4):308-23.

Edwards B. *Drawing on the Right Side of the Brain*. Los Angeles, CA: Tarcher, 1979.

Edwards R, Ibison M, Jessel-Kenyon J, Taylor R. Light emission from the human body. *Comple Med Res.* 1989;3(2):16-19.

Edwards R, Ibison M, Jessel-Kenyon J, Taylor R. Measurements of human bioluminescence. *Acup Elect Res, Intl Jnl,* 1990;15:85-94.

Einstein In Need Of Update? Calculations Show The Speed Of Light Might Change. *Science Daily*. 2001 Feb 12. www.sciencedaily.com/releases/ 2001/02/010212075309.htm. Accessed: 2007 Oct.

Eisenberg MJ. Magnesium deficiency and sudden death. *Am. Heart J.* 1992;124:544-549.

Electromagnetic fields: the biological evidence. *Science.* 1990;249:1378-1381.

Electronic Evidence of Auras, Chakras in UCLA Study. *Brain/Mind Bulletin.* 1978;3:9 Mar 20.

Eltiti S, Wallace D, Ridgewell A, Zougkou K, Russo R, Sepulveda F, et al. Does Short-Term Exposure to Mobile Phone Base Station Signals Increase Symptoms in Individuals who Report Sensitivity to Electromagnetic Fields? A Double-Blind Randomised Provocation Study. *Environ Health Perspect.* 2007;115(11):1603-1608.

Elwood PC. Epidemiology and trace elements. *Clin Endocrinol Metab.* 1985 Aug;14(3):617-28.

Emoto M (Thayne D, transl). *The Hidden Messages in Water.* Hillsboro, OR: Beyond Words, 2004.

Endler P, Pongratz W, van Wijk R, Waltl K, Hilgers H, Brandmaier R. Transmission of hormone information by non-molecular means. *FASEB Jnl.* 1994;8:A400.

Endler PC, Pongratz W, Kastberger G, Wiegant F, Schulte J. The effect of highly diluted agitated thyroxine on the climbing activity of frogs, *J Vet Hum Tox.* 1994;36:56-59.

Endler PC, Pongratz W, Smith CW, Schulte J. Non-molecular information transfer from thyroxine to frogs with regard to 'homoeopathic' toxicology, *J Vet Hum Tox.* 1995:37:259-260.

Endler PC, Pongratz W, Van Wijk R, Kastberger G, Haidvogl M. Effects of highly diluted sucussed thyroxine on metamorphosis of highland frogs, *Berlin J Res Hom.* 1991;1:151-160.

Endler PC, Pongratz W, Van Wijk R, Waltl K, Hilgers H, Brandmaier R. Transmission of hormone information by non-molecular means, *FASEB J.* 1994;8:A400.

Endler PC, Pongratz W, Van Wijk R, Wiegant F, Waltl K, Gehrer M, Hilgers H. A zoological example on ultra high dilution research. In: Endler PC, Schulte J (eds.): *Ultra High Dilution. Physiology and Physics.* Dordrecht: Kluwer Academic Publishers. 1994:39-68.

Endler PC, Pongratz W. *On effects of agitated highly diluted thyroxine (E-30).* Comprehensive report, available at the Institute for Zoology. University of Graz, Universitätsplatz 2, A-8010 Graz, 1994.

Endler PC, Schulte, J. *Ultra High Dilution. Physiology and Physics.* Dordrecht: Kluwer Academic Publ, 1994.

Environmental Protection Agency. *EPA Asbestos Materials Bans: Clarification.* 1999 May 18.

Environmental Working Group. *Human Toxome Project.* 2007. http://www.ewg.org/sites/humantoxome/. Accessed: 2007 Sep.

EPA. *A Brief Guide to Mold, Moisture and Your Home.* Environmental Protection Agency, Office of Air and Radiation/Indoor Environments Division. EPA 2002;402-K-02-003.

Ernst E. A systematic review of systematic reviews of homeopathy. *Br J Clin Pharmacol.* 2002 Dec;54(6):577-82.

Ernst E. Herbal remedies for anxiety - a systematic review of controlled clinical trials. *Phytomedicine.* 2006 Feb;13(3):205-8.

Esch T, Stefano GB. The Neurobiology of Love. *Neuro Endocrinol Lett.* 2005 Jun;26(3):175-92.

Eschenhagen T, Zimmermann WH. Engineering myocardial tissue. *Circ Res.* 2005 Dec 9;97(12):1220-31.

European Union (1980) Council Directive 80/778/EEC of 15 July 1980 relating to the

European Union Council Directive 98/83/EC of 3 November 1998 on the quality of water intended for human consumption. *Off. J. Eur. Commun.* 1998;L330:32-54.

Falcon CT. *Happiness and Personal Problems.* Lafayette, LA: Sensible Psychology, 1992.

Fan X, Zhang D, Zheng J, Gu N, Ding A, Jia X, Qing H, Jin L, Wan M, Li Q. Preparation and characterization of magnetic nano-particles with radiofrequency-induced hyperthermia for cancer treatment. *Sheng Wu Yi Xue Gong Cheng Xue Za Zhi.* 2006 Aug;23(4):809-13.

414

REFERENCES AND BIBLIOGRAPHY

Fawell J, Nieuwenhuijsen MJ. Contaminants in drinking water. *Br Med Bull.* 2003;68:199-208.

Fecher LA, Cummings SD, Keefe MJ, Alani RM. Toward a molecular classification of melanoma. *J Clin Oncol.* 2007 Apr 20;25(12):1606-20.

Feleszko W, Jaworska J, Rha RD, Steinhausen S, Avagyan A, Jaudszus A, Ahrens B, Groneberg DA, Wahn U, Hamelmann E. Probiotic-induced suppression of allergic sensitization and airway inflammation is associated with an increase of T regulatory-dependent mechanisms in a murine model of asthma. *Clin Exp Allergy.* 2007 Apr;37(4):498-505.

Field RW, Krewski D, Lubin JH, Zielinski JM, Alavanja M, Catalan VS, Klotz JB, Letourneau EG, Lynch CF, Lyon JL, Sandler DP, Schoenberg JB, Steck DJ, Stolwijk JA, Weinberg C, Wilcox HB. An overview of the North American residential radon and lung cancer case-control studies. *J Toxicol Environ Health A.* 2006 Apr;69(7):599-631.

Fischer JL, Mihelc EM, Pollok KE, Smith ML. Chemotherapeutic selectivity conferred by selenium: a role for p53-dependent DNA repair. *Mol Cancer Ther.* 2007 Jan;6(1):355-61.

Fisher P. Homeopathy and The Lancet. Evid Based Complement Alternat Med. 2006 March; 3(1):145–147. Folic acid metabolism in human subjects revisited: potential implications for proposed mandatory folic acid fortification in the UK. British Journal of Nutrition. 2007; Oct;98(4).

Frawley D, Lad V. *The Yoga of Herbs.* Sante Fe: Lotus Press, 1986.

Freeman HL, Stansfield SA. Psychosocial effects of urban environments, noise, and crowding. In Lundberg, A. (ed) *Environment and Mental Health.* London: Lawrence Erlbaum. 1998:147-173.

French SD, Cameron M, Walker BF, Reggars JW, Esterman AJ. Superficial heat or cold for low back pain. *Cochrane Database Syst Rev.* 2006 Jan 25;(1):CD004750.

Frey A. Electromagnetic field interactions with biological systems. *FASEB Jnl.* 1993;7:272-28.

Fukada Y, Okano T. Circadian clock system in the pineal gland. *Mol Neurobiol.* 2002 Feb;25(1):19-30.

Galaev, YM. The Measuring of Ether-Drift Velocity and Kinematic Ether Viscosity within Optical Wave Bands. *Spacetime & Substance.* 2002;3(5):207-224.

Gandhi TK, Weingart SN, Borus J, Seger AC, Peterson J, Burdick E, Seger DL, Shu K, Federico F, Leape LL, Bates DW. Adverse drug events in ambulatory care. *N Engl J Med.* 2003 Apr 17;348(16):1556-64.

Gange R. UVA sunbeds - are there longterm hazards. In Cronley-Dillon J, Rosen E, Marshall J (Eds.):*Hazards of Light, Myths and Realities.* Oxford, U.K.: Pergamon Press, 1986.

Ganz PA, Greendale GA, Kahn B, O'Leary JF, Desmond KA. Are older breast carcinoma survivors willing to take hormone replacement therapy? *Cancer.* 1999 Sep 1;86(5):814-20.

García AM, Sisternas A, Hoyos SP. Occupational exposure to extremely low frequency electric and magnetic fields and Alzheimer disease: a meta-analysis. *Int J Epidemiol.* 2008 Apr;37(2):329-40.

Garcia-Lazaro JA, Ahmed B, Schnupp JW. Tuning to natural stimulus dynamics in primary auditory cortex. *Curr Biol.* 2006 Feb 7;16(3):264-71.

Garzon P, Eisenberg MJ. Variation in the mineral content of commercially available bottled waters: implication for health and disease. *Am. J. Med.* 1998;105:125-130.

Gehr P, Im Hof V, Geiser M, Schurch S. The mucociliary system of the lung—role of surfactants. *Schweiz Med Wochenschr.* 2000 May 13;130(19):691-8.

Geldreich EE, Taylor RH, Blannon JC, Reasoner DJ. Bacterial colonization of point-of-use water treatment devices. *J AWWA.* 1985;77:72-80.

Gerber R. *Vibrational Healing.* Sante Fe: Bear, 1988.

Gibbons E. *Stalking the Healthful Herbs.* New York: David McKay, 1966.

Gittleman AL. *Guess What Came to Dinner.* New York: Avery, 2001.

Goedsche K, Forster M, Kroegel C, Uhlemann C. Repeated Cold Stimulations (Hydrotherapy according to Kneipp) in Patients with COPD. *Forsch Komplementarmed.* 2007 Jun;14(3):158-66.

Goldstein N, Arshavskaya TV. Is atmospheric superoxide vitally necessary? Accelerated death of animals in a quasi-neutral electric atmosphere. *Z Naturforsch.* 1997. May-Jun;52(5-6):396-404.

Golub E. *The Limits of Medicine.* New York: Times Books, 1994.

Golubev IM, Zimin VP. On the standard of total hardness in drinking water. *Gig. Sanit.* 1994;(3):22-23.

Gomes A, Fernandes E, Lima JL. Fluorescence probes used for detection of reactive oxygen species. *J Biochem Biophys Methods.* 2005 Dec 31;65(2-3):45-80.

Gomez-Abellan P, Hernandez-Morante JJ, Lujan JA, Madrid JA, Garaulet M. Clock genes are implicated in the human metabolic syndrome. *Int J Obes.* 2007 Jul 24.

Gonzales, et al. 1987. Polysaccharides as antiviral agents: antiviral activity of carrageenan, *Antimicrobial Agents and Chemotherapy.* 31:1388-1393.

Gosner KL. A simplified table for staging anuran embryos and larvae with notes on identification. *Herpetologica.* 1960:16:183-195.

Grad B, Dean E. Independent confirmation of infrared healer effects. In: White R, Broughton R (Eds.): *Research in parapsychology 1983.* Metuchen, NJ: Scarecrow Press, 1984:81-83.

Grad B. A Telekinetic Effect on Plant Growth. *Intl Jnl Parapsy.* 1964;6:473.

Grad B. The 'Laying on of Hands': Implications for Psychotherapy, Gentling, and the Placebo Effect. *Jnl Amer Soc for Psych Res*. 1967 Oct;61(4):286-305.

Grad, B. A telekinetic effect on plant growth: II. Experiments involving treatment of saline in stoppered bottles. *Internl J Parapsychol*. 1964;6:473-478, 484-488.

Grady D, Herrington D, Bittner V, Blumenthal R, Davidson M, Hlatky M, Hsia J, Hulley S, Herd A, Khan S, Newby LK, Waters D, Vittinghoff E, Wenger N. Cardiovascular disease outcomes during 6.8 years of hormone therapy: Heart and Estrogen/progestin Replacement Study follow-up (HERS II). *JAMA*. 2002 Jul 3;288(1):49-57.

Grandjean AC, Grandjean NR. Dehydration and cognitive performance. *J Am Coll Nutr*. 2007 Oct;26(5 Suppl):549S-554S.

Grant WB, Garland CF. The association of solar ultraviolet B (UVB) with reducing risk of cancer: multifactorial ecologic analysis of geographic variation in age-adjusted cancer mortality rates. *Anticancer Res*. 2006 Jul-Aug;26(4A):2687-99.

Grant WB, Holick MF. Benefits and requirements of vitamin D for optimal health: a review. *Altern Med Rev*. 2005 Jun;10(2):94-111.

Grant WB. An estimate of premature cancer mortality in the U.S. due to inadequate doses of solar ultraviolet-B radiation. *Cancer*. 2002 Mar 15;94(6):1867-75.

Grant WB. Solar ultraviolet irradiance and cancer incidence and mortality. *Adv Exp Med Biol*. 2008;624:16-30.

Gray-Davison F. *Ayurvedic Healing*. New York: Keats, 2002.

Greene Lab Scientists Investigating Viral Disease Incidence and Bee Colony Collapse Disorder. At the Frontline: *Columbia University Mailman School of Public Health*. 2007; May:2:2.

Greger M. Bird Flu: Virus of Our Own Hatching. Mother Earth. 2007 Dec-Jan:103-109.

Grissom C. Magnetic field effects in biology: A survey of possible mechanisms with emphasis on radical pair recombination. *Chem. Rev*. 1995;95:3-24.

Groneberg DA, Wahn U, Hamelmann E. Probiotic-induced suppression of allergic sensitization and airway inflammation is associated with an increase of T regulatory-dependent mechanisms in a murine model of asthma. *Clin Exp Allergy*. 2007 Apr;37(4):498-505.

Grzanna R, Lindmark L, Frondoza CG. Ginger—an herbal medicinal product with broad anti-inflammatory actions. *J Med Food*. 2005 Summer;8(2):125-32.

Guerin M, *et al*. 2003. Haematococcus astaxanthin: applications for human health and nutrition. *Trends Biotechnol*. May;21(5):210-6.

Gupta YK, Gupta M, Kohli K. Neuroprotective role of melatonin in oxidative stress vulnerable brain. *Indian J Physiol Pharmacol*. 2003 Oct;47(4):373-86.

Gutmanis J. *Hawaiian Herbal Medicine*. Waipahu, HI: Island Heritage, 2001.

Haarala C, Bergman M, Laine M, Revonsuo A, Koivisto M, Hamalainen H. Electromagnetic field emitted by 902 MHz mobile phones shows no effects on children's cognitive function. *Bioelectromagnetics*. 2005;Suppl 7:S144-50.

Haas M, Cooperstein R, Peterson D. Disentangling manual muscle testing and Applied Kinesiology: critique and reinterpretation of a literature review. Chiropr Osteopat. 2007 Aug 23;15:11.

Hadji L, Arnoux B, Benveniste J. Effect of dilute histamine on coronary flow of guinea-pig isolated heart. Inhibition by a magnetic field. *FASEB Jnl*. 1991;5:A1583.

Hadji L, Arnoux B, Benveniste J. Effect of dilute histamine on coronary flow of guinea-pig isolated heart. *FASEB J*. 1991;5:A1583.

Hagins WA, Penn RD, Yoshikami S. Dark current and photocurrent in retinal rods. *Biophys J*. 1970 May;10(5):380-412.

Hagins WA, Robinson WE, Yoshikami S. Ionic aspects of excitation in rod outer segments. *Ciba Found Symp*. 1975;(31):169-89.

Hagins WA, Yoshikami S. Ionic mechanisms in excitation of photoreceptors. *Ann N Y Acad Sci*. 1975 Dec 30;264:314-25.

Hagins WA, Yoshikami S. Proceedings: A role for Ca2+ in excitation of retinal rods and cones. *Exp Eye Res*. 1974 Mar;18(3):299-305.

Hagins WA. The visual process: Excitatory mechanisms in the primary receptor cells. *Annu Rev Biophys Bioeng*. 1972;1:131-58.

Halliday GM, Agar NS, Barnetson RS, Ananthaswamy HN, Jones AM. UV-A fingerprint mutations in human skin cancer. *Photochem Photobiol*. 2005 Jan-Feb;81(1):3-8.

Halpern G, Miller A. *Medicinal Mushrooms*. New York: M. Evans, 2002.

Halpern S. *Tuning the Human Instrument*. Palo Alto, CA: Spectrum Research Institute, 1978.

Hamel P. *Through Music to the Self: How to Appreciate and Experience Music*. Boulder: Shambala, 1979.

Hammitt WE. The relation between being away and privacy in urban forest recreation environments. *Environment and Behaviour*. 2000;32 (4):521-540.

Handwerk B. Are Earthquakes Encouraged by High Tides? *National Geographic News*. 2004 Oct 22.

Hans J. *The Structure and Dynamics of Waves and Vibrations*. New York:.Schocken and Co., 1975.

Hantusch B, Knittelfelder R, Wallmann J, Krieger S, Szalai K, Untersmayr E, Vogel M, Stadler BM, Scheiner O, Boltz-Nitulescu G, Jensen-Jarolim E. Internal images: human anti-idiotypic Fab antibodies mimic the IgE epitopes of grass pollen allergen Phl p 5a. *Mol Immunol.* 2006 Jul;43(14):2180-7.

Hardin P. Transcription regulation within the circadian clock: the E-box and beyond. *J Biol Rhythms.* 2004 Oct;19(5):348-60.

Harding OG. The healing power of intercessory prayer. *West Indian Med J.* 2001 Dec;50(4):269-72.

Haring BS, Van Delft W. Changes in the mineral composition of food as a result of cooking in "hard" and "soft" waters. *Arch. Environ. Health.* 1981;36:33-35.

Harkins T, Grissom C. Magnetic Field Effects on B12 Ethanolamine Ammonia Lyase: Evidence for a Radical Mechanism. *Science.* 1994;263:958-960.

Harkins T, Grissom C. The Magnetic Field Dependent Step in B12 Ethanolamine Ammonia Lyase is Radical-Pair Recombination. *J. Am. Chem. Soc.* 1995;117:566-567.

Haye-Legrand I, Norel X, Labat C, Benveniste J, Brink C. Antigenic contraction of guinea pig tracheal preparations passively sensitized with monoclonal IgE: pharmacological modulation. *Int Arch Allergy Appl Immunol.* 1988;87(4):342-8.

Heaney RP, Dowell MS. Absorbability of the calcium in a high-calcium mineral water. *Osteoporos Int.* 1994 Nov;4(6):323-4.

Heaney RP, Dowell MS. Absorbability of the calcium in a high-calcium mineral water. *Osteoporos Int.* 1994 Nov;4(6):323-4.

Heaney RP. Absorbability and utility of calcium in mineral waters. *Am J Clin Nutr.* 2006 Aug;84(2):371-4.

Hectorne KJ, Fransway AF. Diazolidinyl urea: incidence of sensitivity, patterns of cross-reactivity and clinical relevance. *Contact Dermatitis.* 1994 Jan;30(1):16-9.

Heerwagen JH. The psychological aspects of windows and window design'. In Selby, R. I., Anthony, K. H., Choi, J. and Orland, B. (eds) *Proceedings of 21st Annual Conference of the Environmental Design Research Association.* Champaign-Urbana, Illinois, 1990 April:6-9.

Helms JA, Farnham PJ, Segal E, Chang HY. Functional demarcation of active and silent chromatin domains in human HOX loci by noncoding RNAs. *Cell.* 2007 Jun 29;129(7):1311-23.

Hendel B, Ferreira P. *Water & Salt: The Essence of Life.* Gaithersburg: Natural Resources, 2003.

Hernandez Avila M, Walker AM, Jick H. Use of replacement estrogens and the risk of myocardial infarction. *Epidemiology.* 1990 Mar;1(2):128-33.

Heyers D, Manns M, Luksch H, Gü̈ntü̈rkü̈n O, Mouritsen H. A Visual Pathway Links Brain Structures Active during Magnetic Compass Orientation in Migratory Birds. *PLoS One.* 2007;2(9):e937. 2007.

Hietanen M, Hamalainen AM, Husman T. Hypersensitivity symptoms associated with exposure to cellular telephones: no causal link. *Bioelectromagnetics.* 2002 May;23(4):264-70.

Hillecke T, Nickel A, Bolay HV. Scientific perspectives on music therapy. *Ann N Y Acad Sci.* 2005 Dec;1060:271-82.

Hirayama J, Sahar S, Grimaldi B, Tamaru T, Takamatsu K, Nakahata Y, Sassone-Corsi P. CLOCK-mediated acetylation of BMAL1 controls circadian function. *Nature* 450, 1086-1090 (13 December 2007)

Hirshon J, Barrueto F. Plant Poisoning, Herbs. E-Medicine. http://www.emedicine.com/EMERG/topic449.htm. Accessed: 2007 Jan.

Hobbs C. *Medicinal Mushrooms.* Summertown, TN: Botanica Press, 1986.

Hoffmann D. *Holistic Herbal.* London: Thorsons, 1983-2002.

Hollfoth K. Effect of color therapy on health and wellbeing: colors are more than just

Holman CD, Armstrong BK, Heenan PJ. Relationship of cutaneous malignant melanoma to individual sunlight-exposure habits. *J Natl Cancer Inst.* 1986 Mar;76(3):403-14.

Holmquist G. Susumo Ohno left us January 13, 2000, at the age of 71. *Cytogenet and Cell Genet.* 2000;88:171-172.

Honeyman MK. Vegetation and stress: a comparison study of varying amounts of vegetation in countryside and urban scenes. In Relf, D. (ed) *The Role of Horticulture in Human Well-Being and Social Development: A National Symposium.* Portland: Timber Press. 1992:143-145.

Hope M. *The Psychology of Healing.* Longmead UK: Element Books, 1989.

Hopps HC, Feder GL. Chemical qualities of water that contribute to human health in a positive way. *Sci. Total Environ.* 1986;54:207-216.

Hoskin M.(ed.). *The Cambridge Illustrated History of Astronomy.* Cambridge: Cambridge Press, 1997.

Huang D, Ou B, Prior RL. The chemistry behind antioxidant capacity assays. J Agric Food Chem. 2005 Mar 23;53(6):1841-56.

Huffman C. Archytas of Tarentum: *Pythagorean, philosopher and Mathematician King.* Cambridge: Cambridge University Press, 2005.

Igarashi T, Izumi H, Uchiumi T, Nishio K, Arao T, Tanabe M, Uramoto H, Sugio K, Yasumoto K, Sasaguri Y, Wang KY, Otsuji Y, Kohno K. Clock and ATF4 transcription system regulates drug resistance in human cancer cell lines. *Oncogene.* 2007 Jul 19;26(33):4749-60.

Ikeda M, Toyoshima R, Inoue Y, Yamada N, Mishima K, Nomura M, Ozaki N, Okawa M, Takahashi K, Yamauchi T. Mutation screening of the human Clock gene in circadian rhythm sleep disorders. *Psychiatry Res.* 2002 Mar 15;109(2):121-8.

Ikonomov OC, Stoynev AG. Gene expression in suprachiasmatic nucleus and circadian rhythms. *Neurosci Biobehav Rev.* 1994 Fall;18(3):305-12.

Inaba H. INABA Biophoton. Exploratory Research for Advanced Technology. *Japan Science and Technology Agency.* 1991. http://www.jst.go.jp/erato/project/isf_P/isf_P.html. Accessed: 2006 Nov.

Inbar O, Dotan R, Dlin RA, Neuman I, Bar-Or O. Breathing dry or humid air and exercise-induced asthma during swimming. *Eur J Appl Physiol Occup Physiol.* 1980;44(1):43-50.

Ivry GB, Ogle CA, Shim EK. Role of sun exposure in melanoma. *Dermatol Surg.* 2006 Apr;32(4):481-92.

Iwami O, Watanabe T, Moon CS, Nakatsuka H, Ikeda M. Motor neuron disease on the Kii Peninsula of Japan: excess manganese intake from food coupled with low magnesium in drinking water as a risk factor. *Sci. Total Environ.* 1994;149:121-135.

Izbicki G, Chavko R, Banauch GI, Weiden MD, Berger KI, Aldrich TK, Hall C, Kelly KJ, Prezant DJ. World trade center "sarcoid-like" granulomatous pulmonary disease in New York City fire department rescue workers. *Chest.* 2007 May;131(5):1414-23.

Jacqmin H, Commenges D, Letenneur L, Barberger-Gateau P, Dartigues JF. Components of drinking water and risk of cognitive impairment in the elderly. *Am. J. Epidemiol.* 1994;139, 48-57.

Jagetia GC, Aggarwal BB. "Spicing up" of the immune system by curcumin. *J Clin Immunol.* 2007 Jan;27(1):19-35.

Jahn R, Dunne, B. Margins of Reality: the Role of Consciousness in the Physical World. New York: Harcourt Brace Jovanovich, 1987.

Jahn RG, Dunne BJ, Nelson RG, Dobyns YH, Bradish GJ. Correlations of random binary sequences with pre-stated operator intention: a review of a 12-year program. Explore (NY). 2007 May-Jun;3(3):244-53, 341-3.

Jahn RG, Dunne BJ. The PEAR proposition. Explore (NY). 2007 May-Jun;3(3):205-26, 340-1.

Jahn RG, Dunne BJ. The pertinence of the Princeton Engineering Anomalies (PEAR) Laboratory to the pursuit of global health. Epilogue. Explore (NY). 2007 May-Jun;3(3):339.

Janssen S, Solomon G, Schettler T. Chemical Contaminants and Human Disease:A Summary of Evidence. *The Collaborative on Health and the Environment.* 2006. http://www.healthandenvironment.org. Accessed: 2007 Jul.

Jaroff L. The End of Homeopathy? *Time.* 2005;Oct 4.

Jarvis DC. *Folk Medicine.* Greenwich, CN: Fawcett, 1958.

Jeebhay MF, Quirce S. Occupational asthma in the developing and industrialised world: a review. *Int J Tuberc Lung Dis.* 2007 Feb;11(2):122-33.

Jensen B. *Foods that Heal.* Garden City Park, NY: Avery Publ, 1988, 1993.

Jensen B. *Nature Has a Remedy.* Los Angeles: Keats, 2001.

Jhon MS. *The Water Puzzle and the Hexagonal Key.* Uplifting, 2004.

Johansen C. Electromagnetic fields and health effects—epidemiologic studies of cancer, diseases of the central nervous system and arrhythmia-related heart disease. *Scand J Work Environ Health.* 2004;30 Suppl 1:1-30.

Johansen C. Rehabilitation of cancer patients - research perspectives. *Acta Oncol.* 2007;46(4):441-5.

Johari H. *Ayurvedic Massage: Traditional Indian Techniques for Balancing Body and Mind.* Rochester, VT: Healing Arts, 1996.

Johari H. *Chakras.* Rochester, VT: Destiny, 1987.

Johnston RE. Pheromones, the vomeronasal system, and communication. From hormonal responses to individual recognition. *Ann N Y Acad Sci.* 1998 Nov 30;855:333-48.

Jonas WB, Kaptchuk TJ, Linde K. A critical overview of homeopathy. *Ann Intern Med.* 2003 Mar 4;138(5):393-9.

Jorgenson J. Therapeutic use of companion animals in health care. *Image J Nurs Sch.* 1997;29(3):249-54.

Jovanovic-Ignjatic Z, Rakovic D. A review of current research in microwave resonance therapy: novel opportunities in medical treatment. *Acupunct Electrother Res.* 1999; 24:105-125.

Jovanovic-Ignjatic Z. Microwave Resonant Therapy: Novel Opportunities in Medical Treatment. *Acup. & Electro-Therap. Res., The Int. J.* 1999;24(2):105-125.

Jutte R, Riley D. A review of the use and role of low potencies in homeopathy. *Complement Ther Med.* 2005 Dec;13(4):291-6.

Kahhak L, Roche A, Dubray C, Arnoux C, Benveniste J. Decrease of ciliary beat frequency by platelet activating factor: protective effect of ketotifen. *Inflamm Res.* 1996 May;45(5):234-8.

Kalsbeek A, Perreau-Lenz S, Buijs RM. A network of (autonomic) clock outputs. *Chronobiol Int.* 2006;23(1-2):201-15.

Kamide Y. We reside in the sun's atmosphere. *Biomed Pharmacother.* 2005 Oct;59 Suppl 1:S1-4.

Kandel E, Siegelbaum S, Schwartz J. *Synaptic transmission. Principles of Neural Science.* New York: Elsevier, 1991.

Kandel E, Siegelbaum S, Schwartz J. *Synaptic transmission. Principles of Neural Science.* New York: Elsevier, 1991.

Kaplan R. The psychological benefits of nearby nature. In: Relf, D. (ed) *The Role of Horticulture in Human Well-Being and Social Development: A National Symposium.* Portland: Timber Press. 1992:125-133.

Kaplan S. A model of person - environment compatibility. *Environment and Behaviour* 1983;15:311-332.

Kaplan S. The restorative environment: nature and human experience. In: Relf, D. (ed) *The Role of Horticulture in Human Well-Being and Social Development: A National Symposium.* Portland: Timber Press. 1992:134-142.

Kaptchuk TJ. The placebo effect in alternative medicine: can the performance of a healing ritual have clinical significance? *Ann Intern Med.* 2002 Jun 4;136(11):817-25.

Karis TE, Jhon MS. Flow-induced anisotropy in the susceptibility of a particle suspension. *Proc Natl Acad Sci USA.* 1986 Jul;83(14):4973-4977.

Karnstedt J. Ions and Consciousness. Whole Self. 1991 Spring.

Kataoka M, Tsumura H, Kaku N, Torisu T. Toxic effects of povidone-iodine on synovial cell and articular cartilage. *Clin Rheumatol.* 2006 Sep;25(5):632-8.

Kato Y, Kawamoto T, Honda KK. Circadian rhythms in cartilage. *Clin Calcium.* 2006 May;16(5):838-45.

Keet E. Beauty or Bust. Alternative Medicine. 2007 Jan.

Keil J. New cases in Burma, Thailand, and Turkey: A limited field study replication of some aspects of Ian Stevenson's work. *J. Sci. Exploration.* 1991;5(1):27-59.

Kelder P. *Ancient Secret of the Fountain of Youth: Book 1.* New York: Doubleday, 1998.

Kerr CC, Rennie CJ, Robinson PA. Physiology-based modeling of cortical auditory evoked potentials. Biol Cybern. 2008 Feb;98(2):171-84.

Keville K, Green M. *Aromatherapy: A Complete Guide to the Healing Art.* Freedom, CA: Crossing Press, 1995.

Khan S. Vitamin D deficiency and secondary hyperparathyroidism among patients with chronic kidney disease. *Am J Med Sci.* 2007 Apr;333(4):201-7.

Kheifets L, Monroe J, Vergara X, Mezei G, Afifi AA. Occupational electromagnetic fields and leukemia and brain cancer: an update to two meta-analyses. *J Occup Environ Med.* 2008 Jun;50(6):677-88.

Kirlian SD, Kirlian V. Photography and Visual Observation by Means of High-Frequency Currents. J Sci Appl Photogr. 1963;6(6).

Kiyose C, *et al.* Biodiscrimination of alpha-tocopherol stereoisomers in humans after oral administration. *Am J Clin Nutr.* 1997 Mar; 65 (3):785-9.

Klatz RM, Goldman RM, Cebula C. *Infection Protection.* New York: HarperResource, 2002.

Kleffmann J. Daytime Sources of Nitrous Acid (HONO) in the Atmospheric Boundary Layer. *Chemphyschem.* 2007 Apr 10;8(8):1137-1144.

Klein R, Landau MG. *Healing: The Body Betrayed.* Minneapolis: DCI:Chronimed, 1992.

Kloss J. *Back to Eden.* Twin Oaks, WI: Lotus Press, 1939.

Kniazeva TA, Kuznetsova LN, Otto MP, Nikiforova TI. Efficacy of chromotherapy in patients with hypertension. *Vopr Kurortol Fizioter Lech Fiz Kult.* 2006 Jan-Feb;(1):11-3.

Kohlhammer Y, Döring A, Schäfer T, Wichmann HE, Heinrich J; KORA Study Group. Swimming pool attendance and hay fever rates later in life. *Allergy.* 2006 Nov;61(11):1305-9.

Kondratyuk VA. On the health significance of microelements in low-mineral water. *Gig. Sanit.* 1989;(2):81-82.

Kosenko E, Kaminsky Y, Stavrovskaya I, Sirota T, Kondrashova MN. The stimulatory effect of negative air ions and hydrogen peroxide on the activity of superoxide dismutase. *FEBS Lett.* 1997 Jun 30;410(2-3):309-12.

Kowalchik C, Hylton W (eds). *Rodale's Illustrated Encyclopedia of Herbs.* Emmaus, PA: 1987.

Kowalczyk E, Krzesiński P, Kura M, Niedworok J, Kowalski J, Błaszczyk J. Pharmacological effects of flavonoids from Scutellaria baicalensis. *Przegl Lek.* 2006;63(2):95-6. Review.

Krause R, Buhring M, Hopfenmuller W, Holick MF, Sharma AM. Ultraviolet B and blood pressure. *Lancet.* 1998 Aug 29;352(9129):709-10.

Krebs K. The spiritual aspect of caring—an integral part of health and healing. *Nurs Adm Q.* 2001 Spring;25(3):55-60.

Kreig M. *Black Market Medicine.* New York: Bantam, 1968.

Kren A, Mamnun YM, Bauer BE, Schüller C, Wolfger H, Hatzixanthis K, Mollapour M, Gregori C, Piper P, Kuchler K. War1p, a Novel Transcription Factor Controlling Weak Acid Stress Response in Yeast. Mol and Cell Bio. 2003;23(5):1775-1785.

Krueger AP, Reed EJ. Biological impact of small air ions. *Science.* 1976 Sep 24;193(4259):1209-13.

Küller R, Laike T. The impact of flicker from fluorescent lighting on well-being, performance and physiological arousal. *Ergonomics.* 1998 Apr;41(4):433-47.

Kung HC, Hoyert DL, Xu J, Murphy SL. Deaths: Final Data for 2005. *National Vital Statistics Reports.* 2008;56(10). http://www.cdc.gov/nchs/data/ nvsr/nvsr56/nvsr56_10.pdf. Accessed: 2008 Jun.

Kurz R. Clinical medicine versus homeopathy. *Padiatr Padol.* 1992;27(2):37-41.

Kuuler R, Ballal S, Laike T Mikellides B, Tonello G. The impact of light and colour on psychological mood: a cross-cultral study of indoor work environments. Ergonomics. 2006 Nov 15;49(14):1496.

Kuznetsova TA, *et al.* 2004. Biological activity of fucoidans from brown algae and the prospects of their use in medicine. *Antibiot Khimioter.* 49(5):24-30.

Kwang Y, Cha , Daniel P, Wirth J, Lobo R. Does Prayer Influence the Success of *in Vitro.* Fertilization–Embryo Transfer? Report of a Masked, Randomized Trial. J Reproductive Med. 2001;46(9).

Lad V. Ayurveda: *The Science of Self-Healing.* Twin Lakes, WI: Lotus Press.

Lafrenière, G. The material Universe is made purely out of Aether. *Matter is made of Waves.* 2002. http://www.glafreniere.com/matter.htm. Accessed: 2007 Jun.

LaKind JS, McKenna EA, Hubner RP, Tardiff RG. A Review of the Comparative Mammalian Toxicity of Ethylene Glycol and Propylene Glycol. *Crit Rev Tox.* 1999:29(4);331-365.

Lakin-Thomas PL. Transcriptional feedback oscillators: maybe, maybe not. *J Biol Rhythms.* 2006 Apr;21(2):83-92.

Langhinrichsen-Rohling J, Palarea RE, Cohen J, Rohling ML. Breaking up is hard to do: unwanted pursuit behaviors following the dissolution of a romantic relationship. *Violence Vict.* 2000 Spring;15(1):73-90.

Lappe FM. *Diet for a Small Planet.* New York: Ballantine, 1971.

Lappe JM, Travers-Gustafson D, Davies KM, Recker RR, Heaney RP. Vitamin D and calcium supplementation reduces cancer risk: results of a randomized trial. *Am J Clin Nutr.* 2007 Jun;85(6):1586-91.

Lattin GL, Moore CJ, Moore SL, Weisberg SB, Zellers A. Density of Plastic Particles found in zooplankton trawls from Coastal Waters of California to the North Pacific Central Gyre. Proceedings of the *Plastic Debris Rivers to Sea Conference.* 2005 Sept 8.

Lattin GL, Moore CJ, Zellers AF, Moore SL, Weisberg SB. A comparison of neustonic plastic and zooplankton at different depths near the northern California shore. Mar Pollut Bull. 2004 Aug;49(4):291-4.

LaValle JB. *The Cox-2 Connection.* Rochester, VT: Healing Arts, 2001.

Lean G. US study links more than 200 diseases to pollution. London Independent. 2004 Nov 14.

Leape L. Lucian Leape on patient safety in U.S. hospitals. Interview by Peter I Buerhaus. *J Nurs Scholarsh.* 2004;36(4):366-70.

Leder D. Spooky actions at a distance: physics, psi, and distant healing. *J Altern Complement Med.* 2005 Oct;11(5):923-30.

Lee MY, Dordick JS. Enzyme activation for nonaqueous media. *Curr Opin Biotechnol.* 2002 Aug;13(4):376-84.

Lefort J, Sedivy P, Desquand S, Randon J, Coeffier E, Maridonneau-Parini I, Floch A, Benveniste J, Vargaftig BB. Pharmacological profile of 48740 R.P., a PAF-acether antagonist. *Eur J Pharmacol.* 1988 Jun 10;150(3):257-68.

Lehmann B. The vitamin D3 pathway in human skin and its role for regulation of biological processes. Photochem Photobiol. 2005 Nov-Dec;81(6):1246-51.

Lenn NJ, Beebe B, Moore RY (1977) Postnatal development of the suprachiasmatic nucleus of the rat. Cell Tissue Res 178:463-475.

Lennihan B. Homeopathy: natural mind-body healing. J Psychosoc Nurs Ment Health Serv. 2004 Jul;42(7):30-40.

Levin AI, Novikov JV, Plitman SI, Noarov JA, Lastochkina KO. Effect of water of varying degrees of hardness on the cardiovascular system. *Gig. Sanit.* 1981;(10):16-19.

Lewis A. Rescue remedy. *Nurs Times.* 1999 May 26-Jun 1;95(21):27.

Lewis WH, Elvin-Lewis MPF. *Medical Botany: Plants Affecting Man's Health.* New York: Wiley, 1977.

Leyel CF. *Culpeper's English Physician & Complete Herbal.* Hollywood, CA: Wilshire, 1971.

Li Q, Gandhi OP. Calculation of magnetic field-induced current densities for humans from EAS countertop activation/deactivation devices that use ferromagnetic cores. *Phys Med Biol.* 2005 Jan 21;50(2):373-85.

Lin PW, Chan WC, Ng BF, Lam LC. Efficacy of aromatherapy (Lavandula angustifolia) as an intervention for agitated behaviours in Chinese older persons with dementia: a cross-over randomized trial. *Int J Geriatr Psychiatry.* 2007 May;22(5):405-10.

Linde K, Clausius N, Ramirez G, Melchart D, Eitel F, Hedges LV, Jonas WB. Are the clinical effects of homeopathy placebo effects? A meta-analysis of placebo-controlled trials. *Lancet.* 1997 Sep 20;350(9081):834-43.

Lininger S, Gaby A, Austin S, Brown D, Wright J, Duncan A. *The Natural Pharmacy.* New York: Three Rivers, 1999.

Lipski E. *Digestive Wellness.* Los Angeles, CA: Keats, 2000.

Litime M, Aïssa J, Benveniste J. Antigen signaling at high dilution. *FASEB Jnl.* 1993;7:A602.

Livanova L, Levshina I, Nozdracheva L, Elbakidze MG, Airapetiants MG. The protective action of negative air ions in acute stress in rats with different typological behavioral characteristics. *Zh Vyssh Nerv Deiat Im I P Pavlova.* 1998 May-Jun;48(3):554-7.

Lloyd D and Murray D. Redox rhythmicity: clocks at the core of temporal coherence. *BioEssays.* 2007;29(5):465-473.

Lloyd JU. *American Materia Medica, Therapeutics and Pharmacognosy.* Portland, OR: Eclectic Medical Publications, 1989-1983.

Lorenz I, Schneider EM, Stolz P, Brack A, Strube J. Sensitive flow cytometric method to test basophil activation influenced by homeopathic histamine dilutions. *Forsch Komplementarmed Klass Naturheilkd.* 2003 Dec;10(6):316-24.

Lovejoy S, Pecknold S, Schertzer D. Stratified multifractal magnetization and surface geomagnetic fields—I. Spectral analysis and modeling. *Geophysical Journal International.* 2001;145(1);112-126.

Lydic R, Schoene WC, Czeisler CA, Moore-Ede MC. Suprachiasmatic region of the human hypothalamus: homolog to the primate circadian pacemaker? *Sleep.* 1980;2(3):355-61.

Lythgoe JN. Visual pigments and environmental light. *Vision Res.* 1984;24(11):1539-50.

Maas J, Jayson, J. K.. & Kleiber, D. A. Effects of spectral differences in illumination on fatigue. *J Appl Psychol.* 1974;59:524-526.

Macdessi JS, Randell TL, Donaghue KC, Ambler GR, van Asperen PP, Mellis CM. Adrenal crises in children treated with high-dose inhaled corticosteroids for asthma. *Med J Aust.* 2003 Mar 3;178(5):214-6.

Magnusson A, Stefansson JG. Prevalence of seasonal affective disorder in Iceland. *Arch Gen Psychiatry.* 1993 Dec;50(12):941-6.

Mahachoklertwattana P, Sudkronrayudh K, Direkwattanachai C, Choubtum L, Okascharoen C. Decreased cortisol response to insulin induced hypoglycaemia in asthmatics treated with inhaled fluticasone propionate. *Arch Dis Child.* 2004 Nov;89(11):1055-8.

Maier R, Greter SE, Maier N. Effects of pulsed electromagnetic fields on cognitive processes - a pilot study on pulsed field interference with cognitive regeneration. *Acta Neurol Scand.* 2004 Jul;110(1):46-52.

Maier T, Korting HC. Sunscreens - which and what for? *Skin Pharmacol Physiol.* 2005 Nov-Dec;18(6):253-62.

Makomaski Illing EM, Kaiserman MJ. Mortality attributable to tobacco use in Canada and its regions, 1998. *Can J Public Health.* 2004;95(1):38-44.

Manson JE, *et al.* Estrogen plus progestin and the risk of coronary heart disease. *NE J Med.* 2003; 349(6):523–534.

Mansour HA, Monk TH, Nimgaonkar VL. Circadian genes and bipolar disorder. *Ann Med.* 2005;37(3):196-205.

Manz F. Hydration and disease. *J Am Coll Nutr.* 2007 Oct;26(5 Suppl):535S-541S.

Marie PJ. Strontium ranelate: a physiological approach for optimizing bone formation and resorption. *Bone.* 2006 Feb;38(2 Suppl 1):S10-4.

Mathie RT. The research evidence base for homeopathy: a fresh assessment of the literature. *Homeopathy.* 2003 Apr;92(2):84-91.

Mattix KD, Winchester PD, Scherer LR. Incidence of abdominal wall defects is related to surface water atrazine and nitrate levels. *J Pediatr Surg.* 2007 Jun;42(6):947-9.

Matutinovic Z, Galic M. Relative magnetic hearing threshold. *Laryngol Rhinol Otol.* 1982 Jan;61(1):38-41.

Mayaux MJ, Guihard-Moscato ML, Schwartz D, Benveniste J, Coquin Y, Crapanne JB, Poiterin B, Rodary M, Chevrel JP, Mollet M. Controlled clinical trial of homoeopathy in postoperative ileus. *Lancet.* 1988 Mar 5;1(8584):528-9.

Mayron L, Ott J, Nations R, Mayron E. Light, radiation and academic behaviour: Initial studies on the effects of full-spectrum lighting and radiation shielding on behaviour and academic performance of school children. *Acad Ther.* 1974;10, 33-47.

Mayron L. Hyperactivity from fluorescent lighting - fact or fancy: A commentary on the report by O'Leary, Rosenbaum and Hughes. *J Abnorm Child Psychol.* 1978;6:291-294.

McCauley B. *Achieving Great Health.* Lansing, MI: Spartan, 2005.

McColl SL, Veitch JA. Full-spectrum fluorescent lighting: a review of its effects on physiology and health. *Psychol Med.* 2001 Aug;31(6):949-64.

McConnaughey E. *Sea Vegetables.* Happy Camp, CA: Naturegraph, 1985.

McDougall J, McDougall M. *The McDougal Plan.* Clinton, NJ: New Win, 1983.

McTaggart L. *The Field.* New York: Quill, 2003.

Mee L. Reviving Dead Zones. *Sci Amer.* 2006 Nov.

Melles Z, Kiss SA. Influence of the magnesium content of drinking water and of magnesium therapy on the occurrence of preeclampsia. *Magnes. Res.* 1992;5:277-279.

Mendoza J. Circadian clocks: setting time by food. *J Neuroendocrinol.* 2007 Feb;19(2):127-37.

Meolie AL, Rosen C, Kristo D, Kohrman M, Gooneratne N, Aguillard RN, Fayle R, Troell R, Townsend D, Claman D, Hoban T, Mahowald M. Oral nonprescription treatment for insomnia: an evaluation of products withn limited evidence. *J Clin Sleep Med.* 2005 Apr 15;1(2):173-87.

Merchant RE and Andre CA. 2001. A review of recent clinical trials of the nutritional supplement Chlorella pyrenoidosa in the treatment of fibromyalgia, hypertension, and ulcerative colitis. *Altern Ther Health Med.* May-Jun;7(3):79-91.

Merrell WC, Shalts E. Homeopathy. *Med Clin North Am.* 2002 Jan;86(1):47-62.

Milgrom LR. Is homeopathy possible? *J R Soc Health.* 2006 Sep;126(5):211-8.

Miller GT. *Living in the Environment.* Belmont, CA: Wadsworth, 1996.

Miller JD, Morin LP, Schwartz WJ, Moore RY. New insights into the mammalian circadian clock. Sleep. 1996 Oct;19(8):641-67.

Mindell E, Hopkins V. *Prescription Alternatives.* New Canaan CT: Keats, 1998.

Mineev VN, Bulatova NIu, Fedoseev GB. Erythrocyte insulin-reactive system and carbohydrate metabolism in bronchial asthma. Ter Arkh. 2002;74(3):14-7.

Mocchegiani E, Giacconi R, Cipriano C, Costarelli L, Muti E, Tesei S, Giuli C, Papa R, Marcellini F, Mariani E, Rink L, Herbein G, Varin A, Fulop T, Monti D, Jajte J, Dedoussis G, Gonos ES, Trougakos IP, Malavolta M. Zinc, metallothioneins, and longevity—effect of zinc supplementation: zincage study. Ann N Y Acad Sci. 2007 Nov;1119:129-46.

Modern Biology. Austin: Harcourt Brace, 1993.

Mohan JE, Ziska LH, Schlesinger WH, Thomas RB, Sicher RC, George K, Clark JS. Biomass and toxicity responses of poison ivy (Toxicodendron radicans) to elevated atmospheric CO_2. Proc Natl Acad Sci U S A. 2006 Jun 13;103(24):9086-9.

Monarca S, Donato F, Zerbini I, Calderon RL, Craun GF. Review of epidemiological studies on drinking water hardness and cardiovascular diseases. Eur J *Cardiovasc Prev Rehabil.* 2006 Aug;13(4):495-506.

Monarca S. Zerbini I, Simonati C, Gelatti U. Drinking water hardness and chronic degenerative diseases. Part II. Cardiovascular diseases. *Ann. Ig.* 2003;15:41-56.

Montanes P, Goldblum MC, Boller F. The naming impairment of living and nonliving items in Alzheimer's disease. *J Int Neuropsychol Soc.* 1995 Jan;1(1):39-48.

Moore CJ, Moore SL, Leecaster MK, Weisberg SB. A comparison of plastic and plankton in the north Pacific central gyre. Mar Pollut Bull. 2001 Dec;42(12):1297-300.

Moore CJ, Moore SL, Weisberg SB, Lattin GL, Zellers AF. A comparison of neustonic plastic and zooplankton abundance in southern California's coastal waters. Mar Pollut Bull. 2002 Oct;44(10):1035-8. PubMed PMID: 12474963.

Moore CJ. Synthetic polymers in the marine environment: a rapidly increasing, long-term threat. Environ Res. 2008 Oct;108(2):131-9.

Moore KH. Conservative management for urinary incontinence. Baillieres Best Pract Res Clin Obstet Gynaecol. 2000 Apr;14(2):251-89.

Moore R. Circadian Rhythms: A Clock for the Ages. Science 1999 June 25;284(5423):2102 – 2103.

Moore RY, Speh JC. Serotonin innervation of the primate suprachiasmatic nucleus. Brain Res. 2004 Jun 4;1010(1-2):169-73.

Moore RY. Neural control of the pineal gland. Behav Brain Res. 1996;73(1-2):125-30.

Moore RY. Organization and function of a central nervous system circadian oscillator: the suprachiasmatic hypothalamic nucleus. Fed Proc. 1983 Aug;42(11):2783-9.

Moorhead KJ, Morgan HC. *Spirulina: Nature's Superfood.* Kailua-Kona, HI: Nutrex, 1995.

Morell V. Minds of their Own. *Nat Geo.* 2008 Mar:36-61.

Morton C. *Velocity Alters Electric Field.* www.amasci.com/ freenrg/ morton1.html. Accessed 2007 July.

Muhlack S, Lemmer W, Klotz P, Muller T, Lehmann E, Klieser E. Anxiolytic effect of rescue remedy for psychiatric patients: a double-blind, placebo-controlled, randomized trial. *J Clin Psychopharmacol.* 2006 Oct;26(5):541-2.

Municino A, Nicolino A, Milanese M, Gronda E, Andreuzzi B, Oliva F, Chiarella F, Cardio-HKT Study Group. Hydrotherapy in advanced heart failure: the cardio-HKT pilot study. *Monaldi Arch Chest Dis.* 2006 Dec;66(4):247-54.

Murchie G. *The Seven Mysteries of Life.* Boston: Houghton Mifflin Company, 1978.

Murphy R. *Organon Philosophy Workbook.* Blacksburg, VA: HANA, 1994.

Murray M and Pizzorno J. *Encyclopedia of Natural Medicine.* 2nd Edition. Roseville, CA: Prima Publishing, 1998.

Muzzarelli L, Force M, Sebold M. Aromatherapy and reducing preprocedural anxiety: A controlled prospective study. *Gastroenterol Nurs.* 2006 Nov-Dec;29(6):466-71.

Myss C. *Anatomy of the Spirit.* New York: Harmony, 1996.

Nadkarni AK, Nadkarni KM. *Indian Materia Medica.* (Vols 1 and 2). Bombay, India: Popular Pradashan, 1908, 1976.

Nagaoka M, et al. 2000. Anti-ulcer effects and biological activities of polysaccharides from marine algae. Biofactors. 12(1-4):267-74.

Nakatani K, Yau KW. Calcium and light adaptation in retinal rods and cones. Nature. 1988 Jul 7;334(6177):69-71.

Napoli N, Thompson J, Civitelli R, Armamento-Villareal R. Effects of dietary calcium compared with calcium supplements on estrogen metabolism and bone mineral density. Am J Clin Nutr. 2007;85(5):1428-1433.

Nardi G, Donato F, Monarca S, Gelatti U. Drinking water hardness and chronic degenerative diseases. I. Analysis of epidemiological research. Ann Ig. 2003 Jan-Feb;15(1):35-40.

Natarajan E, Grissom C. The Origin of Magnetic Field Dependent Recombination in Alkylcobalamin Radical Pairs. Photochem Photobiol. 1996;64:286-295.

Navarro Silvera SA, Rohan TE. Trace elements and cancer risk: a review of the epidemiologic evidence. Cancer Causes Control. 2007 Feb;18(1):7-27.

Nelson RD, Jahn RG, Dunne BJ, Dobyns YH, Bradish GJ. FieldREG II: consciousness field effects: replications and explorations. Explore (NY). 2007 May-Jun;3(3):279-93, 344.

Neumann von J. (1955): Mathematical Foundations of Quantum Mechanics.Princeton: Princeton University Press. Translated by R. Beyer from Mathematische Grundlagen der Quantenmechanik, Springer, Berlin, 1932.

Neushul. 1990. Antiviral carbohydrates from marine red algae. Hydrobiologia. 204/205:99-104.

Newmark T, Schulick P. Beyond Aspirin. Prescott, AZ: Holm, 2000.

Newton PE. The Effect of Sound on Plant Grwoth. JAES. 1971 Mar;19(3):202-205.

Nickmilder M, Bernard A. Ecological association between childhood asthma and availability of indoor chlorinated swimming pools in Europe. Occup Environ Med. 2007 Jan;64(1):37-46.

Niculescu MD, Wu R, Guo Z, da Costa KA, Zeisel SH. Diethanolamine alters proliferation and choline metabolism in mouse neural precursor cells. Toxicol Sci. 2007 Apr;96(2):321-6.

Nielsen LR, Mosekilde L. Vitamin D and breast cancer. Ugeskr Laeger. 2007 Apr 2;169(14):1299-302.

Nievergelt CM, Kripke DF, Remick RA, Sadovnick AD, McElroy SL, Keck PE Jr, Kelsoe JR. Examination of the clock gene Cryptochrome 1 in bipolar disorder: mutational analysis and absence of evidence for linkage or association. Psychiatr Genet. 2005 Mar;15(1):45-52.

Nishigori C, Hattori Y, Toyokuni S. Role of reactive oxygen species in skin carcinogenesis. Antioxid Redox Signal. 2004 Jun;6(3):561-70.

North J. The Fontana History of Astronomy and Cosmology. London: Fontana Press, 1994.

O'Dwyer JJ. College Physics. Pacific Grove, CA: Brooks/Cole, 1990.

Oehme FW (ed.). Toxicity of heavy metals in the environment. Part 1. New York: M.Dekker, 1979.

Oh CK, Lücker PW, Wetzelsberger N, Kuhlmann F. The determination of magnesium, calcium, sodium and potassium in assorted foods with special attention to the loss of electrolytes after various forms of food preparations. Mag.-Bull. 1986;8:297-302.

Okamura H. Clock genes in cell clocks: roles, actions, and mysteries. J Biol Rhythms. 2004 Oct;19(5):388-99.

Okayama Y, Begishvili TB, Church MK. Comparison of mechanisms of IL-3 induced histamine release and IL-3 priming effect on human basophils. Clin Exp Allergy. 1993 Nov;23(11):901-10.

Ole D. Rughede, On the Theory and Physics of the Aether. Progress in Physics. 2006; (1).

O'Leary KD, Rosenbaum A, Hughes PC. Fluorescent lighting: a purported source of hyperactive behavior. J Abnorm Child Psychol. 1978 Sep;6(3):285-9.

One Hundred Million Americans See Medical Mistakes Directly Touching Them as Patients, Friends, Relatives. National Patient Safety Foundation. Press Release. 1997 Oct 9. http://npsf.org/pr/pressrel/ final-sur.htm. Accessed: 2007 Mar.

Ostrander S, Schroeder L, Ostrander N. Super-Learning. New York: Delta, 1979.

Ozone Hole Healing Gradually. Associated Press. 2005 Sept 16, 03:18 pm ET.

Pacione M. Urban environmental quality and human wellbeing-a social geographical perspective. Landscape and Urban Planning 2003;986:1-12.

Packard CC. Pocket Guide to Ayurvedic Healing. Freedom, CA: Crossing Press, 1996.

Palmqvist C, Wardlaw AJ, Bradding P. Chemokines and their receptors as potential targets for the treatment of asthma. Br J Pharmacol. 2007 Apr 30.

Palumbo A. Gravitational and geomagnetic tidal source of earthquake triggering. Italian Physical Society. 1989 Nov;12(6).

Pardo A, Nevo K, Vigiser D, Lazarov A. The effect of physical and chemical properties of swimming pool water and its close environment on the development of contact dermatitis in hydrotherapists. Am J Ind Med. 2007 Feb;50(2):122-6.

Pavlovic M. Einstein's Theory of Relativity - Scientific Theory or Illusion? http://www.milanrpavlovic.freeservers.com. Accessed 2007 Oct..

Payment P, Franco E, Richardson L, Siemiatyck, J. Gastrointestinal health effects associated with the consumption of drinking water produced by point-of-use domestic reverse-osmosis filtration units. *Appl. Environ. Microbiol.* 1991;57:945-948.

Payment, P. (1989) Bacterial colonization of reverse-osmosis water filtration units. *Can. J. Microbiol.* 1989;35:1065-1067.

PDR. *Physicians' Desk Reference.* Montvale, NJ: Thomson, 2003.

Pendell D. *Plant Powers, Poisons, and Herbcraft.* San Francisco: Mercury House, 1995.

Penn RD, Hagins WA. Kinetics of the photocurrent of retinal rods. *Biophys J.* 1972 Aug;12(8):1073-94.

Penn RD, Hagins WA. Signal transmission along retinal rods and the origin of the electroretinographic a-wave. *Nature.* 1969 Jul 12;223(5202):201-4.

Penson RT, Kyriakou H, Zuckerman D, Chabner BA, Lynch TJ Jr. Teams: communication in multidisciplinary care. *Oncologist.* 2006 May;11(5):520-6.

Peroxisomes from pepper fruits (Capsicum annuum L.): purification, characterisation and antioxidant activity. *J Plant Physiol.* 2003 Dec;160(12):1507-16.

Perreau-Lenz S, Kalsbeek A, Van Der Vliet J, Pevet P, Buijs RM. In vivo evidence for a controlled offset of melatonin synthesis at dawn by the suprachiasmatic nucleus in the rat. *Neuroscience.* 2005;130(3):797-803.

Pert C. *Molecules of Emotion.* New York: Scribner, 1997.

Petiot JF, Sainte-Laudy J, Benveniste J. Interpretation of results on a human basophil degranulation test. *Ann Biol Clin (Paris).* 1981;39(6):355-9.

Pflege Z. *Physics.* 2000 Feb;53(2):111-2.

Piggins HD. Human clock genes. *Ann Med.* 2002;34(5):394-400.

Pilkington K, Kirkwood G, Rampes H, Fisher P, Richardson J. Homeopathy for depression: a systematic review of the research evidence. *Homeopathy.* 2005 Jul;94(3):153-63.

Pilkington K, Kirkwood G, Rampes H, Fisher P, Richardson J. Homeopathy for depression: a systematic review of the research evidence. *Homeopathy.* 2005 Jul;94(3):153-63.

Piluso LG, Moffatt-Smith C. Disinfection using ultraviolet radiation as an antimicrobial agent: a review and synthesis of mechanisms and concerns. PDA J Pharm Sci Technol. 2006 Jan-Feb;60(1):1-16.

Pinckney C. *Callanetics.* New York: Avon, 1984.

Pinto JT, Sinha R, Papp K, Facompre ND, Desai D, El-Bayoumy K. Differential effects of naturally occurring and synthetic organoselenium compounds on biomarkers in androgen responsive and androgen independent human prostate carcinoma cells. *Int J Cancer.* 2007 Apr 1;120(7):1410-7.

Piper PW. Yeast superoxide dismutase mutants reveal a pro-oxidant action of weak organic acid food preservatives. *Free Radic Biol Med.* 1999 Dec;27(11-12):1219-27.

Pitt-Rivers R, Trotter WR. *The Thyroid Gland.* London: Butterworth Publisher, 1954.

Plaut T, Jones T. *Asthma Guide for People of All Ages.* Amherst MA: Pedipress, 1999.

Poitevin B, Davenas E, Benveniste J. In vitro immunological degranulation of human basophils is modulated by lung histamine and Apis mellifica. *Br J Clin Pharmacol.* 1988 Apr;25(4):439-44.

Pongratz W, Endler PC, Poitevin B, Kartnig T. Effect of extremely diluted plant hormone on cell culture, *Proc. 1995 AAAS Ann. Meeting,* Atlanta, 1995.

Pont AR, Charron AR, Brand RM. Active ingredients in sunscreens act as topical penetration enhancers for the herbicide 2,4-dichlorophenoxyacetic acid. *Toxicol Appl Pharmacol.* 2004 Mar 15;195(3):348-54.

Pool R. Is there an EMF-Cancer connection? *Science.* 1990;249:1096-1098.

Postlethwait EM. Scavenger receptors clear the air. *J Clin Invest.* 2007 Mar;117(3):601-4.

Poulos LM, Toelle BG, Marks GB. The burden of asthma in children: an Australian perspective. *Paediatr Respir Rev.* 2005 Mar;6(1):20-7.

Preisinger E, Quittan M. Thermo- and hydrotherapy. *Wien Med Wochenschr.* 1994;144(20-21):520-6.

Proctor R. *Human Factors in Web Design.* New York: Routledge, 2005.

Protheroe WM, Captiotti ER, Newsom GH. *Exploring the Universe.* Columbus, OH: Merrill, 1989,

Provalova NV, Suslov NI, Skurikhin EG, Dygaĭ AM. Local mechanisms of the regulatory action of Scutellaria baicalensis and ginseng extracts on the erythropoiesis after paradoxical sleep deprivation. *Eksp Klin Farmakol.* 2006 Sep-Oct;69(5):31-5.

Rachmanin Y, Filippova AV, Michailova R, Belyaeva N, Lamentova T, Robbins DJ, Sly MR. Serum zinc and demineralized water. *Am J Cli. Nutr.* 1981;34:962-963.

Radin D. *The Conscious Universe.* San Francisco: HarperEdge, 1997.

Raloff J. Ill Winds. *Science News:* 2001;160(14):218.

Ramello A, Vitale C, Marangella M. Epidemiology of nephrolithiasis. *J Nephrol.* 2000 Nov-Dec;13 Suppl 3:S45-50.

Rappoport J. Both sides of the pharmaceutical death coin. *Townsend Letter for Doctors and Patients.* 2006 Oct.

Regel SJ, Negovetic S, Roosli M, Berdinas V, Schuderer J, Huss A, *et al.* UMTS base station-like exposure, well-being, and cognitive performance. *Environ Health Perspect.* 2006 Aug;114(8):1270-5.

Reger D, Goode S, Mercer E. *Chemistry: Principles & Practice*. Fort Worth, TX: Harcourt Brace, 1993.

Regis E. *Virus Ground Zero*. New York: Pocket, 1996.

Reichrath J. The challenge resulting from positive and negative effects of sunlight: how much solar UV exposure is appropriate to balance between risks of vitamin D deficiency and skin cancer? *Prog Biophys Mol Biol.* 2006 Sep;92(1):9-16.

Reilly D, Taylor M, Beattie N, Campbell J, McSharry C, Aitchison T, Carter R, Stevenson R. Is evidence for homoeopathy reproducible? *Lancet*, 1994;344:1601-1606.

Reilly D. The puzzle of homeopathy. *J Altern Complement Med.* 2001;7 Suppl 1:S103-9.

Reilly T, Stevenson I. An investigation of the effects of negative air ions on responses to submaximal exercise at different times of day. *J Hum Ergol.* 1993 Jun;22(1):1-9.

Reilly T, Taylor M, McSharry C, Aitchison T. Is homoeopathy a placebo response? Controlled trial of homoeopathic potency, with pollen in hayfever as model. *Lancet.* 1986;II:881-886.

Reiter RJ, Garcia JJ, Pie J. Oxidative toxicity in models of neurodegeneration: responses to melatonin. *Restor Neurol Neurosci.* 1998 Jun;12(2-3):135-42.

Reiter RJ, Tan DX, Korkmaz A, Erren TC, Piekarski C, Tamura H, Manchester LC. Light at night, chrono-disruption, melatonin suppression, and cancer risk: a review. *Crit Rev Oncog.* 2007;13(4):303-28.

Reiter RJ, Tan DX, Manchester LC, Qi W. Biochemical reactivity of melatonin with reactive oxygen and nitrogen species: a review of the evidence. *Cell Biochem Biophys.* 2001;34(2):237-56.

Retallack D. *The Sound of Music and Plants*. Marina Del Rey, CA: Devorss, 1973.

Ring K. *Life at Death: A Scientific Investigation of the Near-Death Experience*. New York: Quill, 1982.

Rinn JL, Kertesz M, Wang JK, Squazzo SL, Xu X, Brugmann SA, Goodnough LH,

Roberts JE. Light and immunomodulation. *Ann N Y Acad Sci.* 2000;917:435-45.

Rodale R. *Our Next Frontier*. Emmaus, PA: Rodale, 1981.

Rodermel SR, Smith-Sonneborn J. Age-correlated changes in expression of micronuclear damage and repair in Paramecium tetraurelia. *Genetics.* 1977 Oct;87(2):259-74.

Rodgers JT, Puigserver P. Fasting-dependent glucose and lipid metabolic response through hepatic sirtuin 1. *Proc Natl Acad Sci USA.* 2007 Jul 31;104(31):12861-6.

Rodriguez E, Valbuena MC, Rey M, Porras de Quintana L. Causal agents of photoallergic contact dermatitis diagnosed in the national institute of dermatology of Colombia. *Photodermatol Photoimmunol Photomed.* 2006 Aug;22(4):189-92.

Rosenlund M, Picciotto S, Forastiere F, Stafoggia M, Perucci CA. Traffic-related air pollution in relation to incidence and prognosis of coronary heart disease. *Epidemiology.* 2008 Jan;19(1):121-8.

Rosenthal N, Blehar M (Eds.). *Seasonal affective disorders and phototherapy*. New York: Guildford Press, 1989.

Rossouw JE, Prentice RL, Manson JE, Wu L, Barad D, Barnabei VM, Ko M, LaCroix AZ, Margolis KL, Stefanick ML. Postmenopausal hormone therapy and risk of cardiovascular disease by age and years since menopause. *JAMA.* 2007 Apr 4;297(13):1465-77.

Routasalo P, Isola A. The right to touch and be touched. Nurs Ethics. 1996 Jun;3(2):165-76.

Roybal K, Theobold D, Graham A, DiNieri JA, Russo SJ, Krishnan V, Chakravarty S, Peevey J, Oehrlein N, Birnbaum S, Vitaterna MH, Orsulak P, Takahashi JS, Nestler EJ, Carlezon WA Jr, McClung CA. Mania-like behavior induced by disruption of CLOCK. Proc Natl Acad Sci USA 2007;104(15):6406-6411.

Rubenowitz E, Molin I, Axelsson G, Rylander R. (2000) Magnesium in drinking water in relation to morbidity and mortality from acute myocardial infarction. Epidemiology. 2000;11:416-421.

Rubin GJ, Hahn G, Everitt BS, Cleare AJ, Wessely S. Are some people sensitive to mobile phone signals? Within participants double blind randomised provocation study. *BMJ.* 2006 Apr 15;332(7546):886-91.

Russek LG, Schwartz GE. Narrative descriptions of parental love and caring predict health status in midlife: a 35-year follow-up of the Harvard Mastery of Stress Study. *Altern Ther Health Med.* 1996 Nov;2(6):55-62.

Russo PA, Halliday GM. Inhibition of nitric oxide and reactive oxygen species production improves the ability of a sunscreen to protect from sunburn, immunosuppression and photocarcinogenesis. *Br J Dermatol.* 2006 Aug;155(2):408-15.

Saarijarvi S, Lauerma H, Helenius H, Saarilehto S. Seasonal affective disorders among rural Finns and Lapps. *Acta Psychiatr Scand.* 1999 Feb;99(2):95-101.

Sainte-Laudy J, Belon P. Analysis of immunosuppressive activity of serial dilutions of histamine on human basophil activation by flow cytometry. *Inflam Rsrch.* 1996 Suppl. 1:S33-S34.

Sakugawa H, Cape JN. Harmful effects of atmospheric nitrous acid on the physiological status of

Salpeter SR, Buckley NS, Ormiston TM, Salpeter EE. Meta-analysis: effect of long-acting beta-agonists on severe asthma exacerbations and asthma-related deaths. *Intern Med.* 2006 Jun 20;144(12):904-12.

Sanders R. Slow brain waves play key role in coordinating complex activity. UC Berkeley News. 2006 Sep 14.

Sarah Janssen S, Solomon G, Schettler T. Chemical Contaminants and Human Disease: A Summary of Evidence. The Collaborative on Health and the Environment. 2006. http://www.healthand-environment.org. Accessed: 2007 Jul.

Sarveiya V, Risk S, Benson HA. Liquid chromatographic assay for common sunscreen agents: application to in vivo assessment of skin penetration and systemic absorption in human volunteers. *J Chromatogr B Analyt Technol Biomed Life Sci.* 2004 Apr 25;803(2):225-31.

Sato TK, Yamada RG, Ukai H, Baggs JE, Miraglia LJ, Kobayashi TJ, Welsh DK, Kay SA, Ueda HR, Hogenesch JB. Feedback repression is required for mammalian circadian clock function. *Nat Genet.* 2006 Mar;38(3):312-9.

Satyanarayana S, Sushruta K, Sarma GS, Srinivas N, Subba Raju GV. Antioxidant activity of the aqueous extracts of spicy food additives—evaluation and comparison with ascorbic acid in in-vitro systems. *J Herb Pharmacother.* 2004;4(2):1-10.

Sauvant M, Pepin D. Drinking water and cardiovascular disease. *Food Chem Toxicol.* 2002;40:1311-1325.

Schlumpf M, Cotton B, Conscience M, Haller V, Steinmann B, Lichtensteiger W. In vitro and in vivo estrogenicity of UV screens. *Environ Health Perspect.* 2001 Mar;109(3):239-44.

Schmidt H, Quantum processes predicted? *New Sci.* 1969 Oct 16.

Schmitt B, Frölich L. Creative therapy options for patients with dementia—a systematic review. *Fortschr Neurol Psychiatr.* 2007 Dec;75(12):699-707.

Schulz T, Zarse K, Voigt A, Urban N, Birringer M, Ristow M. Glucose Restriction Extends Caenorhabditis elegans Life Span by Inducing Mitochondrial Respiration and Increasing Oxidative Stress. *Cell Metabolism.* 2007 Oct 3;6:280-293.

Schumacher P. *Biophysical Therapy Of Allergies.* Stuttgart: Thieme, 2005.

Schumann K, Elsenhans B, Reichl F, Pfob H, Wurster K. Does intake of highly demineralized water damage the rat gastrointestinal tract? *Vet Hum Toxicol.* 1993;35:28-31.

Schwartz GG, Skinner HG. Vitamin D status and cancer: new insights. *Curr Opin Clin Nutr Metab Care.* 2007 Jan;10(1):6-11.

Schwartz S, De Mattei R, Brame E, Spottiswoode S. Infrared spectra alteration in water proximate to the palms of therapeutic practitioners. In: Wiener D, Nelson R (Eds.): *Research in parapsychology 1986.* Metuchen, NJ: Scarecrow Press, 1987:24-29.

Schwellenbach LJ, Olson KL. McConnell KJ, Stolepart RS, Nash JD, Merenich JA. The triglyceride-lowering effects of a modest dose of docosahexaenoic acid alone versus in combination with low dose eicosapentaenoic acid in patients with coronary artery disease and elevated triglycerides. *J Am Coll Nutr.* 2006;25(6):480-485.

Scots pine trees. *Environ Pollut.* 2007 Jun;147(3):532-4.

Senekowitsch F, Endler P, Pongratz W, Smith C. Hormone effects by CD record/replay. *FASEB J.* 1995;9:A392.

Senior F. Fallout. *New York Magazine.* Fall: 2003.

Serra-Valls A. Electromagnetic Industion and the Conservation of Momentum in the Spiral Paradox. *Cornell University Library.* http://arxiv.org/ftp/physics/papers/0012/0012009.pdf. Accessed: 2007 Jul.

Serway R. *Physics For Scientists & Engineers.* Philadelphia: Harcourt Brace, 1992.

Shaffer D. *Developmental Psychology: Theory, Research and Applications.* Monterey, CA: Brooks/Cole, 1985.

Shafik A. Role of warm-water bath in anorectal conditions. The "thermosphincteric reflex". *J Clin Gastroenterol.* 1993 Jun;16(4):304-8.

Shang A, Huwiler-Müntener K, Nartey L, Juni P, Dorig S, Sterne JA, *et al.* Are the clinical effects of homoeopathy placebo effects? Comparative study of placebo-controlled trials of homoeopathy and allopathy. Lancet. 2005;366:726–32.

Shankar R. My Music, My Life. New York: Simon & Schuster, 1968.

Shearman LP, Zylka MJ, Weaver DR, Kolakowski LF Jr, Reppert SM. Two period homologs: circadian expression and photic regulation in the suprachiasmatic nuclei. *Neuron.* 1997 Dec;19(6):1261-9.

Shevelev IA, Kostelianetz NB, Kamenkovich VM, Sharaev GA. EEG alpha-wave in the visual cortex: check of the hypothesis of the scanning process. *Int J Psychophysiol.* 1991 Aug;11(2):195-201.

Shui-Yin Lo. Anomalous State of Ice. *Mod Phys Lttrs.* 1996;10(19):909-919.

Simon H. A.D.A.M. University of Maryland Medical Center. March, 2007. http://www.umm.edu/patiented/articles/which_adults_at_risk_asthma_000004_5.htm. Accessed: 2007 Jun.

Sin DD, Man J, Sharpe H, Gan WQ, Man SF. Pharmacological management to reduce exacerbations in adults with asthma: a systematic review and meta-analysis. *JAMA.* 2004 Jul 21;292(3):367-76.

Skwerer RG, Jacobsen FM, Duncan CC, Kelly KA, Sack DA, Tamarkin L, Gaist PA, Kasper S, Rosenthal NE. Neurobiology of Seasonal Affective Disorder and Phototherapy. *J Biolog Rhyth.* 1988;3(2):135-154.

Smith CW. Coherence in living biological systems. *Neural Network World.* 1994:4(3):379-388.

Smith MJ. "Effect of Magnetic Fields on Enzyme Reactivity" in Barnothy M.(ed.), *Biological Effects of Magnetic Fields.* New York: Plenum Press, 1969.

Smith MJ. *The Influence on Enzyme Growth By the 'Laying on of Hands: Dimenensions of Healing.* Los Altos, California: Academy of Parapsychology and Medicine, 1973.

Smith T. *Homeopathic Medicine: A Doctor's Guide.* Rochester, VT: Healing Arts, 1989.

Smith-Sonneborn J. Age-correlated effects of caffeine on non-irradiated and UV-irradiated Paramecium Aurelia. *J Gerontol.* 1974 May;29(3):256-60.

Smith-Sonneborn J. DNA repair and longevity assurance in Paramecium tetraurelia. *Science.* 1979 Mar 16;203(4385):1115-7.

Smits MG, Williams A, Skene DJ, Von Schantz M. The 3111 Clock gene polymorphism is not associated with sleep and circadian rhythmicity in phenotypically characterized human subjects. *J Sleep Res.* 2002 Dec;11(4):305-12.

Snow WB. *The Therapeutics of Radiant Light and Heat and Convective Heat.* New York: Scientific Authors Publishing Company. 1909.

Snyder K. Researchers Produce Firsts with Bursts of Light: Team generates most energetic terahertz pulses yet, observes useful optical phenomena. *Press Release: Brookhaven National Laboratory.* 2007 July 24.

Soler M, Chandra S, Ruiz D, Davidson E, Hendrickson D, Christou G. A third isolated oxidation state for the Mn12 family of singl molecule magnets. *ChemComm;* 2000; Nov 22.

Soni MG, Carabin IG, Burdock GA. Safety assessment of esters of p-hydroxybenzoic acid (parabens). *Food Chem Toxicol.* 2005 Jul;43(7):985-1015.

Spanagel R, Rosenwasser AM, Schumann G, Sarkar DK. Alcohol consumption and the body's biological clock. *Alcohol Clin Exp Res.* 2005 Aug;29(8):1550-7.

Speed Of Light May Not Be Constant, Physicist Suggests. Science Daily. 1999 Oct 6. www.sciencedaily.com/releases/1999/10/991005114024.htm. Accessed: 2007 Jun.

Spence A. *Basic Human Anatomy.* Menlo Park, CA: Benjamin/Commings, 1986.

Spillane M. Good Vibrations, A Sound 'Diet' for Plants. *The Growing Edge.* 1991 Spring.

Spiller G. *The Super Pyramid.* New York: HRS Press, 1993.

Stampfer MJ, Willett WC, Colditz GA, Rosner B, Speizer FE, Hennekens CH. A prospective study of postmenopausal estrogen therapy and coronary heart disease. N Engl J Med. 1985 Oct 24;313(17):1044-9.

Steck B. Effects of optical radiation on man. *Light Resch Techn.* 1982;14:130-141.

Steiner R. *Agriculture.* Kimberton, PA: Bio-Dynamic Farming, 1924-1993.

Stevenson I. *Reincarnation and Biology: A Contribution to the Etiology of Birthmarks and Birth Defects.* (2 volumes). Westport, CN: Praeger Publishers, 1997.

Stoebner-Delbarre A, Thezenas S, Kuntz C, Nguyen C, Giordanella JP, Sancho-Garnier H, Guillot B; Le Groupe EPI-CES. Sun exposure and sun protection behavior and attitudes among the French population. *Ann Dermatol Venereol.* 2005 Aug-Sep;132(8-9 Pt 1):652-7.

Straus, *et al.* 1984. Suppression of frequently recurring gential herpes. *N Eng J of Medicine.* 24:1545-50.

Sugarman E. *Warning, The Electricity Around You May be Hazardous To Your Health.* New York: Simon & Schuster. 1992.

Sulman FG, Levy D, Lunkan L, Pfeifer Y, Tal E. New methods in the treatment of weather sensitivity. *Fortschr Med.* 1977 Mar 17;95(11):746-52.

Sulman FG. Migraine and headache due to weather and allied causes and its specific treatment. *Ups J Med Sci Suppl.* 1980;31:41-4.

Suppes P, Han B, Epelboim J, Lu ZL. Invariance of brain-wave representations of simple visual images and their names. Proceedings of the National Academy of Sciences Psychology-BS. 1999;96(25):14658-14663.

Tahvanainen K, Nino J, Halonen P, Kuusela T, Alanko T, Laitinen T, Lansimies E, Hietanen M, Lindholm H. Effects of cellular phone use on ear canal temperature measured by NTC thermistors. *Clin Physiol Funct Imaging.* 2007 May;27(3):162-72.

Tan DX, Manchester LC, Reiter RJ, Qi WB, Karbownik M, Calvo JR. Significance of melatonin in antioxidative defense system: reactions and products. *Biol Signals Recept.* 2000 May-Aug;9(3-4):137-59.

Taoka S, Padmakumar R, Grissom C, Banerjee R. Magnetic Field Effects on Coenzyme B-12 Dependent Enzymes: Validation of Ethanolamine Ammonia Lyase Results and Extension to Human Methylmalonyl CoA Mutase. *Bioelectromagnetics.* 1997;18:506-513.

Taraban M, Leshina T, Anderson M, Grissom C. Magnetic Field Dependence and the Role of electron spin in Heme Enzymes: Horseradish Peroxidase. *J. Am. Chem. Soc.* 1997;119:5768-5769.

Teitelbaum J. From Fatigue to Fantastic. New York: Avery, 2001.

Tevini M, ed. UV-B Radiation and Ozone Depletion: Effects on humans, animals, plants, microorganisms and materials. Boca Raton: Lewis Pub, 1993.

Thaut MH. The future of music in therapy and medicine. Ann N Y Acad Sci. 2005 Dec;1060:303-8.

The Timechart Company. Timetables of Medicine. New York: Black Dog & Leventhal, 2000.

Thie J. *Touch for Health.* Marina del Rey, CA: Devorss Publications, 1973-1994.

Thomas Y, Litime H, Benveniste J. Modulation of human neutrophil activation by "electronic" phorbol myristate acetate (PMA). *FASEB Jnl.* 1996;10:A1479.

Thomas Y, Schiff M, Belkadi L, Jurgens P, Kahhak L, Benveniste J. Activation of human neutrophils by electronically transmitted phorbol-myristate acetate. *Med Hypoth.* 2000;54:33-39.

Thomas Y, Schiff M, Litime M, Belkadi L, Benveniste J. Direct transmission to cells of a molecular signal (phorbol myristate acetate, PMA) via an electronic device. *FASEB Jnl.* 1995;9:A227.

Thompson D. *On Growth and Form.* Cambridge: Cambridge University Press, 1992.

Tierra L. *The Herbs of Life.* Freedom, CA: Crossing Press, 1992.

Tierra M. *The Way of Herbs.* New York: Pocket Books, 1990.

Timofeev I, Steriade M. Low-frequency rhythms in the thalamus of intact-cortex and decorticated cats. *J Neurophysiol.* 1996 Dec;76(6):4152-68.

Ting W, Schultz K, Cac NN, Peterson M, Walling HW. Tanning bed exposure increases the risk of malignant melanoma. Int J Dermatol. 2007 Dec;46(12):1253-7.

Tisler T, Zagorc-Koncan J. Aquatic toxicity of selected chemicals as a basic criterion for environmental classification. *Arh Hig Rada Toksikol.* 2003 Sep;54(3):207-13.

Tišler T, Zagorc-Koncan J. The 'whole-effluent' toxicity approach. *Internl J Environ Poll.* 2007, 31, 3-12.

Tisserand R. *The Art of Aromatherapy.* New York: Inner Traditions, 1979.

Tiwari M. Ayurveda: *A Life of Balance.* Rochester, VT: Healing Arts, 1995.

Todd GR, Acerini CL, Ross-Russell R, Zahra S, Warner JT, McCance D. Survey of adrenal crisis associated with inhaled corticosteroids in the United Kingdom. *Arch Dis Child.* 2002 Dec;87(6):457-61.

Tomasek L, Rogel A, Tirmarche M, Mitton N, Laurier D. Lung cancer in French and Czech uranium miners: Radon-associated risk at low exposure rates and modifying effects of time since exposure and age at exposure. *Radiat Res.* 2008 Feb;169(2):125-37.

Tompkins, P, Bird C. *The Secret Life of Plants.* New York: Harper & Row, 1973.

Toomer G. "Ptolemy". *The Dictionary of Scientific Biography.* New York: Gale Cengage, 1970.

Traber MG, *et al.* Synthetic as compared with natural vitamin E is preferentially excreted as a-CEHC in human urine: studies using deuterated a-tocopheryl acetate. *FEBS Letters.* 1998 Oct 16; 437:145-8.

Trivedi B. Magnetic Map Found to Guide Animal Migration. *National Geographic Today.* 2001 Oct 12.

Trowbridge FL, Hand KE, Nichaman MZ. Findings relating to goiter and iodine in the Ten-State Nutrition Survey. *Am J Clin Nutr.* 1975 Jul;28(7):712-6.

Tsinkalovsky O, Smaaland R, Rosenlund B, Sothern RB, Hirt A, Steine S, Badiee A, Abrahamsen JF, Eiken HG, Laerum OD. Circadian variations in clock gene expression of human bone marrow CD34+ cells. *J Biol Rhythms.* 2007 Apr;22(2):140-50.

Tsong T. Deciphering the language of cells. *Trends in Biochem Sci.* 1989;14:89-92.

Tubek S. Role of trace elements in primary arterial hypertension: is mineral water style or prophylaxis? *Biol Trace Elem Res.* 2006 Winter;114(1-3):1-5.

Udermann H, Fischer G. Studies on the influence of positive or negative small ions on the catechol amine content in the brain of the mouse following shorttime or prolonged exposure. *Zentralbl Bakteriol Mikrobiol Hyg.* 1982 Apr;176(1):72-8.

Ullman D. Controlled clinical trials evaluating the homeopathic treatment of people with human immunodeficiency virus or acquired immune deficiency syndrome. *J Altern Complement Med.* 2003 Feb;9(1):133-41.

Ullman D. *Discovering Homeopathy.* Berkeley, CA: North Atlantic, 1991.

Ulrich RS. Aesthetic and affective response to natural environment. In Altman, I. and Wohlwill, J. F. (eds) *Human Behaviour and Environment: Advances in Theory and Research. Volume 6: Behaviour and the Natural Environment.* New York: Plenum Press: 1983:85-125.

Ulrich RS. Influences of passive experiences with plants on individual wellbeing and health. In Relf, D. (ed) *The Role of Horticulture in Human Well-Being and Social Development: A National Symposium.* Portland: Timber Press, Portland. 1992:93 -105.

Ulrich RS. Natural versus urban scenes: some psychophysiological effects. *Environment and Behaviour.* 1981:523-556.

Ulrich RS. View through window may influence recovery from surgery. *Science.* 1984;224:420 - 421.

Ulrich RS. Visual landscapes and psychological well being. *Landscape Research.* 1979;4:17-23.

Vallance A. Can biological activity be maintained at ultra-high dilution? An overview of homeopathy, evidence, and Bayesian philosophy. *J Altern Complement Med.* 1998 Spring;4(1):49-76.

van Wijk K, Haney M, Scales JA. 1D energy transport in a strongly scattering laboratory model. Phys Rev E Stat Nonlin Soft Matter Phys. 2004 Mar;69(3 Pt 2):036611.

Van Wijk R, Wiegant FAC. *Cultured mammalian cells in homeopathy research: the similia principle in self-recovery.* Utrecht, University Utrecht Publisher, 1994.

Vescelius E. *Music and Health.* New York: Goodyear Book Shop, 1918.

Vickers AJ, Smith C. Homoeopathic Oscillococcinum for preventing and treating influenza and influenza-like syndromes. *Cochrane Database Syst Rev.* 2004;(1):CD001957.

Vigny P, Duquesne M. *On the fluorescence properties of nucleotides and polynucleotides at room temperature.* In. Birks J (ed.). Excited states of biological molecules. London-NY: J Wiley, 1976:167-177.

von Schantz M, Archer SN. *Clocks, genes and sleep. J R Soc Med.* 2003 Oct;96(10):486-9.

Vyasadeva S. *Srimad Bhagavatam.* Approx rec 4000 BCE.

Wachiuli M, Koyama M, Utsuyama M, Bittman BB, Kitagawa M, Hirokawa K. Recreational music-making modulates natural killer cell activity, cytokines, and mood states in corporate employees. *Med Sci Monit.* 2007 Feb;13(2):CR57-70.

Walach H, Jonas WB, Ives J, van Wijk R, Weingartner O. Research on homeopathy: state of the art. *J Altern Complement Med.* 2005 Oct;11(5):813-29.

Walach H. Is homeopathy accessible to research? *Schweiz Rundsch Med Prax.* 1994 Dec 20;83(51-52):1439-47.

Walch JM, Rabin BS, Day R, Williams JN, Choi K, Kang JD. The effect of sunlight on postoperative analgesic medication use: a prospective study of patients undergoing spinal surgery. *Psychosom Med.* 2005 Jan-Feb;67(1):156-63.

Walker M. *The Power of Color.* New Delhi: B. Jain Publishers. 2002.

Waser M, *et al.* PARSIFAL Study team. Inverse association of farm milk consumption with asthma and allergy in rural and suburban populations across Europe. *Clin Exp Allergy.* 2007 May;37(5):661-70.

Watson L. Beyond Supernature. New York: Bantam, 1987.

Watson L. Supernature. New York: Bantam, 1973.

Wayne R. Chemistry of the Atmospheres. Oxford Press, 1991.

Weatherley-Jones E, Thompson E, Thomas K. The placebo-controlled trial as a test of complementary and alternative medicine: observations from research experience of individualised homeopathic treatment. *Homeopathy.* 2004 Oct;93(4):186-9.

Weaver J, Astumian R. The response of living cells to very weak electric fields: the thermal noise limit. *Science.* 1990;247:459-462.

Wee K, Rogers T, Altan BS, Hackney SA, Hamm C. Engineering and medical applications of diatoms. *J Nanosci Nanotechnol.* 2005 Jan;5(1):88-91.

Weinberger P, Measures M. The effect of two audible sound frequencies on the germination and growth of a spring and winter wheat. *Can. J. Bot.* 1968;46(9):1151-1158.

Weiner MA. *Secrets of Fijian Medicine.* Berkeley, CA: Univ. of Calif., 1969.

Weiss RF. *Herbal Medicine.* Gothenburg, Sweden: Beaconsfield, 1988.

Welsh D, Yoo SH, Liu A, Takahashi J, Kay S. Bioluminescence Imaging of Individual Fibroblasts Reveals Persistent, Independently Phased Circadian Rhythms of Clock Gene Expression. *Current Biology.* 2004;14:2289-2295.

Werbach M. *Nutritional Influences on Illness.* Tarzana, CA: Third Line Press, 1996.

West P. *Surf Your Biowaves.* London: Quantum, 1999.

Wheeler FJ. *The Bach Remedies Repertory.* New Canaan, CN: Keats, 1997.

Whittaker E. *History of the Theories of Aether and Electricity.* New York: Nelson LTD, 1953.

WHO. *Guidelines for Drinking-water Quality.* 2nd ed, vol. 2. Geneva: World Health Organization, 1996.

WHO. *Guidelines on health aspects of water desalination.* ETS/80.4. Geneva: World Health Organization, 1980.

WHO. Health effects of the removal of substances occurring naturally in drinking water, with special reference to demineralized and desalinated water. Report on a working group (Brussels, 20-23 March 1978). *EURO Reports and Studies.* 1979;16.

WHO. How trace elements in water contribute to health. *WHO Chronicle.* 1978;32:382-385.

Wilen J, Hornsten R, Sandstrom M, Bjerle P, Wiklund U, Stensson O, Lyskov E, Mild KH. Electromagnetic field exposure and health among RF plastic sealer operators. *Bioelectromagnetics.* 2004 Jan;25(1):5-15.

Wilkinson SM, Love SB, Westcombe AM, Gambles MA, Burgess CC, Cargill A, Young T, Maher EJ, Ramirez AJ. Effectiveness of aromatherapy massage in the management of anxiety and depression in patients with cancer: a multicenter randomized controlled trial. *J Clin Oncol.* 2007 Feb 10;25(5):532-9.

Williams A. Electron microscopic changes associated with water absorption in the jejunum. *Gut.* 1963;4:1-7.

Williams A. Increased Concentration of Chlorine in Swimming Pool Water Causes Exercise-Induced Bronchochonstriction (EIB). Presented at the American College of Sports Medicine 51st Annual Meeting, Indianapolis, June 2-5, 2004. *News release, American College of Sports Medicine.*

Wilson L. *Nutritional Balancing and Hair Mineral Analysis.* Prescott, AZ: LD Wilson Cons., 1998.

Winchester AM. *Biology and its Relation to Mankind.* New York: Van Nostrand Reinhold, 1969.

Winfree AT. *The Timing of Biological Clocks.* New York: Scientific American, 1987.

Wittenberg JS. *The Rebellious Body.* New York: Insight, 1996.

Wolf, M. Beyond the Point Particle - *A Wave Structure for the Electron. Galilean Electrodynamics.* 1995 Oct;6(5):83-91.

Wolverton BC. How to Grow Fresh Air: 50 House Plants that Purify Your Home or Office. New York: Penguin, 1997.

Wood M. *The Book of Herbal Wisdom.* Berkeley, CA: North Atlantic, 1997.

Wood YA, Fenn M, Meixner T, Shouse PJ, Breiner J, Allen E, Wu L. Smog nitrogen and the rapid acidification of forest soil, san Bernardino mountains, southern california. ScientificWorld J. 2007 Mar 21;7:175-80.

Worwood VA. *The Complete Book of Essential Oils & Aromatherapy.* San Rafael, CA: New World, 1991.

Yadav H, Jain S, Sinha PR. Antidiabetic effect of probiotic dahi containing Lactobacillus acidophilus and Lactobacillus casei in high fructose fed rats. *Nutrition.* 2007 Jan;23(1):62-8.

Yang CY, Cheng MF, Tsai SS, Hsieh YL. Calcium, magnesium, and nitrate in drinking water and gastric cancer mortality. *Jpn J Cancer Res.* 1998;89:124-130.

Yang CY, Chiu H, Chiu J, Tsai SS., Cheng MF. Calcium and magnesium in drinking water and risk of death from colon cancer. *Jpn J Cancer Res.* 1997;88:928-933

Yang CY, Chiu HF, Cheng MF, Hsu TY, Cheng MF, Wu TN. Calcium and magnesium in drinking water and the risk of death from breast cancer. *J Toxicol Environ Health.* 2000;60:231-241.

Yang CY, Chiu HF, Cheng MF, Tsai SS, Hung CF, Chiu HF. Rectal cancer mortality and total hardness levels in Taiwan's drinking water. *Environ Research.* 1999;80:311-316.

Yang CY, Chiu HF, Cheng MF, Tsai SS, Hung CF, Tseng YT. Pancreatic cancer mortality and total hardness levels in Taiwan's drinking water. *J Toxicol. Environ. Health.* 1999;56, 361-369.

Yang CY, Chiu HF, Cheng,MF, Tsai SS, Hung C, Lin M. Esophageal cancer mortality and total hardness levels in Taiwan's drinking water. *Environ Research.* 1999;A;81:302-308.

Yarows SA, Fusilier WB, Weder AB. Sodium concentration from water softeners. *Arch Intern Med.* 1997 Jan 27;157(2):218-22.

Yeager S. *The Doctor's Book of Food Remedies.* Emmaus, PA: Rodale Press, 1998.

Yellen G. The voltage-gated potassium channels and their relatives. *Nature* 2002 Sept 5;419:35-42.

Yokoi S, Ikeya M, Yagi T, Nagai K. Mouse circadian rhythm before the Kobe earthquake in 1995. *Bioelectromagnetics.* 2003 May;24(4):289-91.

Youbicier-Simo BJ, Boudard F, Meckaouche M, Bastide M, Baylé JD. The effects of embryonic bursectomy and in ovo administration of highly diluted bursin on adrenocorticotropic and immune response of chicken, *Int. J. Immunother.* 1993;9:169-190.

Zaets VN, Karpov PA, Smertenko PS, Blium IaB. Molecular mechanisms of the repair of UV-induced DNA damages in plants. *Tsitol Genet.* 2006 Sep-Oct;40(5):40-68. Review.

Zaks A, Klibanov AM. The effect of water on enzyme action in organic media. *J Biol Chem.* 1988 Jun 15;263(17):8017-21.

Zamora JL. Chemical and microbiologic characteristics and toxicity of povidone-iodine solutions. *Am J Surg.* 1986 Mar;151(3):400-6.

Zhang C, Popp, F., Bischof, M.(eds.). *Electromagnetic standing waves as background of acupuncture system. Current Development in Biophysics - the Stage from an Ugly Duckling to a Beautiful Swan.* Hangzhou: Hangzhou University Press, 1996.

Zi N. *The Art of Breathing.* New York: Bantam, 1986.

Zou Z, Li F, Buck L. Odor maps in the olfactory cortex. *Proc Natl Acad of Sci.* 2005;102(May 24):7724-7729.

Index

holography, 82, 83, 84
homeopathy, 74, 75, 76, 79, 150, 328
homeostasis, 101, 162, 197
homunculi, 82
humidity, 145, 201, 203, 230, 242
humility, 331, 388
hurricanes, 63, 152, 405
hydrocarbons, 18, 116, 233, 242, 306, 308
hydrogenation, 311, 312
hydrotherapy, 187, 188, 189
hyperpolarization, 339
hypertension, 157, 185, 211, 321
hypnosis, 105
hypoxia, 181
identity, 89, 375, 384, 388, 406, 407
imidacloprid, 309
immunity, 190, 310, 321, 372
indole, 228
infrared, 28, 49, 54, 89, 90, 100, 101, 104, 106, 107, 108, 129, 143, 152, 246, 247, 255, 271, 272, 273, 274
ingestion, 160
inhalation, 236
iodopsin, 339
isoflavones, 253, 326, 345
isoiocyanates, 345
isomer, 313
isopropylbenzene, 213
isotope, 25, 31
kapha, 407
keratinocytes, 257
keratoses, 257
kerogen, 115
kinesiology, 126
krypton, 195, 247
lactobacillus, 229
larynx, 203, 204, 392

limonenes, 213, 276
liquefaction, 112
lithium, 160
liver, 85, 102, 119, 158, 167, 170, 177, 178, 179, 191, 204, 219, 232, 251, 308, 309, 310, 312, 313, 314, 321, 322, 323, 345
lobelia, 228
lutein, 185, 187
lycopene, 270, 345
lymph, 88, 89, 123, 224
macrocosm, 5
macrophage, 185
magnesium, 12, 36, 91, 111, 112, 146, 160, 161, 162, 163, 178, 186, 190, 246
mango, 345
mantras, 240
meadowsweet, 321, 322, 323
melanin, 251, 254, 255, 256, 270, 272
melanocytes, 257, 270
melanoma, 253, 256, 257, 258
memory, 73, 74, 75, 76, 77, 78, 79, 149, 150, 151, 152, 167, 177, 202, 210, 211, 341, 348, 370, 371, 375, 376
mercury, 39, 97, 180, 220, 306
methylation, 101
miasm, 328
microalgae, 184, 187
microcirculation, 189
microcosm, 25
microorganism, 115
microrhythms, 366
microvilli, 177
minerals, 68, 88, 111, 112, 144, 146, 148, 159, 160, 161, 162, 163, 168, 172, 176, 178, 179, 184, 185, 186, 187, 193, 276, 314, 328

soy, 326
starch, 192
steam, 27, 190, 192
steel, 124
steroid, 186
stratosphere, 196, 197, 198
strontium, 12, 25
styrene, 165, 213, 305
subluxation, 126
sulfate, 183, 184, 187, 190
sulfide, 146, 228
sulforaphanes, 345
sulphur, 141, 159, 164, 212, 218, 219, 243, 332
sunrise, 254, 274, 275, 373
sunscreen, 253, 256, 259, 260, 275
sunset, 254, 274, 275, 340, 346
sunspot, 199, 261, 263
superbug, 3
suprachiasmatic, 249, 268, 271, 402
tannins, 276, 323
tantra, 240
terephthalate, 165
testosterone, 271
tetrachloride, 197, 306
thalamus, 350
theophylline, 226, 228
thermals, 6, 15, 18, 24, 27, 28, 87, 88, 89, 95, 99, 101, 102, 103, 111, 113, 114, 115, 122, 125, 128, 129, 130, 139, 147, 188, 195, 231, 234, 247, 261, 262, 263, 269, 271, 272, 275, 344, 365, 390, 391
thermoreceptors, 118
thermoregulation, 129, 366, 370
thermosphere, 198
thromboxane, 322
thymus, 342

thyroid, 102, 119, 161, 162, 202, 220, 263, 271, 309, 343
thyroxine, 202
tinnitus, 322, 361, 369
titanium, 141
tobacco, 213
toluene, 212, 310
tomato, 270, 345, 372
tongue, 237, 391
toothpaste, 119
topaz, 140
tornado, 63
torque, 64
toxins, 166, 171, 179, 180, 184, 185, 216, 220, 225, 248, 250, 313, 322
trachea, 204, 223
tradewinds, 196
transcendental, 379, 388
transcription, 84, 405
transducers, 151
transducin, 339
trans-esterification, 166
trans-fat, 270, 312
transist, 16, 18
transpiration, 145
trauma, 330, 361
trichloroacetic, 170
trichloroethane, 197
tricholoethylenes, 244
triclinic, 91, 140
triflualin, 120
trihalomethane, 170
trimethylbenzene, 213
trombone, 355
troposphere, 195, 196, 198, 209
tryptophan, 228
tsunami, 123
tuberculosis, 250, 252
tungsten, 13, 247
turquoise, 140, 247, 338